HPLC in
Nucleic Acid Research

CHROMATOGRAPHIC SCIENCE

A Series of Monographs

Editor: JACK CAZES
Fairfield, Connecticut

Volume 1: Dynamics of Chromatography (out of print)
J. Calvin Giddings

Volume 2: Gas Chromatographic Analysis of Drugs and Pesticides
Benjamin J. Gudzinowicz

Volume 3: Principles of Adsorption Chromatography: The Separation of Nonionic Organic Compounds (out of print)
Lloyd R. Snyder

Volume 4: Multicomponent Chromatography: Theory of Interference (out of print)
Friedrich Helfferich and Gerhard Klein

Volume 5: Quantitative Analysis by Gas Chromatography
Joseph Novák

Volume 6: High-Speed Liquid Chromatography
Peter M. Rajcsanyi and Elisabeth Rajcsanyi

Volume 7: Fundamentals of Integrated GC-MS (in three parts)
Benjamin J. Gudzinowicz, Michael J. Gudzinowicz, and Horace F. Martin

Volume 8: Liquid Chromatography of Polymers and Related Materials
Jack Cazes

Volume 9: GLC and HPLC Determination of Therapeutic Agents (in three parts)
Part 1 edited by Kiyoshi Tsuji and Walter Morozowich
Part 2 and 3 edited by Kiyoshi Tsuji

Volume 10: Biological/Biomedical Applications of Liquid Chromatography
Edited by Gerald L. Hawk

Volume 11: Chromatography in Petroleum Analysis
Edited by Klaus H. Altgelt and T. H. Gouw

Volume 12: Biological/Biomedical Applications of Liquid Chromatography II
Edited by Gerald L. Hawk

Volume 13: Liquid Chromatography of Polymers and Related Materials II
Edited by Jack Cazes and Xavier Delamare

Volume 14: Introduction to Analytical Gas Chromatography: History, Principles, and Practice
John A. Perry

Volume 15:	Applications of Glass Capillary Gas Chromatography *Edited by Walter G. Jennings*
Volume 16:	Steroid Analysis by HPLC: Recent Applications *Edited by Marie P. Kautsky*
Volume 17:	Thin-Layer Chromatography: Techniques and Applications *Bernard Fried and Joseph Sherma*
Volume 18:	Biological/Biomedical Applications of Liquid Chromatography III *Edited by Gerald L. Hawk*
Volume 19:	Liquid Chromatography of Polymers and Related Materials III *Edited by Jack Cazes*
Volume 20:	Biological/Biomedical Applications of Liquid Chromatography IV *Edited by Gerald L. Hawk*
Volume 21:	Chromatographic Separation and Extraction with Foamed Plastics and Rubbers *G. J. Moody and J. D. R. Thomas*
Volume 22:	Analytical Pyrolysis: A Comprehensive Guide *William J. Irwin*
Volume 23:	Liquid Chromatography Detectors *Edited by Thomas M. Vickrey*
Volume 24:	High-Performance Liquid Chromatography in Forensic Chemistry *Edited by Ira S. Lurie and John D. Wittwer, Jr.*
Volume 25:	Steric Exclusion Liquid Chromatography of Polymers *Edited by Josef Janča*
Volume 26:	HPLC Analysis of Biological Compounds: A Laboratory Guide *William S. Hancock and James T. Sparrow*
Volume 27:	Affinity Chromatography: Template Chromatography of Nucleic Acids and Proteins *Herbert Schott*
Volume 28:	HPLC in Nucleic Acid Research: Methods and Applications *Edited by Phyllis R. Brown*

Other Volumes in Preparation

HPLC in Nucleic Acid Research
METHODS AND APPLICATIONS

Edited by

Phyllis R. Brown
Department of Chemistry
University of Rhode Island
Kingston, Rhode Island

MARCEL DEKKER, INC. New York and Basel

Library of Congress Cataloging in Publication Data
Main entry under title:

HPLC in nucleic acid research.

(Chromatographic science ; v. 28)
Includes index.
1. Nucleic acids--Analysis. 2. High performance
liquid chromatography. I. Brown, Phyllis R.
II. Title: H.P.L.C. in nucleic acid research.
III. Series.
QP620.H65 1984   574.87'328   84-19916
ISBN 0-8247-7236-9

COPYRIGHT © 1984 by MARCEL DEKKER, INC. ALL RIGHTS RESERVED.

Neither this book nor any part may be reproduced or transmitted in any
form or by any means, electronic or mechanical, including photocopying,
microfilming, and recording, or by any information storage and retrieval
system, without permission in writing from the publisher.

MARCEL DEKKER, INC.
270 Madison Avenue, New York, New York 10016

Current printing (last digit):
10 9 8 7 6 5 4 3 2 1

PRINTED IN THE UNITED STATES OF AMERICA

This book is dedicated to Bertram M. Brown—University of Rhode Island, 1936 and 1938—for his vital interest in this research and in all the students involved in it. His support and encouragement has been a valuable part of any successes we have achieved.

It is also dedicated to the University of Rhode Island, and all the people at the university who provided the opportunities for research and graduate education in chemistry.

Preface

The mysteries of life and living organisms have intrigued many people over the centuries. Only within the last four decades, however, has great progress been made in determining fundamental features of the human organism, such as the structure and function of the nucleic acids. Progress in biochemical research has been made possible mainly by the tremendous strides in the development of new techniques and instrumentation. It is hard to believe that only 40 years ago, most of the elegant, sophisticated instrumentation that is so necessary to biochemical research did not exist. Where it existed, it was used only by a handful of physicists or a selected small group of chemists. Little of this instrumentation was adapted to the needs of the biochemist or biomedical researcher before World War II.

One of the most difficult problems usually facing the chemist is the separation, purification, and identification of compounds in a complex mixture. Since living organisms are composed of intricate structures containing vast numbers of complicated molecules, this problem is magnified in biochemical or biomedical research. In addition, the constituents of cells or physiological fluids can be difficult to handle, since they may be ionic, nonvolatile, thermally labile, or of high molecular weight. Often it is necessary to deal with extremely low concentrations, in the nano- to picomole range, and even in femtomole concentrations. In addition, other compounds which can interfere with the analysis can be present in significantly higher amounts.

In 1971, when I wrote my first book on "high pressure liquid chromatography" (HPLC),* I predicted that the development of commercially available liquid chromatographs would open "new horizons for the analysis of nonvolatile reactive compounds" and would "prove to be an invaluable aid in all biochemical and biomedical research." Also, I predicted that HPLC would be "useful not only in biochemistry but also in inorganic, organic and clinical chemistry." These predictions proved to be true beyond my wildest dreams. Today, because of the explosive growth and development of instrumentation and technology, HPLC is one of the most widely used techniques, not only in analytical chemistry and biochemistry but also in any discipline that requires separation methods. HPLC has brought a new dimension to the separation and quantitative analysis of submicrogram quantities of biologically active molecules. In modern liquid chromatographs, there is sophisticated microprocessor technology for operative control and data handling combined with high-efficiency columns; thus rapid and efficient separations of all types of compounds, including those that are nonvolatile and thermally labile, can be achieved with great sensitivity, selectivity, and convenience.

Liquid chromatographic techniques have become especially important in the areas of protein and nucleic acid research. It is interesting to note that it was the discovery of DNA and the need to know the composition of the nucleic acids that originally gave great impetus to the development of HPLC.

Since 1969 I have been involved in the development of HPLC assays for nucleotides, nucleosides, their bases, and the enzymes which catalyze their reactions. Presently, the research in my laboratory is divided into three areas:

1. Development and refinement of HPLC methods for the analysis of biological samples for nucleic acid constituents and the enzymes active in their metabolism;
2. The use of these methods in biomedical investigations and especially in studies of disease processes;
3. Basic research on retention processes of these compounds using both the reversed-phase and ion-exchange modes.

Over the years, each member of my research group has concentrated on a particular phase of the overall research plan. This wonderful group of young people has collaborated with me in planning and writing this book. For me, this project was a labor of love and the culmination of more than a decade's work. In this book we used a unified approach

---

*High Pressure Liquid Chromatography: Biochemical and Biomedical Applications, Phyllis R. Brown, Academic Press, New York, 1973.

Preface

to the use of HPLC in nucleic acid research. We have used the expertise gained over the years. For this book we distilled the work done in our laboratory and integrated it with published work from other laboratories. We have organized the material so that the book would be easy to read and would be useful as a reference book containing both theoretical and practical information. We hope that our work will aid chromatographers, biochemists, biomedical researchers, and clinical chemists by making readily available to them not only HPLC information on theory and instrumentation, but also applications of this technique. These applications range from investigations of nucleic acid metabolism in normal subjects and disease processes to biogenetic engineering.

It has been exciting to work with my students—past and present— and it has been a joy to see them mature—both personally and professionally. This book is a testimony to everyone who has worked in my laboratory—to their talents, their hard work, and their achievements.

I would like to acknowledge the many people who have helped make this book possible. I especially thank the Gillette Corp., the Teknor-Apex Corp., and Perkin-Elmer Co., who provided academic-year or summer research fellowships for my students; the Perkin-Elmer Co., Waters Associates, Whatman Inc., Kratos Inc., Schoeffel Instrument Division, and Farrand Optical Co. for use of equipment and supplies; and my many colleagues in both chromatography and biomedical research for their advice and encouragement. I would also like to thank the faculty and staff of the Chemistry Department and the Graduate School at the University of Rhode Island for their support, and especially Dr. Douglas M. Rosie, in whose summer course at URI I first learned about chromatography.

Phyllis R. Brown

Contents

Preface		v
Contributors		xi
List of Abbreviations of Purine and Pyrimidine Bases, Their Nucleosides and Nucleotides		xiii

Part I OVERVIEW

    1 Nucleic Acid Research   3
        Phyllis R. Brown

    2 Structures, Properties, and Chromatographic Behavior of Nucleic Acid Constituents   17
        Hubert A. Scoble

Part II METHODOLOGY

    3 Sample Preparation   31
        Phyllis R. Brown

    4 Chromatography   49
        Phyllis R. Brown

    5 Peak Identification   81
        Mary Jo Wojtusik

    6 Quantitative Analysis in Liquid Chromatography   99
        Hubert A. Scoble

7	Microbore Columns *Derek Dezaro and Richard A. Hartwick*	113
8	Detection Systems *Sebastian P. Assenza*	139
9	Mobile Phase Control *Pamela A. Perrone*	161

Part III APPLICATIONS

10	Nucleic Acids *Richard C. Simpson*	181
11	Oligonucleotides *Jonathan B. Crowther and Richard A. Hartwick*	195
12	Free Nucleotides in Biological Samples *Malcolm McKeag*	215
13	Free Nucleosides and Their Bases in Physiological Fluids *Katsuyuki Nakano*	247
14	Cyclic Nucleotides *Robert F. Burgoyne*	267
15	Enzyme Assays by HPLC *Anne P. Halfpenny*	285
16	Nucleotide Coenzymes *Mary E. Dwyer*	303
17	HPLC Analysis of Pyrimidine and Purine Antimetabolite Drugs *Joy R. Miksic*	317
18	Central Nervous System Drugs (Methylxanthine Drugs) *Katsuyuki Nakano*	339
19	Nucleic Acid Constituents in Disease Processes *Mona Zakaria*	365

Selected Recent Publications 389
Index 393

Contributors

Sebastian P. Assenza*   Department of Chemistry, University of Rhode Island, Kingston, Rhode Island

Phyllis R. Brown   Department of Chemistry, University of Rhode Island, Kingston, Rhode Island

Robert F. Burgoyne   Life Science Marketing, Waters Associates, Milford, Massachusetts

Jonathan B. Crowther[†]   Department of Chemistry, Rutgers, The State University of New Jersey, Piscataway, New Jersey

Derek Dezaro   Department of Chemistry, Rutgers, The State University of New Jersey, New Brunswick, New Jersey

Mary E. Dwyer   Department of Chemistry, University of Rhode Island, Kingston, Rhode Island

---

Current affiliations:

*Pharmaceutical Development Department, Analytical Development, Stuart Pharmaceuticals, Division of ICI Americas, Wilmington, Delaware.

†Department of Veterinary Science, Cornell University, Ithaca, New York.

Anne P. Halfpenny* Department of Chemistry, University of Rhode Island, Kingston, Rhode Island

Richard A. Hartwick Department of Chemistry, Rutgers, The State University of New Jersey, New Brunswick, New Jersey

Malcolm McKeag Department of Chemistry, University of Rhode Island, Kingston, Rhode Island

Joy R. Miksic Research and Development, Biodecision Laboratories, Inc., Pittsburgh, Pennsylvania

Katsuyuki Nakano PL Medical Data Center, Tondabayashi, Osaka, Japan

Pamela A. Perrone[†] Department of Chemistry, University of Rhode Island, Kingston, Rhode Island

Hubert A. Scoble Department of Chemistry, Massachusetts Institute of Technology, Cambridge, Massachusetts

Richard C. Simpson Department of Chemistry, University of Rhode Island, Kingston, Rhode Island

Mary Jo Wojtusik Department of Chemistry, University of Rhode Island, Kingston, Rhode Island

Mona Zakaria Quality Control, Hoffmann-La Roche, Inc., Nutley, New Jersey

---

Current affiliations:

*Department of Clinical Chemistry, Memorial Sloan-Kettering Cancer Center, New York, New York.

†Liquid Chromatography Product Department, The Perkin-Elmer Corporation, Norwalk, Connecticut.

List of Abbreviations of Purine and Pyrimidine Bases, Their Nucleosides and Nucleotides

Ade	adenine	GDP	guanosine 5'-diphosphate
Ado	adenosine	GTP	guanosine 5'-triphosphate
AMP	adenosine 5'-phosphate	C-GMP	guanosine 3',5'-cyclicphosphate
ADP	adenosine 5'-diphosphate		
ATP	adenosine 5'-triphosphate	Xan	xanthine
C-AMP	adenosine 3',5'-cyclic phosphate	Xao	xanthosine
		XMP	xanthosine 5'-phosphate
Hyp	hypoxanthine	XDP	xanthosine 5'-diphosphate
Ino	inosine	XTP	xanthosine 5'-triphosphate
IMP	inosine 5'-phosphate	Cyt	cytosine
IDP	inosine 5'-diphosphate	Cyd	cytidine
ITP	inosine 5'-triphosphate	CMP	cytidine 5'-phosphate
C-IMP	inosine 3',5'-cyclicphosphate	CDP	cytidine 5'-diphosphate
		CTP	cytidine 5'-triphosphate
Gua	guanine	Ura	uracil
Guo	guanosine	Urd	uridine
GMP	guanosine 5'-phosphate		

UMP	uridine 5'-phosphate	Thd	thymidine
UDP	uridine 5'-diphosphate	TMP	thymidine 5'-phosphate
UTP	uridine 5'-triphosphate	TDP	thymidine 5'-diphosphate
Thy	thymine	TTP	thymidine 5'-triphosphate

Unless specifically noted:

1. All the nucleosides and nucleotides referred to in this book are ribonucleosides and ribonucleotides, except for thymidine, which occurs naturally as the 2'-deoxyribonucleoside and 2'-deoxyribonucleotide.
2. In free nucleotides, the phosphate is usually on the 5' position of the sugar. However, in hydrolysates of nucleic acids, the phosphate will be on the 2' or 3' position. The position of the phosphate is designated by inserting the position number between the nucleoside and the phosphate (e.g., adenosine 5'-phosphate).
3. Substitution on the purine or pyrimidine ring is designated by the position on the ring followed by the substituent (e.g., 1-methyladenine). Substitution on an amine group is designated by N with a superscript, which shows the position of the amine on the ring, followed by the group (e.g., N^6,N^6-dimethyladenine or N^2-methylguanine).

HPLC in
Nucleic Acid Research

Part I
OVERVIEW

1
Nucleic Acid Research

PHYLLIS R. BROWN University of Rhode Island, Kingston, Rhode Island

   I. History 3
  II. Functions of Nucleotides and Nucleic Acids 6
     A. Nucleotides 6
     B. Nucleic acids 7
 III. Biochemical Pathways 8
     A. Nucleotides, nucleosides, and their bases 8
     B. Nucleic acids 13
     References 14

I. HISTORY

The study of nucleic acid metabolism began in 1776 with the identification of uric acid by Scheele. Thus the chemistry and metabolism of nucleotides and their metabolites is one of the earliest investigations in biochemistry. It was not until 1868-1869 that Miescher discovered a material which contained phosphorus in pus cell nuclei that was soluble in alkali and precipitated in acid. Altman named this material "nucleic acid" [1]. By 1909 it was evident that the nucleic acid obtained from yeast contained the purine bases adenine, guanine, uracil, and cytosine, as well as phosphoric acid and the sugar D-ribose. It was not until 1928, however, that it was shown that a different nucleic acid, then called thymonucleic acid, contained 2-deoxy-D-ribose and thymine instead of D-ribose and uracil. The term nucleotide was introduced by Levene and applied to the phosphoric ester of the thymine nucleoside [2]. Although the four nucleotides subunits of RNA were isolated in the pure form from yeast RNA, the position of the phosphate group was not identified until 1925, when the structure of inosine 5'-phosphate (5'-IMP) was confirmed. Prior to 1936, the purine nucleotides were thought to have the glycosidic linkage at the N-7 position, the position assigned by Fischer. In 1936, however, Gulland, on the basis of absorption spectra, determined that this glycosidic linkage was at the 9 position on the purine. During the 1940s the β configuration as well as the furanose structure of the ribose was established [1-6].

During the post-World War II period, major advances in our knowledge of nucleic acids were made possible by the development of chromatographic and spectroscopic techniques. Cohn [7,8] from Oak Ridge was the first to use ion-exchange chromatography for the separation of nucleotides, and Potter and his coworkers [9-12] were the first to isolate nucleotides from the acid-soluble fraction of tissue. They demonstrated that, in addition to the adenine nucleotides, there was a complex mixture of free nucleotides present in tissue. These mixtures included the mono-, di-, and triphosphate nucleotides as well as nucleotide coenzymes.

The next big quantum leap in nucleic acid research came with the development of a technique called high performance liquid chromatography (HPLC). In 1962 Anderson at Oak Ridge [13,14] developed chromatographic instrumentation using ion-exchange resins to analyze urine for all the low-molecular-weight, ultraviolet-absorbing compounds. Among the hundreds of compounds were purines and pyrimidines, their nucleosides and nucleotides and nucleotide coenzymes. Although these analyses were very time-consuming, it was the start of the utilization of HPLC in nucleic acid research. The tremendous importance of DNA and RNA in biochemical research spurred the development of this technique. It became increasingly evident that the structures of nucleic acids were of vital interest not only in studies of normal growth but also of disease processes. Therefore efficient methods were needed not only to isolate each nucleic acid but also to determine the composition of each DNA or RNA molecule, i.e., the number and kind of the purine and pyrimidine bases present. With the work of Burtis et al. [15-17], Uziel [18], and Horvath and Lipsky [19,20], it became possible to analyze by HPLC the hydrolysates of the nucleic acids on the nucleotide, nucleoside, or base level rapidly, reproducibily, and efficiently. Thus with the technological developments of HPLC in the 1970s, a new era dawned in the analytical capabilities for research in the nucleic acid field. Analyses that previously were tedious and time-consuming could now be done routinely with a precision undreamed of earlier. The tremendous power of HPLC ushered in a new era in research, and this technique became a major tool in genetic engineering, in investigations of disease processes, in studies of normal cell growth and metabolism, in pharmacology, in physiology, and in clinical chemistry.

Because HPLC can be used for so many different assays, it has had a great impact on nucleic acid research. The following analyses are some of the ways HPLC can be used in studies dealing with nucleic acids, their structures, their metabolism, and their synthesis.

1. Separation, isolation, and purification of nucleic acids in biological samples
2. Separation and identifcation of hydrolysates of RNA and DNA on the nucleotide, nucleoside, or base level.
3. Determination of structures and purity of oligonucleotides in the synthesis of nucleic acids
4. Analysis of metabolic end products and intermediates of nucleic acids (e.g., nucleotides, nucleosides, and bases).
5. Determination of activities of enzymes in nucleic acid metabolic reactions
6. Qualitative and quantitative analysis of coenzymes and cofactors involved in nucleic acid metabolism.

HPLC is also playing a vital role in studies of the effects of disease on nucleic acid metabolism. For these studies, rapid and efficient HPLC assays are now available for the qualitative and quantitative analysis of the following:

1. Free nucleotides; total profiles of all nucleotides or assays for a single nucleotide (e.g., ATP)
2. Purines and pyrimidines and their nucleosides; total profiles in physiological fluids or assays for a single nucleoside (e.g., adenosine)
3. Cyclic nucleotides, especially cyclic AMP and cyclic GMP, which regulate important biological functions
4. Enzymes and coenzymes needed to catalyze metabolic reactions

In addition, HPLC is a powerful technique for therapeutic drug monitoring. Assays are available to determine levels of purine and pyrimidine drugs in serum or other physiological fluids rapidly, reliably, and with great sensitivity. With these assays it is also possible to determine the effects of these drugs on nucleic acid metabolism as well as to determine their effectiveness and/or the onset of their toxicity. We believe that HPLC will continue to be a tremendously important tool in biochemical research, and that it will play a vital role in nucleic acid research and biotechnology.

Therefore we wrote this book so that the pertinent information on the HPLC analyses of all the compounds of interest in nucleic acid metabolism would be readily available to the researcher and the clinical chemist.

II. FUNCTIONS OF NUCLEOTIDES AND NUCLEIC ACIDS

A. Nucleotides

Nucleotides in biological systems have many functions, which can be classified in five general categories:

1. Nucleic acid structural units; as precursors in the biosynthesis of DNA and RNA or as catabolites in nucleic acid metabolism
2. Energy functions; for transporting metabolically available energy and supporting energetically unfavorable reactions
3. Donor functions; for the transfer of a wide variety of groups to appropriate acceptors
4. Physiological functions; as mediators of reactions on the organism, organ, or cell level
5. Regulatory functions; to regulate enzyme reactions and metabolic processes

1. Structural units

Nucleotides are the building blocks of nucleic acids. In addition, three vitamins also have structures in which heterocyclic rings are derived from nucleotide or purine bases. One example is folic acid, in which the 2-amino-4-hydroxy-6-methyldihydropterin moiety of folic acid comes from guanosine 5'-triphosphate (GTP); the synthesis of this moiety involves several steps. In riboflavin biosynthesis, purines are precursors; and in thiamine biosynthesis, the pyrimidine ring is provided by an intermediate in the de novo pathway of purine. In some cells guanine is involved in thiamine biosynthesis, whereas in other cells the purine is hypoxanthine.

2. Energy metabolism

ATP is the nucleotide mainly involved in energy metabolism. The free energy derived from biological oxidations is usually converted into chemical energy in the form of ATP. There are many types of oxidative and photophosphorylation reactions. One of the most important reactions is the hydrolysis of pyrophosphate bonds of ATP or other high-energy triphosphate nucleotides. This reaction generates free energy which helps drive energetically unfavorable reactions. There are many reactions in which a phosphate group is transferred to another molecule. These reactions are classified according to the type of receptor molecule. The energy released in bond breakage is used to

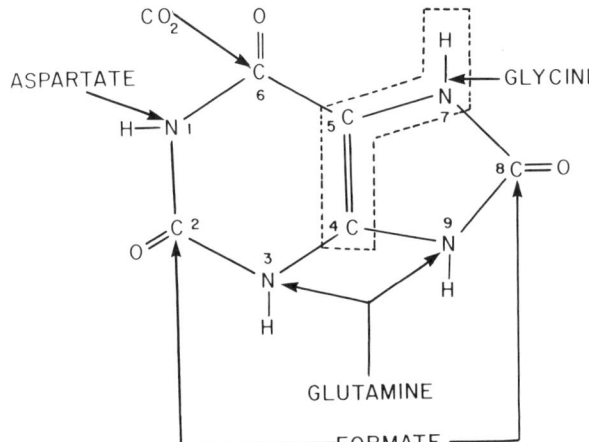

Figure 1. Source atoms in purines in de novo synthesis of nucleotides: 1 from aspartate; 2 and 8 from formate; 3 and 9 from glutamine; 4, 5, and 7 from glycine; 6 from CO_2 and HCO_3^-. The ribose and phosphate groups come from the PRPP in the first step of the biosynthesis.

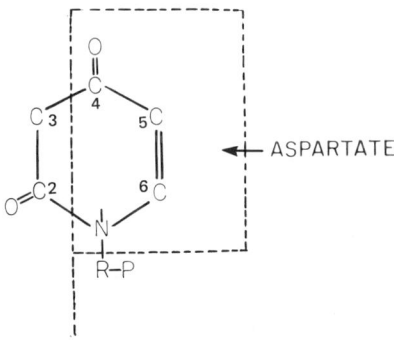

Figure 2. Source of atoms in pyrimidines in the de novo biosynthesis of nucleotides: 1, 4, 5, and 6 from aspartate; 2 from CO_2 and HCO_3^-; 3 from ammonia. The ribose and phosphate groups are from PRPP, which reacts with orotidylate in the fifth step of the biosynthesis, just before the decarboxylation in which the orotate is converted to uridylate.

Table 1 Enzymes in the de Novo Synthesis of Purine Nucleotides

Reactions	Enzymes (trivial name)	Enzyme commission number
Amide transfers	PRPP amidotransferase	2.4.2.14
	PR glycineamide synthetase	6.3.1.3
	PR formylglycine amide synthetase	6.35.3
One-carbon transfers	PR glycineamidine formyltransferase	2.1.22
	PR aminoimidazole carboxamide transferase	
Cyclization	PR aminoimidazole synthetase	6.3.3.1
	IMP 1,2-hydrolase	3.5.4.10
Displacement	Adenylosuccinateylase	4.3.22
Aspartate amino transfer	PR aminoimidazole succinocarboxamide synthetase	6.3.2.6
Carboxylate transfer	PR aminoimidazole carboxylase	4.1.1.21

Table 2 Enzymes in the de Novo Synthesis of Pyrimidine Nucleotides

Type of reaction	Enzyme (trivial name)	Enzyme commission number
Phosphate transfers	Carbonylphosphate synthetase I	2.7.2.5
	Carbonate kinase	2.7.2.2
Deamination	CTP synthetase	6.3.42
Decarboxylation	Orotidine 5'-phosphate decarboxylase	4.1.1.23
PPRP transfer	Orotate phosphoribosyl transferase	2.4.2.10
Dehydrogenase	Dihydrorotate dehydrogenase	1.3.3.1
Hydrogen transfers	Dihydroorotase	3.5.23
Transfer of aspartate carbonyl group	Aspartate carbonyl transferase	2.1.3.2

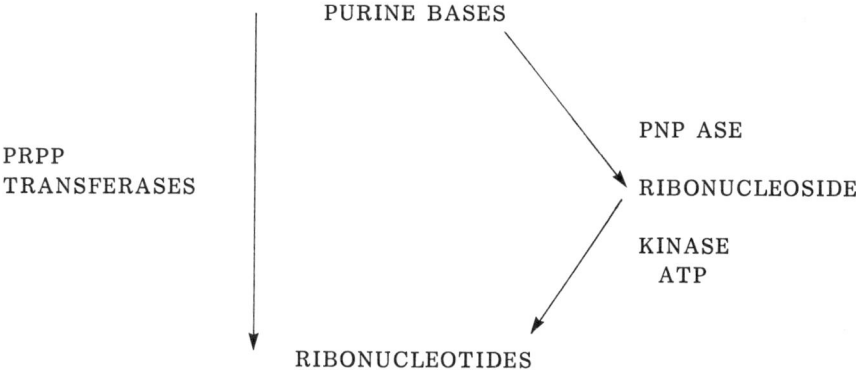

Figure 3. Basic schematic diagram of the salvage pathway of biosynthesis of nucleotides.

Salvage pathways: Purine and pyrimidine nucleotides can also be synthesized via the salvage pathway; i.e., from preformed bases or nucleosides using relatively few enzymes. The nucleosides can be converted to nucleotides by nucleoside kinases in the presence of ATP, while the bases can be converted to ribonucleotides via phosphoribosyl transferases in the presence of pyrophosphoribosyl-phosphate (RPRP). Another possible route of the biosynthesis of ribonucleotides is a more indirect route; i.e., the conversion of bases to nucleosides via the general enzyme, purine nucleoside phosphorylase (PNPase, EC 2.4.2.1), followed by conversions of nucleoside to the nucleotide. The latter step is catalyzed by a base-specific kinase such as adenosine kinase. Steps in the salvage pathway are shown in Fig. 3.

2. *Catabolism*

The major pathway of purine catabolism is the conversion of nucleotides to uric acid. Among the types of enzymatic reactions required for the overall conversion are dephosphorylation, deamination, and oxidation reactions, as well as reactions involving the cleavage of glycosidic bonds. Both adenylate and guanylate can be converted to xanthylate via inosate (Fig. 4). In addition conversions to uric acid can be made on both the nucleoside and base levels. Humans and other higher primates excrete uric acid as the end product of purine metabolism. Other groups of animals have specific end products of purine catabolism, such as allantoic acid, ammonia, or urea; the type of end product depends on the specific enzyme present to catalyze the terminal steps of purine catabolism. Thus it can be seen that animal species have distinctive patterns of enzyme specialization.

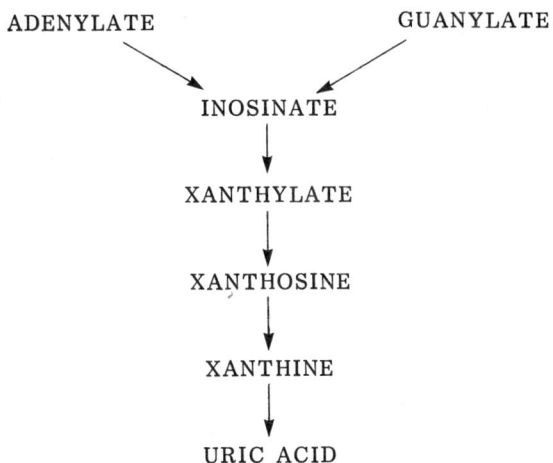

Figure 4. Catabolism of ribonucleotides directly to uric acid.

In contrast to purine metabolism, catabolic pathways of pyrimidines are reductive rather than oxidative in higher animals. The catabolism of pyrimidine nucleotides involves glycosidic bond cleavage, deamination, and dephosphorylation. The end product of the catabolism of uracil is malonyl coenzyme A (CoA), whereas that of thymine is methylmalonyl CoA.

3. *Interconversion*

In the metabolic pathways of purines and pyrimidines, there can be interconversion among the various compounds on the ribonucleotide, ribonucleoside, or base levels. Since inosate is the end product of de novo synthesis, inosate must be converted into guanylate and adenylate for use in nucleic acid biosynthesis. Adenylosuccinate is an important intermediate in the conversion of inosinate to adenylate, and xanthylate in the conversion of inosinate to guanylate. Reverse conversion of guanylate and adenylate to inosate can also occur. All four monophosphates are also precursers for the biosynthesis of their di- and triphosphate nucleotides. Interconversions can also occur on the nucleoside and base levels. For example, the deamination of adenine compounds can occur on the nucleotide, nucleoside, or base level, each catalyzed by a specific enzyme. The conversion of adenosine to inosine has been found to be especially important in certain disease states and especially in severe immunodeficiencies. On the base level, both hypoxanthine and guanine are converted to xanthine prior to oxidative reactions to uric acid. Thus there are several alternative pathways for the catabolism of purine nucleotides to uric acid.

B. Nucleic acids

1. *Biosynthesis*

Deoxynucleic acids: DNA molecules can be synthesized from deoxynucleotides, and the enzyme which catalyzes this reaction is DNA polymerase. In this reaction, the substrates must be on the triphosphate deoxyribonucleotide level. The polymerization reaction will not take place on the diphosphate level. In addition, the substrate structures must include all four of the bases. The reaction is reversible, and depolymerization can take place if high concentrations of pyrophosphate are present. A cofactor that must be present is a divalent metal ion, usually magnesium. A DNA primer molecule must also be present. If all required substrates and cofactors are present, then quantities of DNA which exceed the primer material by 20-fold may be produced.

Ribonucleic acids: There are several enzymes which can catalyze the polymerization of ribonucleotides to RNA. Each enzyme has specific substrate and primer requirements.

Among the polymerases is the enzyme polynucleotide phosphorylase, which catalyzes the formation of RNA from ribonucleotides. The requirements are that the ribonucleotides be in the diphosphate form, that magnesium ions be present, and that primer RNA molecules be present; however, there is no relationship between primer and product structures. The reaction is readily reversible in the presence of large concentrations of organic phosphate, and both homopolymers and heteropolymers can be formed.

RNA polymerase is similar to DNA polymerase in that it catalyzes polymerization of nucleotides on the triphosphate level and that magnesium ions are required. As in its DNA analog, the reaction is reversible. However, it is interesting to note that the primer is a DNA rather than an RNA molecule.

There are also RNA replicases, i.e., RNA-dependent RNA polymerases. Some RNA polymerases can use either RNA or DNA as primer, with the primer functioning as the template.

2. *Inhibition of nucleic acid biosynthesis*

There are many inhibitors of nucleic acid synthesis. These inhibitors are useful in establishing the interdependences of protein synthesis with RNA and DNA synthesis. In addition, these inhibitors can function as antimetabolites and as antitumor agents. Most of the inhibitors of nucleic acid synthesis are related structurally to normally occurring purine and pyrimidine compounds—for example, 6-mercaptopurine or 8-azaguanine. The chromatography of antimetabolites will be discussed in Chap. 17.

3. Hydrolysis of nucleic acids

Nucleic acids may be broken down by several enzymes. An enzyme called exonuclease is relatively nonspecific with respect to the base and catalyzes the hydrolysis of both polyribo- and polydeoxyribonucleotides. In a stepwise manner it liberates nucleoside 5'-phosphates. Endonucleases attack intranucleotide bonds in which the phosphate of the P-O bond must be attached to the 3'-OH of a pyrimidine nucleotide. In this reaction the first step is the formation of a cyclic nucleotide from the polynucleotide. The formation of cyclic nucleotides can also be catalyzed by phosphodiesterases. The endonucleases and phosphodiesterases are specific for RNA, as they require a 2'-OH group.

REFERENCES

1. J. F. Henderson and A. R. P. Paterson, *Nucleotide Metabolism, an Introduction*, Academic Press, New York, 1973.
2. W. Jones, *Nucleic Acids—Their Chemical Properties and Physiological Conduct*, 2nd ed., Longmans, Green, New York, 1970.
3. A. M. Michelson, *The Chemistry of Nucleosides and Nucleotides*, Academic Press, New York, 1963.
4. P. A. Levene and L. W. Bass, *Nucleic Acids*, Chemical Catalog Co. (Tudor), New York, 1963.
5. E. Chargaff and J. N. Davidson (Eds.), *The Nucleic Acids*, vols. 1-3, Academic Press, New York, 1955.
6. J. N. Davidson, *The Biochemistry of Nucleic Acids*, Methuen, London, 1960.
7. W. E. Cohn, Science, *109*, 377 (1949).
8. W. E. Cohn, J. Am. Chem. Soc., 72, 1471 (1950).
9. R. B. Hurlburt, H. Schmitz, A. F. Brumm, and V. R. Potter, J. Biol. Chem., *209*, 23 (1954).
10. H. Schmitz, R. B. Hurlburt, and V. R. Potter, J. Biol. Chem., *209*, 41 (1954).
11. H. Schmitz, V. R. Potter, R. B. Hurlburt, and D. M. White, Cancer Res., *14*, 66 (1954).
12. V. R. Potter, *Nucleic Acid Outlines*, vol. 1, Burgess, Minneapolis, 1960.
13. N. G. Anderson, Anal. Biochem., 4, 269 (1962).
14. N. E. Anderson and C. C. Ladd, Biochem. Biophys. Acta, 55, 275 (1962).
15. C. A. Burtis and D. Gere, *Nucleic Acid Constituents by Liquid Chromatography*, Varian Aerograph, California, 1970.
16. C. A. Burtis, D. Gere, and F. MacDonald, Chromatographia, 3, 116 (1970).

17. C. A. Burtis, M. N. Munk, and R. R. MacDonald, Clin. Chem., 16(8), 1970.
18. M. Uziel, C. K. Koh, and E. E. Cohn, Anal. Biochem., 25 (1978).
19. C. Horvath, B. Preiss, and S. R. Lipsky. Anal. Chem., 39, 1422 (1967).
20. C. Horvath and S. R. Lipsky, Anal. Chem., 41, 1227 (1969).
21. D. E. Atkinson, Regulation of Energy Metabolism: Exploitable Molecular Mechanisms and Neoplasia, Williams & Wilkins, Baltimore, 1969.

2
Structures, Properties, and Chromatographic Behavior of Nucleic Acid Constituents

HUBERT A. SCOBLE Massachusetts Institute of Technology, Cambridge, Massachusetts

   I. Structures and Properties 17
  II. Chromatographic Methods 21
 III. Mechanisms of RPLC Retention 22
  IV. Structure-Retention Relationships 23
   V. Mobile Phase Effects 24
       A. pH effects 24
       B. Organic modifiers 24
       C. Ionic strength 25
       References 26

I. STRUCTURES AND PROPERTIES

The parent compounds of the two classes of nitrogenous bases found in nucleic acid constituents are the heterocyclic compounds, purine and pyrimidine. Purine itself may be thought of as a derivative of pyrimidine. It consists of a pyrimidine ring and an imidazole ring fused together. Three pyrimidine derivatives, uracil, thymine, and cytosine, and two purine derivatives, adenine and guanine, constitute the major nitrogenous bases found in RNA and DNA.

Through x-ray diffraction studies, it has been determined that the pyrimidines are planar and the purines are nearly planar with a slight

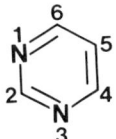

Figure 1. Structure and systematic numbering of the pyrimidine base.

pucker. The important functional groups involved in the formation of hydrogen bonds are the amino groups of adenine, guanine, and cytosine, the ring -NH- group of position 1 of adenine and guanine and position 3 of the pyrimidine bases, and the strongly electronegative oxygen atom at position 2 of the pyrimidines and position 6 of guanine.

The basic structures and numbering systems of pyrimidines are shown in Fig. 1. In the pyrimidine ring, the 2, 4, and 6 positions exhibit a marked π electron deficiency. Since the 1 and 3 positions of the ring nitrogen atoms reinforce this π electron deficiency, the 2, 4, and 6 positions are subject to nucleophilic attack. The 5 position of pyrimidine is only slightly electron-deficient; thus it is much more resistant to nucleophilic attack, and if electrophilic substitution occurs it will be at the 5 position.

When there are electron-releasing substituents on the ring such as in uracil (2,6-dihydroxypyrimidine), the π electron deficiency is counteracted; thus the system is similar to an aromatic ring. Since the 2, 4, and 6 positions are deactivated toward nucleophilic attack, electrophilic substitution is facilitated.

If substituents in the 2, 4, or 6 position are hydroxy, mercapto, or amino groups, the tautomers oxo, thio, or imino can form and the hydrogen is accommodated on a ring nitrogen. Although the hydroxy group substituents prefer the oxo tautomeric form and mercapto groups the thio tautomeric form, the amino group exists mainly as the free -NH_2. Pyrimidine tautomeric forms affect reversed-phase retention behavior, as have been shown by studies of Brown and Gushka [1].

Purine is formed by the fusion of a pyrimidine and an imidazole ring. The structures of purine bases and the numbering used in nomenclature are shown in Fig. 2. Purine ring systems are also π electron deficient. Thus ring carbon atoms are activated toward nucleophilic attack and deactivated toward electrophilic attack. Purines, like pyrimidines, also exist in oxo, thio, and imino tautomeric forms.

The free purine and pyrimidine bases are weakly basic compounds. The tautomeric forms in which they exist depend on the pH. Hypoxanthine, for example, occurs in the lactam and lactim forms; however, at pH 7.0 the lactam form of hypoxanthine predominates (Fig. 3).

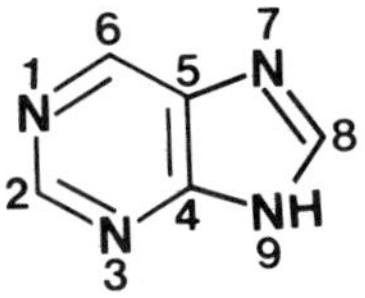

Figure 2. Structure and systematic numbering of the purine base.

In addition to the common bases mentioned above, a large number of purine and pyrimdine derivatives called minor bases occur in small amounts in some nucleic acids. Among the minor pyrimidines are 5-methylcytosine and 5-hydroxymethylcytosine; the minor pyrimidines include 6-methyladenine and 2-methylguanine. Minor bases are especially important in transfer RNAs, which may contain up to 10% of these components.

All purine and pyrimidine derivative bases of nucleic acid strongly absorb ultraviolet radiation in the region of 250 to 280 nm. This property is very useful for the detection of purine- and pyrimdine-containing compounds in open column, paper, thin layer, and high performance liquid chromatography.

The monomeric units of DNA are the deoxyribonucleotides, and those of RNA are ribonucleotides. Each nucleotide contains three characteristic components: (a) a nitrogen heterocyclic base, which is a derivative of either purine or pyrimidine; (b) a pentose; and (c) one or more molecules of phosphoric acid. To describe a nucleotide, not only the specific base, pentose, and number of phosphate groups must be given, but also the site of attachment of the carbohydrate to the base, the configuration of the glycosidic linkage, and the site of attachment of the phosphate group(s) (Fig. 4).

Four different deoxyribonucleotides serve as the major compounds of DNA; they differ from each other only in their nitrogenous based components. The four bases characteristic of DNA are the purine derivatives adenine and guanine and the pyrimidine derivatives cytosine

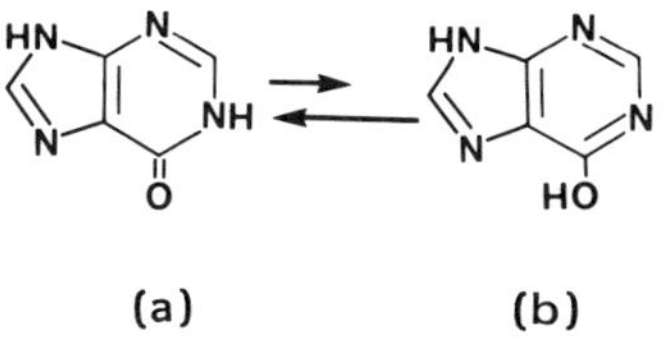

Figure 3. Lactim (a) and Lactam (b) tautomeric forms of the purine base hypoxanthine.

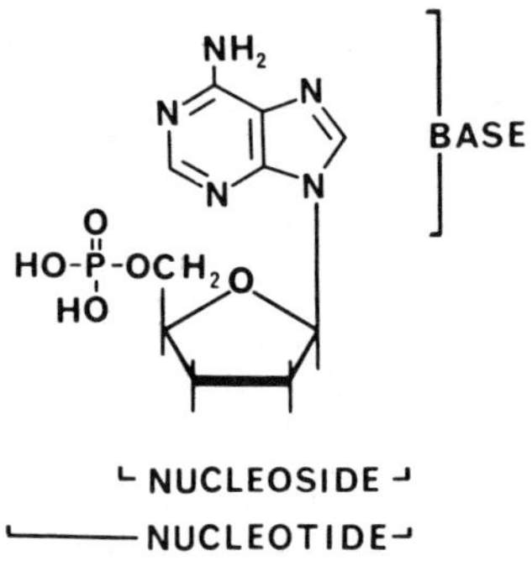

Figure 4. Structure of the nucleotide, 5'-adenosine monophosphate showing the base and nucleoside components of the nucleotide.

and thymine. Thus thymine, which is 5-methyluracil, is characteristically present in DNA but not in RNA, whereas uracil is normally present in RNA but only rarely in DNA.

The other difference in the composition between these two kinds of nucleic acids is that deoxyribonucleotides contain the pentose component 2'-deoxy-D-ribose, whereas ribonucleotides contain D-ribose. The pentose is joined to the base by a β-N-glycosyl bond between carbon atom 1 of the pentose and nitrogen atom 9 of the purine base or nitrogen atom 1 of the pyrimidine base. The phosphate group of nucleotides is in an ester linkage with carbon atom 5 of the pentose. When the phosphate group is removed from the nucleotide by hydrolysis, the structure remaining is called a nucleoside.

All common ribonucleotides and deoxyribonucleotides occur in cells not only as the 5'-monophosphates, but also as the 5'-diphosphates and the 5'-triphosphates. The nucleoside 5'-di- and 5'-triphosphoric acids are relatively strong acids.

In addition to these nucleotides, ribonucleotide 3',5'-cyclic phosphates and ribonucleotide 3'-phosphates occur due to the hydrolytic action of ribonucleases on certain ribonucleotide linkages of RNA. Cyclic nucleotides, which are important regulators of many biochemical reactions and body functions, are discussed in Chap. 14. Two of these nucleotides, adenosine 3',5'-cyclic phosphate (cyclic AMP) and guanosine 3',5'-cyclic phosphate (cyclic GMP), play a key role in regulating hormone reactions.

In addition, many coenzymes contain nucleotide moieties. Important examples of nucleotide coenzymes are the flavin coenzymes (FAD^+ and

FADH) and the nicotinamide coenzymes (NAD$^+$, NADH, NADP$^+$, NADPH). There is a discussion of nucleotide coenzymes in Chap. 16.

II. CHROMATOGRAPHIC METHODS

The earliest chromatographic methods used for the analysis of nucleotides, nucleosides, and their bases was by ion exchange [2]. Nucleotides separated readily upon anion exchange because of the presence of the negatively charged phosphate groups. The development of gradient elution by Crampton et al. [3] made the separation rapid and efficient. The development of microparticle chemically bonded ion-exchange packings further improved the efficiency of the separation and decreased the time required for an analysis [4]. Both anion and cation exchange were used to separate the purines and pyrimidines or the individual members of the nucleosides [4-7]. However, because of poor retention of some of the pyrimidines, it was difficult to separate the nucleosides and bases when they were present together as a complex mixture.

It was not until the development of the reversed-phase mode with microparticle chemically bonded packings that the nucleosides and bases present in biochemical samples could be separated in one analysis [8]. The reversed-phase mode proved ideal for nucleosides and bases, in combination or separately, since these compounds are not usually ionized at the pH of the mobile phase. Thus the reversed-phase mode became the method of choice for nucleosides and/or bases and was adapted for all types of analyses, from a single nucleoside such as adenosine [9] to a complete assay of all the nucleosides and bases in a physiological fluid such as serum or plasma [10].

For the nucleic acids, it was first thought that the only mode of high performance liquid chromatography (HPLC) that could be effectively used for such large molecules was size exclusion (or gel chromatography) [11]. However, it has been found that both the reversed-phase and ion exchange modes can be used to isolate the DNA or RNA molecules [12,13]. Usually, special packings give superior separations to the ones routinely used for other nucleic acid constituents [13].

Coenzymes such as NADH and NAD$^+$ and their metabolites can be readily separated using the reversed-phase mode, as can oligonucleotides. Although oligonucleotides are not normally found in physiological fluids or tissues, the analysis of these compounds is very important in studies of DNA and RNA and in genetic engineering.

New developments in the separation of nucleic acid constituents involve the use of ion pairing [14], mixed beds [15], and column switching [16] to determine nucleotides, nucleosides, and bases in one analysis.

III. MECHANISMS OF RPLC RETENTION

Since the majority of the separations of nucleosides and bases and now even nucleotides are made by reversed-phase liquid chromatography (RPLC), we shall discuss RPLC in greater detail than other modes. At present the mechanism of RPLC retention is not clearly understood. There are many mechanisms postulated, such as partition [17-19]; adsorption [20-28]; dispersive interaction [29]; solubility in the mobile phase [30]; solvophobic effects [31-36]; combined solvophobic and silanophilic interaction [37-39]; and a mechanism based on compulsory absorption [40]. It appears that there is not just a single retention mechanism in RPLC, but that mixed-mode mechanisms are operative. However, solvophobicity (hydrophobicity) appears to be, at present, the primary mechanism for solute retention.

Solvophobicity can be regarded as a reversible association of the solute molecules with the hydrocarbonaceous stationary phase. The association is not based on great attraction of the solute for the stationary phase. Rather, the solute is forced onto or into the stationary phase because of mutual repulsion of the solute and the solvent—thus the term solvophobic. Although there are several important factors in determining the magnitudes of the solvophobic effect, the dielectric constant and surface tension of the solvent appear to play important roles in governing solute retention. There can be forces opposing those of the solvophobic effect, such as the interaction of the solute with the mobile phase. Solutes which have polar substituents interact more strongly with the polar hydroorganic mobile phase; thus there is a decrease in retention compared to similar compounds with no polar moiety. When a solute molecule is ionized under the appropriate mobile phase conditions, there is an increase in electrostatic attraction between solute and eluent. Ultimately this interaction leads to decreased retention. To explain the atypical behavior of some solutes under reversed-phase conditions, Horvath and coworkers [37-39] introduced the concept of a dual binding mechanism. In addition to solvophobic forces, solutes can interact with the free surface silanols of packing materials. To denote a reversible binding mechanism between solute molecules and silanol groups, the term silanophilic interaction is used.

In addition to the hydrocarbonaceous ligates, silanol groups on the stationary phase surface are accessible to solute molecules. The eluite can therefore bind in two different ways to the surface of a stationary phase: the solvophobic interactions and silanophilic interactions.

Such a dual binding mechanism can explain satisfactorily the anomalous retention of many solutes which do not react according to purely solvophobic terms.

IV. STRUCTURE-RETENTION RELATIONSHIPS

It has been difficult to rationalize the retention behavior of purine and pyrimidine compounds based on their chemical structure and physiochemical characteristics; recently, however, Brown and Grushka [1] formulated some rules for predicting retention behavior of these compounds. These rules are summarized as follows:

1. Substituents that cause the compound to exist mainly as the non-aromatic tautomer (the lactam or amine form) decrease the capacity ratio, k', of the compound.
2. Substituents that cause charge formation decrease the capacity ratio of the compound.
3. The group type and position of the substituent on the ring affect the retention characteristics, the order of the effect being OH < H < NH_2 < NHR. Both methyl and ribosyl groups approximately double the k' value when compared with the parent compound.
4. The replacement of a hydroxyl group with a hydrogen in the 2'-position of the ribosyl moiety (2'-deoxy compounds) results in an increase of the k' over the corresponding ribonucleoside.
5. The addition of a linear phosphate group dramatically decreases the k' over the corresponding ribonucleoside.
6. The addition of a 3',5'-cyclized phosphate increases the k' over that of the corresponding ribonucleoside.

A unique phenomenon occurring in aromatic compounds which contain hetero atoms is that of vertical associations through base stacking in aqueous solutions [41-48]. Brown and Grushka investigated this base stacking of purine and pyrimidine compounds in reversed-phase systems [1]. Unlike base pairing in RNA and DNA, where hydrogen bonding is the dominant force, the primary mechanism for base stacking is believed to be π electron overlap [41]. Stacking can be either heterogeneous, i.e., between a purine and a pyrimidine, or homogeneous, i.e., between the same type compounds. The order of stacking (as determined by the free energy and equilibrium constants of stacking) of nucleotides, nucleosides, and bases has been correlated to the order of retention [1]. It is not yet known whether stacking affects retention or whether the same factors that control retention also control stacking. However, since the concentrations of nucleosides and bases in physiological fluids are much lower than those in which the stacking parameters were determined, it appears that stacking does not affect retention but that the solvophobic factors involved in retention are also operative in stacking.

V. MOBILE PHASE EFFECTS

The reversed-phase high performance liquid chromatographic retention behavior of purine and pyrimidine based compounds is affected by many mobile phase characteristics. Some of the mobile phase parameters which have been investigated include pH, ionic strength, and percent organic modifier.

A. pH effects

The retention behavior of purines and pyrimidine bases is a function of pH (Table 1). It has been found that as the pH of the mobile phase is varied and the ratio of neutral to ionized species increases, the retention of that compound also increases.

The presence of an electron-withdrawing ribose moiety causes an increase in retention time; thus there is an increase in retention behavior of nucleosides over that of corresponding bases at any mobile phase pH.

While nucleoside and base retention mechanisms can be adequately explained in terms of solvophobic considerations, nucleotide retention behavior can best be explained in terms of a mixed-mode mechanism [49]. It has been observed that at a low pH, ribonucleotides elute in order of increasing negative charge, an elution pattern atypical for the reversed-phase mode. The elution can perhaps be explained in terms of silanophilic interactions in which nucleotide phosphate moieties interact with free surface silanols of the stationary phase.

B. Organic modifiers

When concentrations of an organic modifier are increased, the retention behavior of nucleosides and bases is typical of that expected based on solvophobic mechanisms; with increasing concentrations of organic modifier, solute retention is decreased. When a stronger eluent is used with reversed-phase conditions (e.g., acetonitrile vis-à-vis methanol) there is also a subsequent decrease in the capacity ratios.

In varying the concentration of organic modifier, the nucleotides again exhibit anomalous retention characteristics. At low pH values and increasing concentrations of organic modifier, the order of retention is monophosphate, diphosphate, and finally triphosphate ribonucleotides. This behavior is not in accordance with the solvophobic mechanism, since at low pH triphosphates, which have three ionized phosphates, elute after the monophosphates, which have one ionized phosphate group. Again this anomalous behavior may be described in terms of the silanophilic mechanism or, more likely, a mixed-mode retention mechanism is in effect.

Table 1 Ultraviolet and Ionization Data for Some Nucleic Acid Components

Compound	pK_1	pK_2	X_{max} (nm) Acidic	Neutral	Basic
Cyt	4.5	12.2	276	267	282
Cyd	4.2	12.5	280	270	273
Thy	–	9.9	265	265	291
Thd	–	9.7	267	267	267
Ade	4.2	9.8	263	261	269
Ado	3.5	12.5	257	260	259
Gua	3.2	9.6	248, 276	246, 276	274
Guo	1.6	9.2	256	254	256, 266
Hyp	2.0	8.9	248	250	263
Ino	1.2	8.8	248	249	253
AMP	3.7	6.6	257	260	259
ADP	3.9	7.2	257	260	259
ATP	4.0	7.7	257	260	259

It is interesting to note that the increase in concentration of organic solvents not only affects retention behavior, but also disrupts the associative process, thus causing destacking of the nucleic acid constituents in aqueous solutions.

C. Ionic strength

Changing the mobile phase ionic strength has little effect on the retention behavior of nucleosides and bases [8]. However, there is a definite effect of mobile phase ionic strength in the case of ribonucleotides, where the compounds are ionized under the mobile phase condition.

With increasing ionic strength the triphosphate ribonucleotides, which have the largest negative charge, have the greatest decrease in retention. In addition, ribonucleotides with protonated bases exhibit a smaller change in k' than those with neutral bases. Thus, it appears that there is a net decrease in the total ribonucleotide charge due to interaction between the protonated base and the negatively charge phosphate group.

REFERENCES

1. P. R. Brown, and E. Grushka, Anal. Chem., 52, 1210 (1980).
2. W. E. Cohn, Science, 109, 377 (1949).
3. C. F. Crompton, F. R. Frankel, A. M. Benson, and A. Wade, Anal. Biochem., 1, 249 (1960).
4. R. A. Hartwick and P. R. Brown, J. Chromatogr., 112, 651 (1975).
5. C. G. Horvath and S. R. Lipsky, Anal. Chem., 41, 1227 (1969).
6. M. Uziel, C. K. Koh, and W. E. Cohn, Anal. Biochem., 25, 77 (1968).
7. C. A. Burtis, M. N. Munk, and F. R. MacDonald, Clin. Chem., 16, 201 (1970).
8. R. A. Hartwick and P. R. Brown, J. Chromatogr., 126, 679 (1976).
9. R. A. Hartwick and P. R. Brown, J. Chromatogr., 143, 383 (1977).
10. R. A. Hartwick, A. M. Krstulovic, and P. R. Brown, J. Chromatogr., 186, 659 (1979).
11. L. Graeve, W. Goemann, P. Foldi, and J. Kruppa, Biochem. Biophys. Res. Commun., 107, 1559 (1982).
12. P. N. Nguyen, J. L. Bradley, and P. M. McGuire, J. Chromatogr., 236, 508 (1982).
13. F. Regnier, Abstract, International Symposium on Column Liquid Chromatography, Cherry Hill, N. J., June 1982.
14. N. E. Hoffman and J. C. Liao, Anal. Chem., 49, 2231 (1977).
15. J. B. Crowther, J. P. Coronia, and R. A. Hartwick, Anal. Biochem., 124, 65 (1982).
16. A. P. Halfpenny, P. R. Brown, and J. A. Korpi, Abstract #470, Pittsburgh Conference on Analytical Chemistry and Applied Spectroscopy, 1983.
17. J. H. Knox and A. Pyrde, J. Chromatogr., 112, 171 (1975).
18. C. H. Lochmuller and D. R. Wilder, J. Chromatogr. Sci., 17, 574 (1979).
19. R. P. W. Scott and P. Kucera, J. Chromatogr., 142, 213 (1979).
20. H. Colin and G. Guiochon, J. Chromatogr., 158, 183 (1978).
21. H. Colin, N. Ward, and G. Guiochon, J. Chromatogr., 149, 169 (1978).
22. T. Hanai and K. Fujmura, J. Chromatogr. Sci., 14, 140 (1976).
23. H. Hemetsberger, M. Kellerman, and H. Ricken, Chromatographia, 10, 276 (1977).
24. H. Hemetsberger, W. Maasfeld, and H. Ricken, Chromatographia, 9, 303 (1976).

25. J. J. Kirkland, J. Chromatogr. Sci., 10, 129 (1972).
26. R. E. Leitch and J. J. DeStafano, J. Chromatogr. Sci., 11, 105 (1975).
27. M. J. Telepchak, Chromatographia, 6, 234 (1973).
28. K. K. Unger, N. Becker, and P. Roumeliotis, J. Chromatogr., 125, 115 (1976).
29. R. Karch. I. Sebestian, I. Halasz, and H. Engelhardt, J. Chromatogr., 122, 171 (1976).
30. D. C. Locke, J. Chromatogr. Sci., 12, 433 (1974).
31. C. Horvath, W. Melander, and I. Molnar, J. Chromatogr., 125, 129 (1976).
32. B. L. Karger, J. R. Grant, A. Hartkopf, and P. H. Weiner, J. Chromatogr., 128, 65 (1976).
33. R. M. McCormick and B. L. Karger, J. Chromatogr., 119, 259 (1980).
34. C. Horvath, W. Melander, I. Molnar, and P. Molnar, Anal. Chem., 49, 2295 (1977).
35. W. R. Melander, D. E. Campbell, and C. Horvath, J. Chromatogr., 158, 215 (1978).
36. C. Horvath, W. Melander, and A. Nahum, J. Chromatogr., 186, 371 (1979).
37. K. E. Bij, C. Horvath, W. R. Melander, and A. Nahum, J. Chromatogr., 203, 65 (1981).
38. W. R. Melander and C. Horvath, J. Chromatogr., 201, 211 (1980).
39. A. Nahum and C. Horvath, J. Chromatogr., 203, 53 (1981).
40. J. A. Baker, R. E. Skeleton, and C. Y. Ma, J. Chromatogr., 168, 417 (1979).
41. A. D. Broom, M. P. Schweizer, and P. O. P. Ts'O, J. Am. Chem. Soc., 89, 3612 (1967).
42. M. P. Schweizer, S. I. Chan, and P. O. P. Ts'O, J. Am. Chem. Soc., 87, 5241 (1965).
43. T. N. Sollie and J. A. Schellman, J. Mol. Biol., 33, 61 (1968).
44. P. O. P. Ts'O, N. S. Kondo, R. K. Robbins, and A. D. Broom, J. Am. Chem. Soc., 91, 5625 (1969).
45. P. O. P. Ts'O, in Molecular Associations in Biology (B. Pullman, Ed.), Academic Press, New York, 1968, p. 59.
46. P. O. P. Ts'O and S. I. Chan, J. Am. Chem. Soc., 86, 4176 (1964).
47. P. O. P. Ts'O, I. S. Melvin, and C. Olsen, J. Am. Chem. Soc., 85, 1289 (1963).
48. C. Bugge, in The Jerusalem Symposium on Quantum Chemistry and Biochemistry, Israel Academy of Sciences and Humanities, Jerusalem, 1972, vol. IV, pp. 194-195.
49. M. Zakaria, P. R. Brown, and E. Grushka, Anal. Chem., 55, 457 (1983).

Part II
METHODOLOGY

3
Sample Preparation

PHYLLIS R. BROWN University of Rhode Island, Kingston, Rhode Island

   I. Introduction 31
  II. Deproteinization Methods 32
      A. Specific methods 33
      B. Effectiveness of methods 36
      C. Matrix effects 37
 III. Chromatographic Methods 39
      A. General discussion 39
      B. Specific methods 39
 IV. Sample Preparation Methods for Specific Biological Matrices 41
      A. Blood fluids 41
      B. Blood cells 41
      C. Urine and other physiological fluids 45
      D. Tissues 46
  V. Conclusions 47
      References 47

I. INTRODUCTION

Several factors must be considered in choosing a sample preparation technique. Since a high performance liquid chromatography (HPLC) analysis is very sensitive, often in the picomole range, the sample preparation technique chosen should be chemically clean and introduce minimal extraneous material. A total analysis is no better than the least accurate steps; thus the method should be simple and involve a minimum number of steps in which the sample is subjected to dilutions or loss of material. The method should also be rapid enough for a reasonable sample throughput and inexpensive enough for routine work. Some of these requirements may be less important than others in a particular analysis; however, the major goal of good sample preparation is to eliminate interfering compounds without greatly diluting the sample or altering the components of analytical interest.

II. DEPROTEINIZATION METHODS

In preparing biological samples prior to HPLC, proteins usually must be removed. Proteins and other macromolecules can interfere with the analysis of nucleic acid constituents and can cause serious problems by clogging the columns, thus decreasing reproducibility and column lifetime. Deproteinization methods commonly utilize the change in dielectric constant, temperature, ionic strength, or pH. In addition, proteins can be removed from a biological matrix by ultrafiltration, adsorption, or specific complexation (Table 1) [1]. Each technique has advantages and disadvantages, and the appropriate deproteinization method must be chosen for the problem at hand.

Table 1 Deproteinization Techniques

	Principles utilized	Examples
1.	Change in pH	Addition to sample of strong acid (e.g., TCA, PCA, HCl)
2.	Change in ionic strength	Addition to sample of salts [e.g., $(NH_4)_2SO_4$]
3.	Change in temperature	Heating of sample (denaturing and precipitating proteins)
4.	Change in dielectric constant	Addition of organic solvent (e.g., CH_3CN or C_2H_5OH)
5.	Specific adsorption	Use of precolumn (using silica)
6.	Specific complexation	Use of specific ligands in column (e.g., borate gels)
7.	Filtration	Use of membranes in ultrafiltration

A. Specific methods

1. Strong acids

The strong acid method for sample preparation is widely used to precipitate proteins in biological samples, especially prior to the chromatographic analysis of nucleotides. In this method, the strong acid is usually perchloric acid (PCA) or trichloracetic acid (TCA); however, other strong acids may also be used. Upon the addition of acid to biological fluids, essentially all the protein is precipitated. The acid must not remain in the solution, or hydrolysis of the nucleotides will result. Studies have been made to determine the best method of removing or neutralizing the acid [1-6]. With TCA, the traditional method has been extraction of the acid with diethyl ether. However, the multiple ether extractions which are required are time-consuming and can cause errors in quantitation. A method reported by Khym [4] has the advantages of simplicity, rapidity, and good recovery of nucleotides from standard solutions. In this method, a solution of trioctylamine in Freon is used to extract the acid from the sample solution. It was found that the concentration of the amine solution is critical, and that concentrated or aged amine solutions can seriously affect the results [3]. Moreover, when the efficiency of extraction of nucleotides from biological samples was investigated, it was found that recovery of nucleotides from the erythrocyte matrix was 20% lower than recovery from the plasma matrix and 40% lower than from aqueous solutions [6] (Table 2, Fig. 1). It was also found that recovery of nucleotides was lower in the presence of excess Freon, and that better recoveries were found at higher acid concentrations (i.e., 12% TCA rather than 6%), suggesting that hydrogen bonding may be involved in extraction efficiency [6].

Recoveries of nucleosides and bases from pooled human serum were poorer than for nucleotides, ranging from 65 to 70% [1]. Although the presence of excess amine in the solution did not affect the recovery of nucleosides, nucleosides were lost because of their solubility into the Freon and coprecipitation with the proteins.

When PCA is used instead of TCA, the acidic solution is usually neutralized by potassium hydroxide. Perchlorate salts then precipitate out. Because of the solubility of the perchlorate, the salt does not always precipitate completely, and the high salt concentration causes problems in the chromatographic step [7]. In addition, coprecipitation can cause loss in the recovery of the nucleic acid constituents [3,8].

2. Organic solvents

A traditional method of deproteinizing a sample prior to chromatographic analysis is by the addition of an organic solvent such as methanol,

Table 2 Matrix and Solute Structure Effects in Recoveries of Nucleotides

Matrix	Percentage recovery		
	CMP	AMP	GMP
Water	85.1 ± 2.9	78.9 ± 2.4	81.0 ± 3.4
Plasma	68.7 ± 1.4	66.2 ± 1.9	59.1 ± 1.5
Erythrocytes	43.0 ± 2.1	37.1 ± 2.5	39.0 ± 2.7

Matrix	Percentage recovery		
	GMP	GDP	GTP
Water	81.0 ± 3.4	77.6 ± 5.0	72.2 ± 6.1
Plasma	59.1 ± 1.5	64.4 ± 1.4	60.4 ± 3.5
Erythrocytes	39.0 ± 2.7	32.9 ± 3.2	23.0 ± 2.3

Source: Ref. 6.

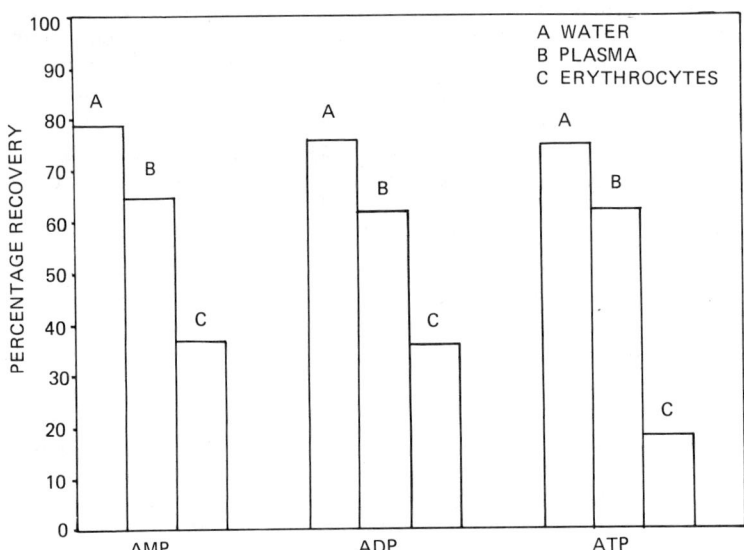

Figure 1. Recoveries of nucleotides (AMP, ADP, ATP) from water, plasma, and erythrocytes (6% TCA). Bar graph of the percent recoveries of added AMP, ADP, and ATP to water (bar A), plasma (bar B), and erythrocytes (bar C), showing the effect of the number of phosphate groups on the percent recovery. (From Ref. 6.)

ethanol, or acetonitrile [9,10]. Because of the presence of the organic solvent, the dielectric constant of the solution decreases, causing the larger proteins to agglomerate. This method is especially useful if the compounds of interest are preferentially soluble in organic solvents; for example, for simple and rapid analyses of certain drugs, proteins are precipitated by an organic solvent in which the drug is very soluble. Thus, extraction of the drug into an organic solvent occurs simultaneously with protein precipitation. However, for many nucleosides and bases, and especially for nucleotides, this method is not useful because of low solubility of nucleic acid constituents in organic solvents.

3. *Salts*

Use of ammonium sulfate, or the "salting out" technique, has been used over the years to remove proteins from biological samples [9,10]. The technique is chemically mild and is the method used when recovery and/or reactivation of the enzymes precipitated is necessary. Essentially 100% recoveries of the nucleosides and bases are obtained with serum samples. A disadvantage of this method is that there is the possibility of contamination of the sample by the added salts. These salts can cause loss of resolution. In addition, sample dilution can cause a significant decrease in sensitivity.

For this procedure, an equal volume of saturated ammonium sulfate solution is added to the serum sample. The samples are vortexed for 2 min, centrifuged at 1140 RCF for 10 min and then filtered through a millipore membrane filter, Type GS (Millipore Corp., Bedford, Mass). The filtered samples are placed in plastic vials and frozen until HPLC analysis [1].

4. *Temperature*

An increase in temperature can cause rupture of the tertiary structure of the proteins and subsequent precipitation of proteins in biological samples. However, heat is not commonly used because a gel is formed in many biological samples, particularly serum. Thus the serum sample is not in a suitable form for quantitative analysis. Although deproteinization by heating in a water bath can be a useful sample preparation technique for the analysis of blood cells, the strong acid technique is more versatile. In addition, since many biological compounds are thermally labile, inaccurate results may be obtained unless proper precautions are taken by analyzing parallel blanks.

5. *Ultrafiltration*

Ultrafiltration is a sample procedure which provides excellent recovery of free nucleosides and their bases [1,11-13]. With this method, only concentrations of free constituents are determined; thus total concentration (free plus protein bound) of the compound cannot be evaluated. For example, the concentrations of tryptophan in serum samples prepared by ultrafiltration are much lower than those obtained with strong acid, since a large proportion of the tryptophan in serum is bound to proteins. In addition, it was found that the recovery of a compound such as theophylline is low because of adsorption onto the ultrafiltration cones. The advantages of ultrafiltration are that the sample is not diluted, and compounds which might interfere with the HPLC analysis are not added to the sample. In addition, multiple samples can be processed rapidly and efficiently. In this method about 1-ml samples of serum or plasma are filtered directly through ultrafiltration cones (with a nominal molecular weight of 25,000) and centrifuged for 15 min at a speed of 500 RCF [1].

6. Precolumns

A method now gaining popularity is concentration of solutes by the use of precolumns or cartridges specially designed to retain the nucleotides and/or nucleosides and bases and to elute the proteins. On changing the solvent, the compounds of interest can then be eluted and separated. For example, prior to a reversed-phase liquid chromatography (RPLC) determination of nucleosides and bases in plasma, the sample can be loaded onto a C_{18} cartridge (e.g., Sep-Pak C_{18}[4] cartridge) and washed with a primary wash of water. The cartridge is washed twice with about 0.5 ml of 2.5% CH_3OH to remove all proteins and other nonretained compounds. The nucleosides and bases are then removed by rinsing the cartridge twice with 100% methanol. The samples are freeze-dried and reconstituted in the RPLC buffer immediately before analysis. An advantage of this method is that the solutes of interest are concentrated by selective adsorption, thus increasing sensitivity. In addition, no interferences are introduced. The technique is rapid, and there are no volume changes. However, protein-bound components are ot recovered unless a low-pH solution is used, since only highly acidic solutions will hydrolyze a protein-substrate bond and free the substrates.

B. Effectiveness of methods

The effectiveness of the sample preparation procedures is evaluated in terms of: (a) efficiency of protein removal; and (b) efficiency of recovery of the compounds of interest from the biological matrix.

Hartwick et al. [1] examined the relative efficiency of the use of strong acid, ultrafiltration, sample cartridges, and ammonium sulfate for protein removal from serum. Except for the ammonium sulfate procedure, in which only 65% of the protein was removed, the other methods examined removed essentially all the serum protein (Table 3).

Table 3 Efficiency of Deproteinization Method[a]

Method	Percent efficiency
TCA	100
$(NH_4)_2SO_4$	(65.5 ± 2.1)[b]
Ultrafiltration	100
Precolumn concentration	100[c]

[a] Burret method of protein assay.
[b] 95% confidence level.
[c] Dependent on elution conditions.
Source: Ref. 11.

Upon evaluating the recoveries of compounds added to pooled human serum, it was found that the efficiency of recovery for the majority of nucleosides and bases was between 70 and 100% for the ammonium sulfate, ultrafiltration, and precolumn methods (Table 4). However, with ultrafiltration and the precolumn methods, it was found that the recovery of xanthosine and theophylline was pH-dependent. Using the strong acid technique, the recoveries of nucleosides and bases were only between 50 and 59% using 6% TCA and between 60 and 69% using 12% TCA [1].

C. Matrix effects

In assays of nucleic acid constituents in biological samples, it is extremely important to know how the steps prior to the HPLC of the sample affect the accuracy of the measurements. In other words, is one accurately measuring the concentration of the constituent or constituents of interest, or is there a loss in recovery during the deproteinization and/or extraction steps? The traditional determination is to add known concentrations of the compounds of interest to the sample (prior to sample preparation) and determine the percentage of recovery. This procedure will give an indication of efficiency of recovery and matrix effects. However, it is possible to determine the recovery of *only* the added compound; the percentage of recovery of the endogenous compound cannot really be determined because the endogenous concentrations or factors affecting the removal of endogenous concentrations are not actually known.

Although exogenous nucleosides and bases can be recovered almost quantitatively from serum [1], it was found by Van Haverbeke and Brown [6] that the type of matrix made a tremendous difference in

Table 4 Recoveries of Compounds Added to Pooled Human Serum[a]

Compound	TCA at: 6%	TCA at: 12%	(NH$_4$)$_2$SO$_4$	Ultrafiltration at: pH 7.8	Ultrafiltration at: pH 5.1	Precolumn
Xao	58.8 ± 6.2	66.4 ± 1.0	97.4 ± 7.7	99.3 ± 2.5	74.8 ± 4.2	b
Ino	59.4 ± 7.7	65.9 ± 2.7	95.6 ± 7.6	98.9 ± 2.2	97.7 ± 1.7	101. ± 2.85
Guo	54.0 ± 4.7	67.6 ± 0.3	84.8 ± 9.6	73.6 ± 3.9	85.9 ± 6.4	92.0 ± 5.32
Trp	—	—	102. ± 12.6	12.1 ± 2.6	96.2 ± 3.7	43.3 ± 2.15
Thp	—	—	82.7 ± 5.1	40.7 ± 5.0	85.5 ± 4.7	91.2 ± 4.8
Thb	—	—	72.1 ± 7.5	81.3 ± 3.2	83.2 ± 4.3	90.2 ± 5.4
Caf	—	—	88.9 ± 6.7	83.0 ± 9.1	87.7 ± 4.5	92.8 ± 5.7

[a] All ranges are reported at the 95% confidence level.
[b] Highly variable recovery, pH-dependent.
Source: Ref. 11.

recovery of nucleotides. Using the strong acid method, 80% of the nucleotides were recovered from an aqueous solution, while the recovery from plasma was only about 60% and from erythrocytes less than 40%. In addition to the matrix effect, it was shown that the structures of the solute also affected efficiency of recovery and that percent recovery decreased with increasing number of phosphate groups (mono > di > triphosphate nucleotides) (Fig. 1). Base structure also affected efficiency of recovery although to a lesser degree. Riss et al. [8] investigated the recovery of 12 nucleotides from liver samples and isolated hepocytes using three different extraction procedures. Although they did not calculate percent recovery, they found that the amount of material recovered was similar using the TCA-amine Freon and the PCA-KOH methods and was very much lower when a charcoal-adsorption sample preparation technique was used. They found a significant amount of breakdown of the triphosphate to the diphosphate nucleotide and a minor amount of breakdown of the diphosphate to the monophosphate nucleotide using all three procedures.

III. CHROMATOGRAPHIC METHODS

A. General discussion

Chromatographic techniques may also be used to prepare biological samples for HPLC analysis. If relatively large amounts of material are needed, e.g., for further identification of the peaks by other methods, the chromatography may be carried out on large open columns. A newer technique is the use of column switching within an HPLC system. In these methods, either the group of compounds of interest is retained while all other compounds are eluted, or, the compounds of interest are eluted and all other compounds are retained. Although any mode of chromatography can be used in sample preparation, the modes that have been particularly useful in the analysis of biological samples for nucleic acid constituents are exclusion, ion-exchange, partition, and affinity chromatography. Since the use of partition has been discussed in the section on deproteinization techniques under precolumns, only the other three chromatographic modes will be discussed in this section.

B. Specific methods

1. Exclusion chromatography

In exclusion chromatography, the molecules are separated according to size (and sometimes shape). Thus in crude biological samples containing molecules in a large range of sizes, the compounds can be separated according to molecular weight with columns filled with gel permeation packings (e.g., Bondagel, Styrogel) [14]. Since pore sizes can vary greatly, it is possible to separate by groups the nucleic acids and other large-molecular-weight compounds such as proteins, polypeptides, and oligonucleotides from the nucleotides, nucleosides, and their bases along with other low-molecular-weight compounds. In this case, the larger molecules will be eluted first and the smaller ones last. Each fraction is collected and can then be analyzed by the appropriate mode (partition or ion exchange) to separate the individual compounds present in that group. This method is gentle and does not alter either the compounds retained or those eluted. Therefore, the biological activities of any of the compounds can be determined by the appropriate bioassays. In addition, the recycling technique can be used to improve the separation of the peaks which overlap or are poorly resolved [14].

2. Ion-exchange chromatography

Ion-exchange packings can be used both in open columns outside the chromatograph or with a precolumn to remove the proteins in the chromatograph using the column switching technique. Charged compounds such as nucleotides are selectively retained, while the non-charge compounds such as nucleosides and bases are eluted without retention. The nucleosides and bases are then separated using the reversed-phase mode. The retained nucleotides can then be eluted by changing the mobile phase, e.g., using an eluent of stronger ionic strength, or the individual nucleotides in the collected fraction can then be separated using gradient elution [7,15].

3. Affinity chromatography

Affinity chromatography is a type of adsorption chromatography in which the stationary phase or adsorbent has a selective affinity for a solute or class of solutes [16,17]. This principle can be used to retain selectively a compound or compounds of interest. Affinity chromatography includes both bioselective adsorption as well as chemically selective adsorption. Bioselective adsorption, a technique used widely in enzymology, utilizes an immobilized bioligand. A crude extract is passed through a column support to which the bioligand is attached. The biologically active molecules are retained on the column as substrate-ligand complexes. Other compounds which do not have an affinity for the ligands will be eluted with buffer. Subsequently, the desired substrates are eluted by changing the conditions (pH, ionic strength, temperature, etc.) [17].

In nucleic acid research, proteins which recognize and bond with nucleotides or nucleic acids can be used as ligands to selectively retain nucleotides in cell mixtures. Conversely nucleotides can be the immobilized as ligands to retain certain kinases [17]. In addition, coenzymes containing nucleotides such as NAD^+ + $NADP^+$ are also used widely as ligands for the separation of dehydrogenases [17].

An important chemically selective adsorption method used in the analyses of nucleosides is the borate column developed by Uziel et al. [18]. They utilized the selective complexation of cis-diols with $B(OH)_3$ molecules in polyacrylamideborate gel columns for the specific retention of ribonucleosides. All other nucleic acid fragments as well as other constituents of the sample are eluted. Gehrke and coworkers used this gel extensively to determine very low concentrations of ribonucleosides and their methylated analogs in urine and other biological fluids [16,79].

Group-specific sample preparation methods reduce the demands made on the chromatographic system. In addition to protecting the analytical column, the solute of interest can be concentrated, thus increasing the sensitivity of the assay.

IV. SAMPLE PREPARATION METHODS FOR SPECIFIC BIOLOGICAL MATRICES

A. Blood fluids

In the preparation of blood fluids (serum or plasma) for HPLC, the fluids must first be separated from the blood cells. In the case of serum, where no anticoagulant is used, the blood is allowed to stand for 20 min and clot naturally. The sample is then centrifuged at 1145 RCF for 15 min and the supernatant removed rapidly to prevent leakage of nucleosides or bases into the serum. The serum is then deproteinized and frozen until HPLC analysis.

For plasma, it was found that the best anticoagulant to use was heparin, since ethylenediamine tetraacetate (EDTA) and acid-citrate dextrose (ACD) solutions both contained ultraviolet-absorbing impurities which interferred with the nucleoside and base analysis (Figs. 2 and 3). However, the heparinized plasma must be separated immediately from the red blood cells to prevent increases in hypoxanthine and/or xanthine levels caused by a breakdown of ATP in the erythrocytes (Fig. 4) and subsequent transport of the bases across the cell membrane into the plasma [20]. The procedure involves drawing blood samples into vacutainer tubes containing heparin. After the samples are allowed to stand for 20 min at room temperature, they are centrifuged for 20 min at 1145 RCF. The plasma is removed from the intact blood cells and centrifuged through ultrafiltration cones for 20 min at 1000 RCF. The deproteinized ultrafiltrates are stored at -20°C until HPLC analysis [20].

B. Blood cells

To determine the nucleotide concentrations in the formed elements of blood, the total cell population must first be separated from the plasma and then each type of cell cleanly separated from the others. The

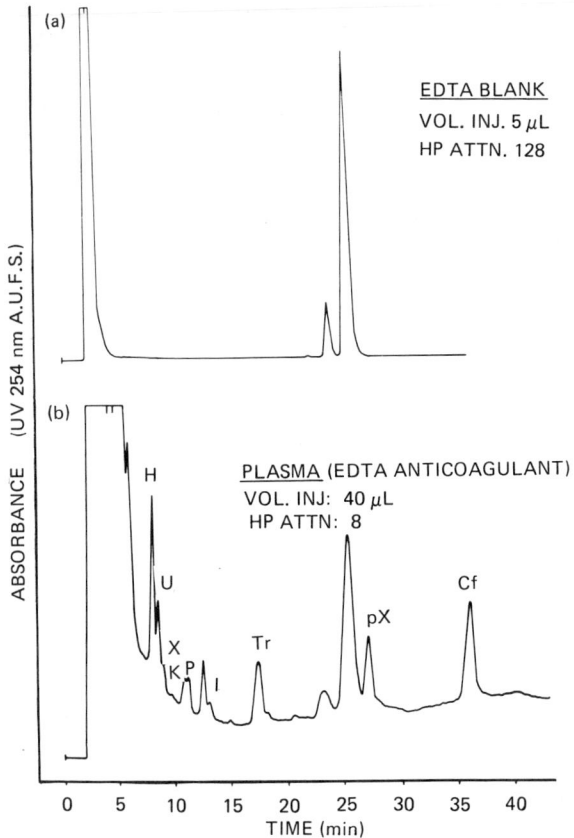

Figure 2. Impurities in the anticoagulant, EDTA. (a) Chromatogram of the EDTA blank. Injection volume 5 μl, integrator attenuation 128. (b) Chromatogram of a plasma sample obtained from blood collected in EDTA. Injection volume 40 μl; integrator attenuation 8; UV detector at 254 nm; column, Whatman ODS-3 (10 μm); guard column, Whatman CoPell ODS packing. Eluents: A = 0.02 M KH_2PO_4, pH 5.7; B = 3:2 methanol:water. Gradient: linear, 0-40% B in 35 min. Flow rate 1.5 ml/min. H, hypoxanthine; U. uridine; X, xanthine; K. L-kynurenine; P, L-phenylalanine; I, inosine; Tr, L-tryptophan; pX, paraxanthine; Cf, caffeine. (From Ref. 20.)

nucleotide concentrations of each type of cell (i.e., erythrocytes, leukocytes, and platelets) are unique [21]. Erroneous determinations of nucleotide concentrations may be made if there is contamination in any one fraction by other cells. Although the amounts of nucleosides and bases in blood fluids are commonly stated in molar terms (or

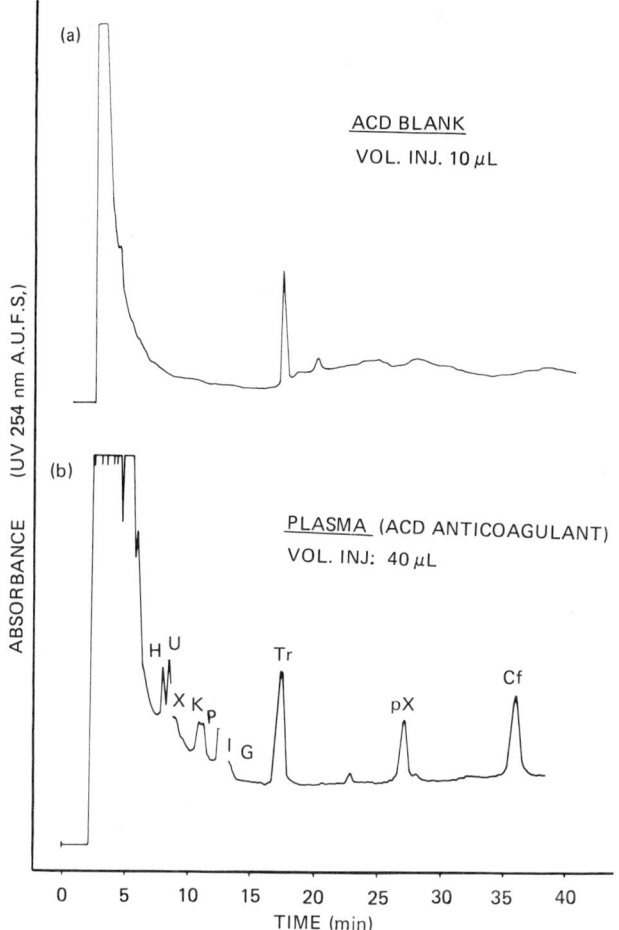

Figure 3. Impurities in the anticoagulant, ACD. (a) Chromatogram of the ACD blank. Injection volume 5 µl, integrator attenuation 128. (b) Chromatogram of a plasma sample obtained from blood collected in ACD. Injection volume 40 µl; integrator attenuation 8; UV detector at 254 nm; column, Whatman ODS-3 (10 µm); guard column, Whatman CoPell ODS packing. Eluents: A = 0.02 M KH_2PO_4, pH 5.7; B = 3/2 methanol/water. Gradient: linear, 0-40% B in 35 min. Flow rate: 1.5 ml/min. Peaks: H, hypoxanthine; U, uridine; X, xanthine; K, L-kynurenine; P, L-phenylalanine; I, inosine; Tr, L-tryptophan; pX, paraxanthine; Cf, caffeine. (From Ref. 20.)

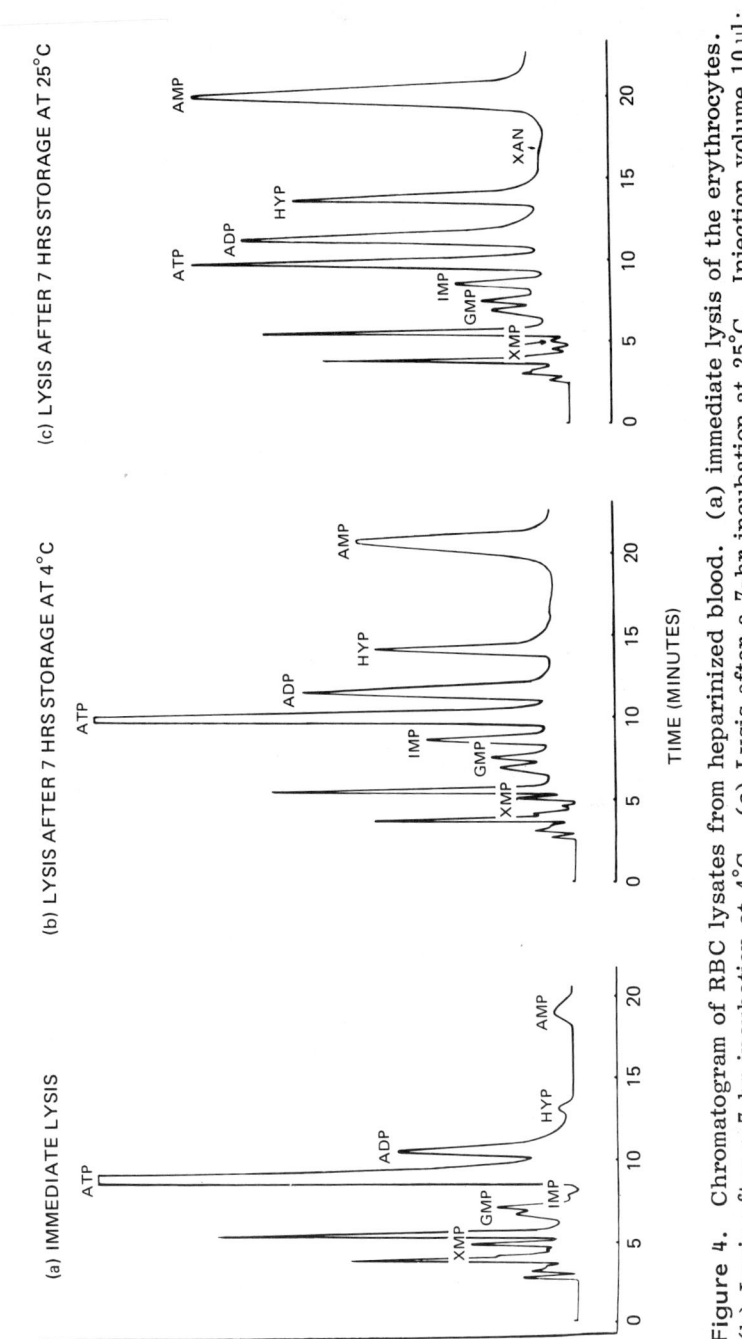

Figure 4. Chromatogram of RBC lysates from heparinized blood. (a) immediate lysis of the erythrocytes. (b) Lysis after a 7-hr incubation at 4°C. (c) Lysis after a 7-hr incubation at 25°C. Injection volume 10 μl; UV detector at 254 nm; column, Whatman ODS-3 (10 μm); guard column, Whatman CoPell ODS packing. Eluent: 0.06 M K_2HPO_4 (pH 6.0) isocratic. Flow rate: 1.0 ml/min. (From Ref. 20.)

picomolar or nanomolar, etc.), amounts of nucleotides in cells are generally given in weight per number of cells, e.g., grams per number of leukocytes. Thus, a careful count of the number of cells is necessary. For a total leukocytic determination this may be difficult, because leukocytes are a mixed population of cells of different sizes. In cells, the nucleotides are predominant and their bases and nucleosides are present proportionally in very small amounts. The opposite is true in blood fluids. Thus, the general procedure is to analyze blood fluids for their nucleosides and bases [11,12,22,23] and blood cells for their nucleotides [7,8,24-26]. However, improved procedures are making possible the determination of the low levels of nucleotides in serum and plasma [27] as well as of nucleosides and bases in the formed elements of blood [17].

C. Urine and other physiological fluids

Since urine is the "garbage can" of the body and material not used by the body is excreted in it, the urine contains, along with proteins and other high-molecular-weight compounds, literally hundreds of low-molecular-weight, ultraviolet-absorbing compounds.

Thus the selective removal of ribonucleosides is particularly useful when urine is the matrix [16,19]. Precolumn concentration can also be utilized for other groups of compounds because large volumes of urine are generally available (24-hr samples), and the compounds of interest may be present in very low concentrations. Therefore, with precolumn concentration not only can substituents which interfere with the HPLC assay be removed, but also the concentration of the compound(s) of interest be increased, thus improving the sensitivity of the analysis.

Although patients with severe kidney disease produce little or no urine, they do retain large quantities of water-containing compounds, which are normally excreted, and a wide variety of toxins. Patients with little or no kidney function are treated by hemodialysis. In studies of kidney disease, hemodialysates both before and after dialysis have been investigated using gas chromatography and HPLC. The sample preparation prior to HPLC is simple and fast. Veening et al. [28,29] collected 200 ml of "used" hemodialysate and filtered the sample through a 0.2-µm filter. The sample was then concentrated by distillation at 25°C and 1 mm pressure to a volume of 25 ml.

Saliva is another physiological fluid which has been used recently to monitor the purine analogs, caffeine [30] and theophylline [31-33], and their metabolites by HPLC. Nakano et al. [34] found that saliva can be injected directly without any sample preparation if a precolumn is used. However, the system must be cleaned frequently with 100% methanol, and the precolumn must be changed regularly. Therefore, if a large number of samples are to be analyzed, they suggested that the samples be centrifuged first and aliquots from the upper layers injected into the HPLC. There is little difference in the percent recovery when saliva is analyzed by direct injection, injection of the upper layer, or injection after ultrafiltration (procedure as described under blood fluids).

D. Tissues

Solid samples from tissues, tumors, or tissue culture must be carefully collected to prevent rapid degradation of nucleotides by the enzymes present. Rapid in vitro freezing of the samples is usually used to keep the nucleotide pools intact. Another problem in working with tissues is that of completely extracting the nucleotides from the solid sample.

In order to obtain a large surface area, Juengling and Kammerheier [35] used a precooled mortar to pulverize samples of cardiac tissue. They then extracted the nucleotides with PCA. After centrifugation, the supernatant fluid was neutralized with KOH and the perchlorate salts filtered off.

To avoid anoxia, Riss et al. [5] used the freeze-clamp method of Bucher and Swaffield [36] to obtain liver tissue. Using a mortar and pestle, they ground the sample to a powder under liquid nitrogen and kept the pulverized sample in the liquid nitrogen while extracting 0.5 to 1.0 g of the powder with PCA. The solution was neutralized with KOH. They also took the same amount of frozen liver powder or 20 to 30 million hepatocytes obtained from frozen liver slices and homogenized them with a Potter-Elvensem homogenizer in three volumes of 10% TCA at 4°C. They reextracted the resulting pellet twice with 20% TCA and removed the TCA from the combined supernatants with an alamine-Freon solution.

Theen et al. [37] studied procedures for processing liver tissue and developed small liquid nitrogen-cooled clamps to obtain liver samples of about 300 mg. Their procedures minimized phosphorylase changes.

Tissue cultured cells can be extracted with cold acid, or solutions of cells can be loaded directly onto precolumns or preparative cartridges. Another possible sample preparation technique is the sonication of the sample in strong acid, followed by neutralization or removal of the acid [38].

In bacteria, Lunden and Thore [39] compared 10 methods of extraction of adenine nucleotides. These methods included the following: Tris-EDTA, KOH, Tris-arsenate-EDTA-N-butanol solution (TAEB), ethanol, butanol, chloroform, formic acid, sulfuric acid, TCA, and PCA. They found that TCA gave the highest recovery of the nucleotides. This method was reproducible and the nucleotides were found to be stable in the neutralized TCA extract to which EDTA had been added.

V. CONCLUSIONS

In conclusion, each sample preparation method has advantages and disadvantages. Presently, either the precolumn cartridge or the ultrafiltration method appears to be the most useful for the routine analysis of nucleosides together with their bases in serum or plasma. In urine, Gehrke's selective method for nucleosides has advantages because of the tremendous number of low-molecular-weight, ultraviolet-absorbing compounds in this matrix.

For nucleotides in blood cells or solid samples, the sample preparation procedure must be chosen and optimized for each specific problem. The factors to be considered are the prompt deactivation of the enzymes, efficiency of deproteinization, and percent of the nucleotides of interest extracted. The use of strong acid is still the method most commonly used and best meets, at present, the necessary criteria for optimal sample preparation.

REFERENCES

1. R. A. Hartwick, D. Van Haverbeke, M. McKeag, and P. R. Brown, J. Liq. Chromatogr., 2(5), 125 (1979).
2. P. R. Brown and R. P. Miech, Anal. Chem., 44, 1072 (1972).
3. S. C. Chen, P. R. Brown, and D. M. Rosie, J. Chromatogr. Sci., 15, 218 (1977).
4. J. X. Khym, Clin. Chem., 21, 1245 (1975).
5. T. L. Riss, N. L. Zorich, M. D. Williams, and A. Richardson, J. Liq. Chromatogr., 3(1), 133 (1980).
6. D. A. Van Haverbeke and P. R. Brown, J. Liq. Chromatogr., 1(4), 507 (1978).
7. P. R. Brown, J. Chromatogr., 52, 257 (1970).
8. G. H. Rao, J. D. Peller, J. G. White, J. Chromatogr., 226(2), 466 (1981).
9. W. C. Hutchinson and H. N. Munro, Analyst, 86, 768 (1961).
10. H. N. Munro and A. Fleck, Methods Biochem. Anal., 14, 113 (1964).
11. R. A. Hartwick, S. P. Assenza, and P. R. Brown, J. Chromatogr., 186, 647 (1979).
12. R. A. Hartwick, A. M. Krstulovic, and P. R. Brown, J. Chromatogr., 186, 659 (1978).
13. R. A. Hartwick and P. R. Brown, CRC Rev., 10, 279 (1981).
14. L. R. Snyder and J. J. Kirkland, *Introduction to Modern Liquid Chromatography*, Wiley-Interscience, New York, 1974.
15. C. G. Horvath, B. A. Preiss, and S. R. Lipsky, Anal. Chem., 39, 1422 (1970).

16. C. W. Gehrke, K. C. Kuo, G. E. Davis, R. D. Suits, P. T. Waalkes, and E. Borek, J. Chromatogr., *150*, 445 (1978).
17. W. H. Scouten, *Affinity Chromatography: Bioselective Adsorption on Inert Matrices*, Wiley-Interscience, New York, 1981.
18. M. Uziel, L. H. Smith, and S. A. Taylor, Clin. Chem., *22*, 1451 (1976).
19. G. E. David, R. D. Suits, K. C. Kuo, C. W. Gehrke, T. P. Waalkes, and E. Borek, Clin. Chem., *23*(8), 1427 (1977).
20. M. Zakaria and P. R. Brown, Anal. Biochem., *120*, 25 (1982).
21. E. S. Scholar, P. R. Brown, R. E. Parks, Jr., and P. Calabresi, Blood, *41*, 927 (1973).
22. P. R. Brown, S. Bobick, and F. L. Hanley, J. Chromatogr., *99*, 587 (1974).
23. R. A. Hartwick and P. R. Brown, J. Chromatogr., *126*, 679 (1976).
24. R. A. DeAbreu, J. M. Van Baal, J. A. Bakkeren, C. H. DeBruyn, and D. A. Egbert, J. Chromatogr., *227*, 45 (1982).
25. R. A. Hartwick and P. R. Brown, J. Chromatogr., *112*, 651 (1975).
26. G. H. Rao, J. D. Peller, K. L. Richards, J. McCullough, and J. G. White, J. Chromatogr., *229*(1), 205 (1982).
27. R. A. DeAbreu, J. M. Van Baal, C. H. DeBruyn, J. A. Bakkeren, and D. A. Egbert, J. Chromatogr., *229*(1), 67 (1982).
28. H. Veening, in *Biological/Biomedical Applications of Liquid Chromatography II*, *12* (G. L. Hawk et al., Eds.) 93 (1979).
29. E. J. Knudson, Y. C. Lau, H. Veening, and D. A. Dayton, Clin. Chem., *24*, 686 (1978).
30. C. E. Cook, C. R. Tallent, E. W. Amerson, M. W. Meyers, J. A. Kepler, G. F. Taylor, and H. D. Christensen, J. Pharmacol. Exp. Ther., *199*, 679 (1976).
31. G. Levy, E. F. Ellis, and R. Koysvoko, Pediatrics, *53*, 873 (1974).
32. J. J. Orcutt, P. P. Kozak, S. A. Gillman, and L. H. Cummins, Clin. Chem., *23*, 599 (1977).
33. J. R. Miksic and B. Hodes, J. Pharm. Sci., *68*, 1200 (1979).
34. K. Nakano, S. P. Assenza, and P. R. Brown, J. Chromatogr. Biomed. Appl., *233*, 51 (1982).
35. J. Juengling and H. Kammerheier, Anal. Biochem., *102*, 358-361 (1980).
36. N. L. Bucher and M. N. Swaffield, Biochem. Biophys. Acta, *129*, 445 (1966).
37. J. Theen, D. P. Gilboe, and F. Q. Nuttall, Am. J. Physiol., *243*(3), E182 (1982).
38. R. J. Simmonds and R. A. Harkness, J. Chromatogr., *226*(2), 369 (1981).
39. A. Lundin and A. Thore, Appl. Microbiol., *30*(5), 713 (1975).

4
Chromatography

PHYLLIS R. BROWN University of Rhode Island, Kingston, Rhode Island

I. Basic Principles 49
   A. Description 49
   B. Types of HPLC separation 50
   C. Terminology and equations 54
II. Instrumentation 60
   A. Injectors 61
   B. Pumps 62
   C. Automation 62
III. Operating Conditions 63
   A. Stationary phase 63
   B. Mobile phase 64
   C. Temperature 67
IV. Optimization and Problem Solving 70
   A. Choice of chromatographic mode 73
   B. Choice of stationary and mobile phases 76
   References 79

I. BASIC PRINCIPLES

A. Description

Chromatography is a separation technique in which components of a mixture are selectively retained by a stationary phase. The process is based on differential rates of migration over a bed or through a column,

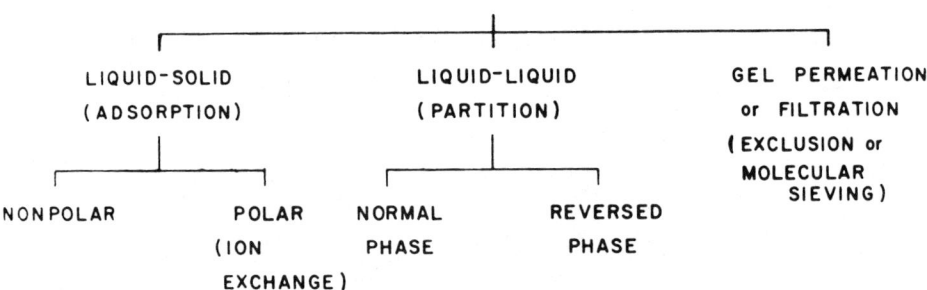

Figure 1. Classification of chromatographic modes.

since each substituent in mixture interacts differently with its environment under the same conditions. Thus separations of solutes are the result of differences in equilibrium distribution between two immiscible phases; a moving (or mobile) phase and a nonmoving (or stationary) phase. In order to move through the stationary phase, the sample constituents must be in the mobile phase. When the substituents favor the stationary phase, i.e., are more strongly "held" by the column, they move slowly. Those substituents which are not strongly attracted to the stationary phase favor the mobile phase and move more rapidly through the column. Thus the velocities of the substituents are different and compounds in the mixture are separated.

In liquid chromatography the mobile phase is a liquid, and high performance (or high pressure, high speed, high flow, etc.) liquid chromatography (HPLC) is defined as that process in which the stationary phase is in a column and the mobile phase is pumped through the column at inlet pressures up to 6000 psi.

B. Types of HPLC separations

In chromatography, methods are classified according to the types of mechanism of retention: adsorption, partition, ion exchange, or molecular sieving. The methods are also classified according to the stationary and mobile phases, with the mobile phase always being named first (Fig. 1). Thus, there is gas-solid (GS), gas-liquid (GL), liquid-solid (LS) and liquid-liquid (LL) chromatography. In this book, we will be concerned only with the two latter types of chromatography, which are both used in HPLC.

Whenever the stationary phase is a solid, the mechanism of retention is ostensibly adsorption. When it is a liquid, the mechanism is thought to be partition; however, because the liquid stationary phase is bonded to a solid support, which is not completely covered, adsorption (and possible exclusion) may be operative in addition to the partition mechanism of retention.

1. Liquid-solid chromatography (LSC)

Although liquid-solid chromatography using a silica stationary phase was used extensively in the early days of HPLC [1-5], this type of HPLC separation is not commonly used at present. However, for the HPLC of nucleotides, ion exchange (IE) a subclass of LS chromatography, is used in which an ionic material or polyelectrolyte is bonded to a silica solid support. The ion-exchange material has fixed charges. Associated with each fixed charge is a mobile counterion of the opposite charge. There are two major types of ion exchangers: cation and anion exchangers. In cation exchange, the fixed ion is an anion such as a sulfonic acid group with an associated positive counterion such as a sodium or potassium ion. In anion exchange, the fixed ion is a cation, generally a quarternary ammonium ion, with a negative counterion such as chloride ion. The ionic solutes to be separated take the place of the counterion of the same charge on the column and are retained by the ions of the opposite charge which are fixed on the stationary phase via electrostatic forces (Fig. 2). Solutes pass through the column at different rates since each type of solute has a unique interaction with the fixed ion. Although the major mechanism of retention is by ionic forces, adsorption on the matrix and molecular sieving may play a role in the separation. This technique is used mainly for ionic or ionizable solutes such as nucleotides, since compounds which either are not ionic or cannot be converted into ionic moieties are poorly retained. Thus, in the analysis of nucleic acid constituents, mainly the nucleotides [6-8], oligonucleotides, and nucleic acids [9] are separated by ion exchange. However, using the appropriate pH, nucleosides and bases can also be separated by ion exchange [10]. The advantages of ion exchange are as follows:

1. Ionized or ionizable compounds are well retained and can be readily separated.
2. Good ion-exchange packings are commericially available which are stable and do not quickly deteriorate.

The limitations are these:

1. Solutes must have an ionizable functional group.
2. Salts are used in the mobile phase which can be difficult to remove.

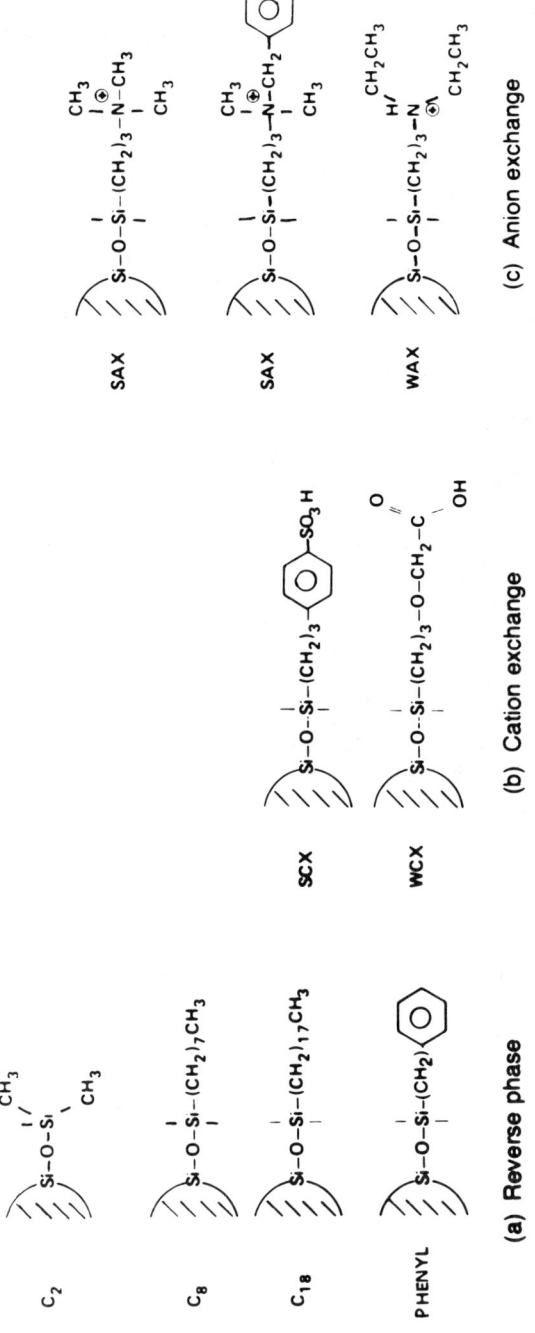

Figure 2. Packings for HPLC. (a) Reversed-phase packings. (b) Cation-exchange packings. (c) Anion-exchange packings. (Courtesy of Varian Associates, Walnut Creek, Calif.)

2. Liquid-liquid chromatography (LLC)

In LLC the solutes are partitioned between two liquid phases: the "liquid" stationary phase and the liquid mobile phase. In HPLC, the "liquid" stationary phase is usually chemically bonded to a solid stationary support such as silica, although some polymeric resins are also used. There are two types or modes of LLC. In chromatography, the term the "normal phase" is used when the stationary phase is polar and the mobile phase is nonpolar. For example, a polar "liquid" such as β,β'-oxydiproprionitrile is coated onto or bonded to a solid support. With this type of column, a nonpolar eluent such as n-heptane is used. The other type of LLC separation is called "reversed-phase liquid chromatography" (RPLC), in which the stationary phase, a nonpolar "liquid" stationary phase such as an alkane, is bonded to a support (Fig. 2). In this mode, the moving phase is usually a polar liquid such as water, methanol, or acetonitrile. However, these particular solvents are not essential to a RPLC separation. In recent literature, any separation in which the mobile phase is *more* polar than the stationary phase is referred to as a reversed-phase separation. It is the most universal mode of HPLC and thus widely used. Approximately 80% of all HPLC separations are done by the reversed-phase mode. Among the many advantages of RPLC are the following:

1. Water-soluble solutes can be readily separated by RPLC; thus it is much more useful in biochemical and biomedical studies, since most of the compounds of interest are water-soluble.
2. It is an excellent technique for nonpolar and moderately polar solutes.
3. Highly polar and ionized compounds can also be separated using techniques such as ion suppression, ion pairing, and ligand exchange.
4. A broad spectrum of closely related and widely different compounds can be analyzed simultaneously.
5. It is easy to use, and the columns are stable and highly efficient.
6. It has tremendous versatility of operating conditions, since the mobile phase can be an aqueous solution. Thus small changes in pH, ionic strength, percent and type of organic modifier can cause significant changes in retention and/or resolution.
7. For the mobile phase, inexpensive solvents such as water or methanol are usually used.

Purine and pyrimidine bases and their nucleosides are now analyzed routinely by RPLC [11,12], and separations of some nucleotides [13] and oligonucleotides [14] have also been obtained.

3. Exclusion chromatography

In exclusion chromatography, which is also known as molecular sieving, gel filtration, or gel permeation, compounds are separated according to their size or molecular weight. The mechanism of separation is different from other modes, as the separation depends on mechanical sorting of molecules, not on the differential interaction with either the stationary or mobile phase. In exclusion chromatography, the separation is dependent only on the size of the pores of the stationary phase, which is usually a highly porous, nonionic gel. The large molecules cannot penetrate the gel whereas the smaller ones can. Thus the large molecules are not retained and have a fast retention time, while the small molecules are "trapped" in the pores and have longer retention times (Fig. 3). This technique, which is known as fractionation, is not usually used to separate one unique type of molecule from another, but mainly to determine molecular weight distribution or to separate classes of compounds by size, e.g., nucleic acids from oligonucleotides from nucleotides. Although exclusion chromatography was originally used for high-molecular-weight compounds [15], gels are now available so that small-molecular-weight compounds and even some unique species can be separated [16].

C. Terminology and equations

In chromatography, the mobile phase is also referred to as the moving phase, solvent, carrier, or column eluent. After emerging from the column, the mobile phase, which is then composed of the eluent and a solute (called the eluite), is referred to as the effluent.

As in all scientific fields, each specialty has its own language, terminology, and symbols. Chromatography is no different, and there are some terms and equations which are used regularly that should be defined.

Retention time t_R is the time a solute is retained on a given column. It is also defined as the time from the time of injection of the sample to the maximum of the solute peak (Fig. 4). The hold-up time t_O is the time from injection to the maximum of a peak of a solute that is *not* retained on the column. The adjusted retention time t_A is the difference in retention time of a retained solute minus the hold-up time (Fig. 5):

$$t_A = t_R - t_O \tag{1}$$

To denote the ability to retain a solute on a column, the term capacity factor, which has the symbol k', is used. Note that the k is always lower case to differentiate it from equilibrium or distribution constants, for which a capital K is used. Also, the prime is used with the k to differentiate it from the rate constant, k. The factor k' is related to both adjusted retention time and hold-up time and is defined in Eq. (2).

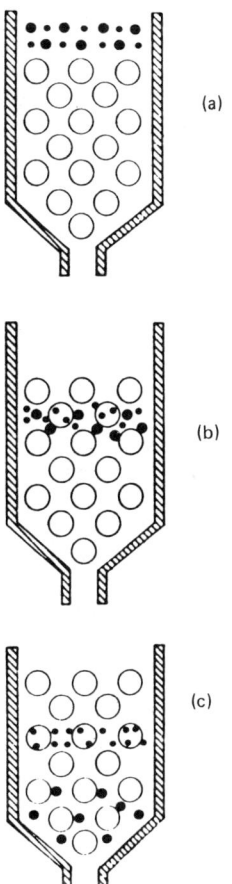

Figure 3. Packings for exclusion chromatography.

$$k' = \frac{t_R - t_O}{t_O} \quad \text{or} \quad \frac{t_A}{t_O} \tag{2}$$

An optimal k' is 2 to 5; however, a k' value of 1 to 10 is acceptable. If k' is too low [17], the components elute too rapidly; whereas when k' is too high, the solute elutes too slowly. This is the only term we will use in which the elution of just *one* component is considered. If all the solutes are resolved, there is a unique k' value for each solute in the mixture.

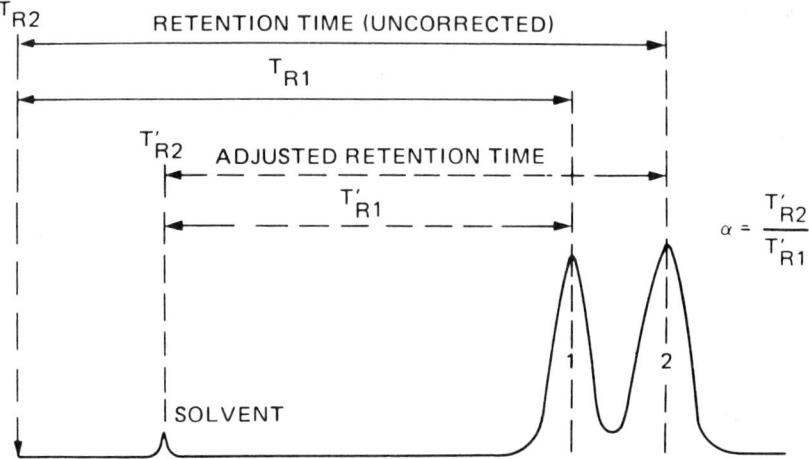

Figure 4. Chromatographic terms. (From E. L. Johnson and R. Stevenson, *Basic Liquid Chromatography*, Varian Associates, Walnut Creek, Calif., 1978, p. 35.)

When we wish to consider the ability of the column to separate two solutes, the term spearation or selectivity, which is denoted by α, is used. Alpha is defined as follows:

$$\alpha = \frac{k'_2}{k'_1} \text{ or } \frac{t_{A2}}{t_{A1}} \tag{3}$$

In k'_2 the 2 refers to the peak with the longer retention time (Fig. 4). Therefore α is always greater than 1. When α is 1, the peaks coincide and there is no separation; but when α is 2, there is good separation (Fig. 4).

The term resolution (R_s) includes both the *separation* of the two peaks and the *width* of the peaks. Therefore it denotes column efficiency, since narrow peaks are most desirable in HPLC. Thus the R_s term includes capacity and selectivity and is defined as follows:

$$R_s = \frac{2(t_2 - t_1)}{(w_2 + w_1)} \tag{4}$$

where w_2 is the width of the base of peak 2 and w_1 is the width of peak 1. It can also be defined using other symbols:

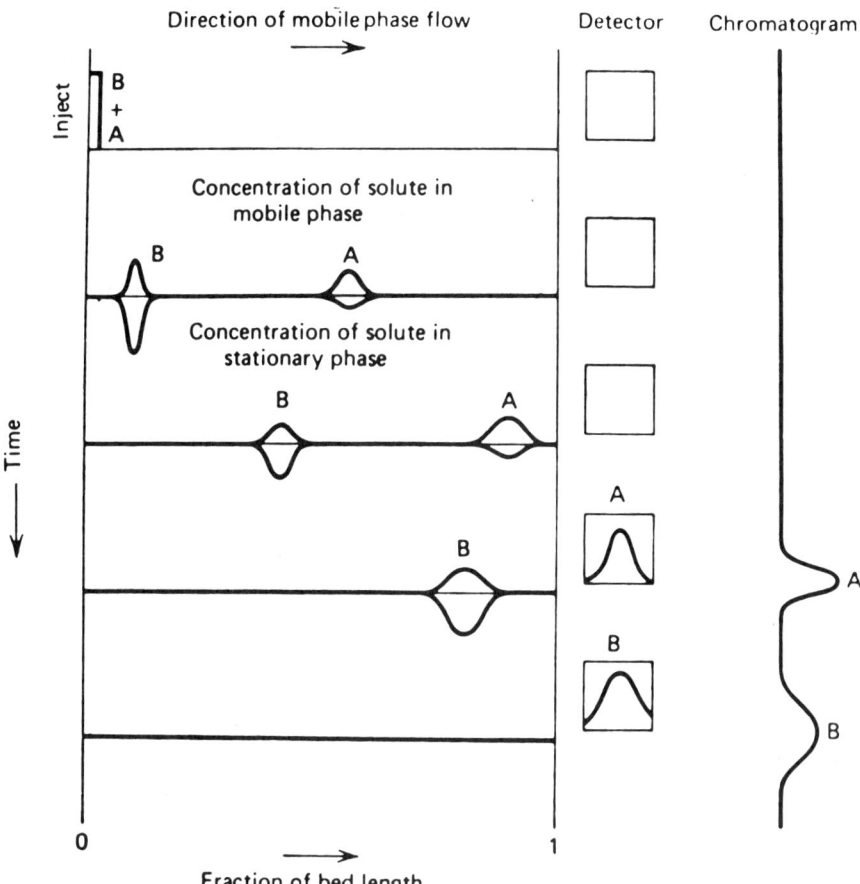

Figure 5. Band spreading. (From James M. Miller, *Separation Methods in Chemical Analysis*, John Wiley & Sons, New York, 1975, p. 10.)

$$R_s = \left(\frac{1}{4}\right)\left(\frac{\alpha - 1}{\alpha}\right)\left(\frac{k'_2}{k'_2 + 1}\right)(N)^{1/2} \tag{5}$$

when N is the number of theoretical plates. The term N, the efficiency of a column, is defined as

$$N = 16\left(\frac{t_R}{w}\right)^2 \tag{6}$$

It is a term used for comparative purposes to describe the efficiency of a column and to compare it with other columns, all other operating parameters being the same (column length, design, packings, solvent, flow rate, etc.). The higher the N value, the more efficient the column. Another term used in denoting column efficiency is "height equivalent to a theoretical plate," symbolized by HETP or H, which was introduced by Martin and Synge [18] and is defined as

$$H = \frac{L}{N} \tag{7}$$

in which L is the length of the column. Since N is inversely proportional to H, the lower the H value, the more efficient is the column. Although the term N is a dimensionless parameter, H is given in terms of unit of length (m, cm, or mm).

In chromatography, a process called band broadening or zone spreading occurs as a solute passes through a column [19] (Fig. 5). When a quantity of solute is introduced into a column, the solute diffuses with time. The longer the time on the column, the broader the chromatographic peak or band; thus the harder it is to separate two adjacent peaks. The diffusion phenomenon is due to many factors. In 1956, Van Deemter et al. [20] proposed a theory of chromatography called the rate theory of chromatography. They proposed that band braodening is caused by several factors which contribute to the efficiency of a chromatographic column. Thus H is equal to the sum of three factors:

$$H = A + \frac{B}{\mu} + C\mu \tag{8}$$

The A factor represents band spreading due to eddy diffusion or nonhomogeneous flow. The B factor is spreading due to diffusion in a longitudinal direction, and C is due to resistance to mass transfer in both the stationary and mobile phases. The term μ is the linear velocity of the mobile phase.

By plotting H vs. μ, the optimal flow rate can be determined in gas chromatography (Fig. 6). This method for obtaining an optimal flow rate does not apply in liquid chromatography, as the longitudinal diffusion term B/μ, which in GC is very large, is very small in LC.

The original Van Deemter equation takes into account the variables affecting efficiency in chromatography. Therefore this equation can be used not only to optimize chromatographic separations, but also to predict the effect of each variable on the efficiency of separation. The original Van Deemter equation [20] is

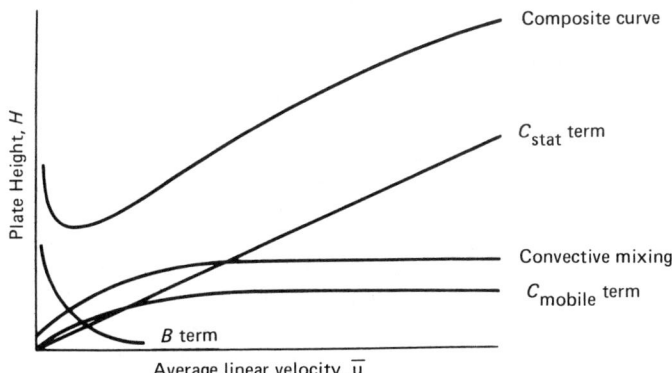

Figure 6. Van Deemter plot. (From Hobart H. Willard, Lynne L. Merritt, Jr., John A. Dean, and Frank A. Settle, Jr., *Instrumental Methods of Analysis*, 6th ed., Copyright © 1981 by Litton Educational Publishing, Inc. Reprinted by permission of Wadsworth Publishing Company, Belmont, CA 94002.)

$$H = 2\lambda d_p + (2\partial D_m)\left(\frac{1}{\mu}\right) + \frac{8}{\pi^2}\frac{k'}{(1+k')^2}\frac{d_f^2}{D_s}(\mu) \qquad (8)$$

where

λ = dimensionless parameter related to packing irregularities
d_p = mean particle diameter of stationary phase.
∂ = parameter related to varying flow
D_m = molecular diffusion coefficient of the solute in the mobile phase
μ = mean mobile phase velocity
k' (K) = partition ratio (amount solute in stationary phase divided amount in mobile phase at any instant)
d_f = stationary liquid film thickness
D_s = molecular diffusion coefficient of solute in stationary phase

Note that the A factor is concerned only with the stationary phase, the B factor only with the mobile phase, and the C factor with both the mobile and stationary phases.

II. INSTRUMENTATION

The basic instrumentation in HPLC is relatively simple. It consists of an injector, a pump, a column, a detector, and a recorder (Fig. 7). Various combinations of these components can be used alone or together with an integrator and/or computer. In this chapter we will not discuss detectors, since they are covered in a separate chapter (Chap. 8). We will also not discuss specific instruments or manufacturers, since the field is changing so rapidly that the material would be out of date before it is published. Each year, the American Association for the Advancement of Science publishes a *Guide to Scientific Instruments* [21] and Analytical Chemistry their Annual Lab Guide, e.g., [22], in which the latest advances in equipment can be found. Also, at the annual Pittsburgh Conference on Analytical Chemistry and Applied Spectroscopy, manufacturers launch their newest developments and all the latest components and instruments are displayed.

The type of instrument to buy depends on many factors. Questions to ask before purchase of an instrument include:

1. Will the analysis be carried out routinely, or is it a one-time research type of analysis?
2. Will the instrument be used frequently?
3. Will it be needed for a large number of analysis?
4. Will the analyses be for the same type of solutes, or for many different types?
5. What sensitivity of detection is required?
6. Is the sample complex; thus is gradient needed?
7. Will the instrument be used for trace analysis? Preparative work? Quantitative work?

Therefore the major question is "Do I need a sophisticated instrument, or will a simple instrument do the job for me?

An inexpensive liquid chromatograph can often be purchased for use as a dedicated instrument for the routine analysis of one or a few compounds in a large volume of samples—e.g., to determine theophylline serum levels in asthmatic patients.

Automation and computerization of HPLC for use in a large clinical or pharmaceutical laboratory or in quality control laboratories will increase the expense; however, a tremendous increase in throughput can be obtained, thus making it economically feasible.

The schematic representation of a basic liquid chromatograph is shown in Fig. 7. Although only one pump is necessary for isocratic work, more than one pump can be used to generate a gradient. Many companies market not only pumps but solvent delivery systems which include programmers to control the type of eluent program one may wish to use.

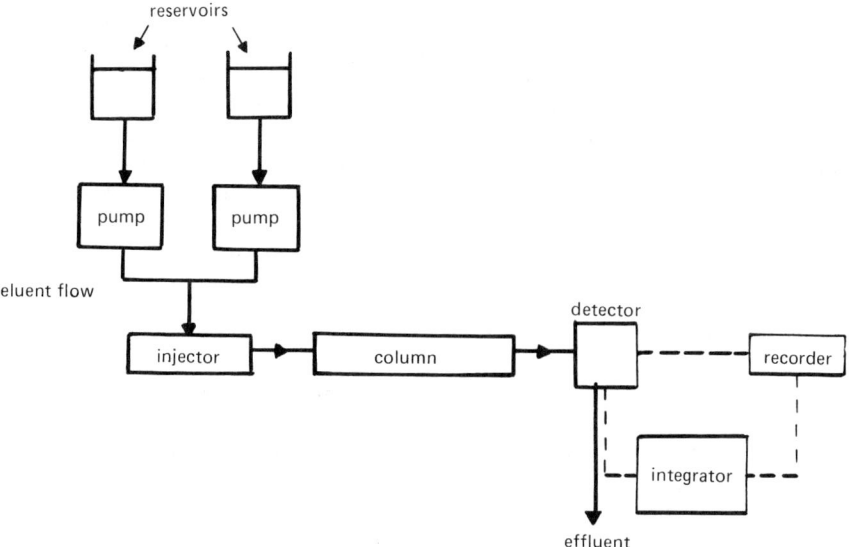

Figure 7. Diagram of a high performance liquid chromatograph.

A. Injectors

There are four types of injectors or sample introduction devices: septum-type syringe injectors, septumless syringe injectors, valve-type injectors, and automatic sample injectors. The simplest and least expensive is the septum-type syringe injector. However, septums tend to deteriorate and then leak with repeated use. In addition, only a limited amount of the sample can be injected because of the high backpressure when injecting samples "on line," i.e., with the mobile phase flowing.

The septumless syringe injectors are more complicated and thus more expensive. However, because they have no septum, deterioration and leakage are not problems. They can be used both on line and with stopped flow. Valve-type injectors are widely used because of their precision. However, the major disadvantage is that only fixed volumes of samples can be injected, 1-10 µl with values with internal sample loops and 8-2000 µl with external sample loops. Automatic sample injectors which are microprocessor-controlled are much more expensive and a sophisticated version of the valve injector. They are a necessity when large numbers of samples are to be analyzed, e.g., in quality control in industry or in the clinical laboratory.

B. Pumps

There are many types of pumps which are commercially available. Syringe pumps which are constant-flow are the type presently used in liquid chromatographs. Of the two types of syringe pumps, piston reciprocating pumps and membrane reciprocating pumps, the piston pumps are most widely used. The advantages of reciprocating pumps are that they can deliver a continuous flow from a large reservoir and there can be a rapid exchange of solvents.

The features that a pump should have for adequate performance in a liquid chromatograph are the following:

1. Operation should be simple.
2. Noise level should be low.
3. The pump should produce maximal pressure with minimal pressure fluctuations and constant flow delivery.
4. Components should be inert to commonly used solvents.

In addition to pumps, a good solvent delivery system will include a solvent degassing system to remove dissolved gases. Gas bubbles in a liquid chromatographic system may cause "spikes" in the chromatogram and/or oscillations of the pressure indicator.

Furthermore, when gradient elution is required for the analysis of complex mixtures, solvent programming is required. In low-pressure gradient systems, the solvent, after being mixed at atmospheric pressure, is pumped into the column by a single high-pressure reciprocating pump. These systems are inexpensive. In addition, they are versatile because more than two solvents can be handled simultaneously. The limitations include problems with ease and speed of changing solvents and reproducibility. Because of these limitations, most chromatographers prefer high-pressure gradient systems. In these systems, only two solvent gradients can be used. Two pumps are used, each delivering a fraction of the total solvent. For a gradient, the delivery of one pump is increased and that of the other pump decreased. The gradient generated by the two high-pressure streams of solvent being mixed in a mixing chamber is then pumped into the column. With this system, linear, concave, and convex gradients can easily be generated.

C. Automation

Microprocessor technology has made possible automation of liquid chromatographs. There are microprocessors capable of controlling any one phase or the complete operation of the chromatograph. For routine work, the operating parameters can be predetermined using automation and will remain constant for repetitive analyses. Automation permits unattended operation of the chromatograph for quality control and in research studies involving large numbers of samples. The microprocessors presently on the market can control the action of the pumps, determine optimal wavelengths when variable-wavelength detectors are used, and optimize conditions for separations; thus methods development can be simplified. Microprocessors are also the basis for the automatic injectors now commercially available. An example of the use of a microprocessor in nucleic acid analyses is the separation of nucleotides, nucleosides, and bases in a single analysis using two columns and two gradients. The microprocessor can control three solvent delivery systems and an external column switching valve to separate the nucleotides by anion exchange. By switching to a reversed-phase column, the bases and nucleosides can then be separated [23]. Microprocessors are also used in data collection, storage, and handling. In sophisticated instruments, all types of mathematical manipulations can be performed, and the data supplied can include retention time, peak areas, and absorbance ratios. In addition, the baseline and/or chromatograms of controls can be subtracted out.

III. OPERATING CONDITIONS

The great advantage of HPLC over GC is its versatility. Not only are there several modes that can be used and stationary phase parameters which can be adjusted slightly or changed, but also, because the mobile phase is a liquid, there are many mobile phase parameters that can influence resolution. To obtain the best resolution, a combination of stationary and mobile phase parameters must be optimized. Although temperature programming is not usually used in HPLC, there can be an optimal temperature for the separation of interest.

In HPLC, solvent programming (also called gradient elution) is the technique most commonly used to achieve separation of complex mixtures.

A. Stationary phase

There are many column packings available for HPLC. A simple instrument with a good column can usually give good separations, but any instrument, no matter how sophisticated, will give inadequate resolution with a poor column. Although there are many parameters that can be varied in the stationary phase of HPLC, once an adequate column is found for the desired separations, optimizing a separation is usually done by varying one or a combination of mobile phase parameters. Stationary phase parameters that can be varied include column length, column diameter, particle size and shape, type of solid support, and type of bonded phase and surface coverage. The effect of many of these parameters on efficiency can be predicted from the Van Deemter equation; e.g., before microparticulate packings were used, it was predicted that greater efficiency could be obtained by decreasing the particle size (H α dp). At present, the majority of the columns which are commercially available range in length from 5 to 30 cm and in diameter from 2.0 to 4.6 mm. At present, longer columns (25-30 cm) are used routinely; however, it has recently been shown that shorter columns (5-10 cm) filled with 5-μm packings give faster separations (i.e., 10 min instead of 35) with excellent resolution (Fig. 8). Generally a silica support is used to which a stationary phase of a long-chain alkane (C_2-C_{18}) is bonded for reversed phase, and an ionic material (-N^+R_3 or -S_3^{2-}) is bonded for ion exchange. Since microbore columns hold great promist for the future, these columns will be discussed in detail in Chap. 7.

B. Mobile phase

Because the mobile phase is a liquid, HPLC has great power for high resolution. Not only are there many parameters that can be varied to achieve the desired separation, but solute-solvent interactions, as well as solute-stationary phase interactions, can be used to achieve optimal selectivity.

In HPLC, the mobile phase can be organic or aqueous. If organic, the solvent can be highly polar, such as methanol or acetonitrile; nonpolar, such as heptane; or somewhere in between, such as methylene chloride [17]. If the solvent is aqueous, then pH, ionic strength, types of cations, types of anions, and type and percent of organic modifier can be varied. A slight change in any of these parameters can cause a significant difference in k' and/or α values. For example, in the RPLC separation of nucleosides and bases, ionic strength has little influence on k' or t_R values, whereas pH affects only those compounds that have a pK_a in the range of pH values of the eluents studied (Fig. 9). On the other hand, an increase in the percentage of methanol in the eluent will decrease the retention of these compounds (Fig. 10). Also, nucleotides, because of the presence of the ionized phosphate group, do not act as consistently in reversed-phase chromatography. For example, an increase in ionic strength of the buffer will dramatically decrease the k' values of the majority of the di- and triphosphate nucleotides. However, there is only a minimal decrease in retention of the monophosphates (Fig. 11). It is interesting to note that when the percentage of methanol is decreased in an eluent of pH 5.5, the decrease in ln k' on nucleotides is linear, as is expected for a reversed phase system (Fig. 12). However, when the eluent is at a pH of 2.95, the plot of ln k' of nucleotides vs. percent methanol is curved, indicating that mixed mechanisms are operative at the lower pH (Fig. 13).

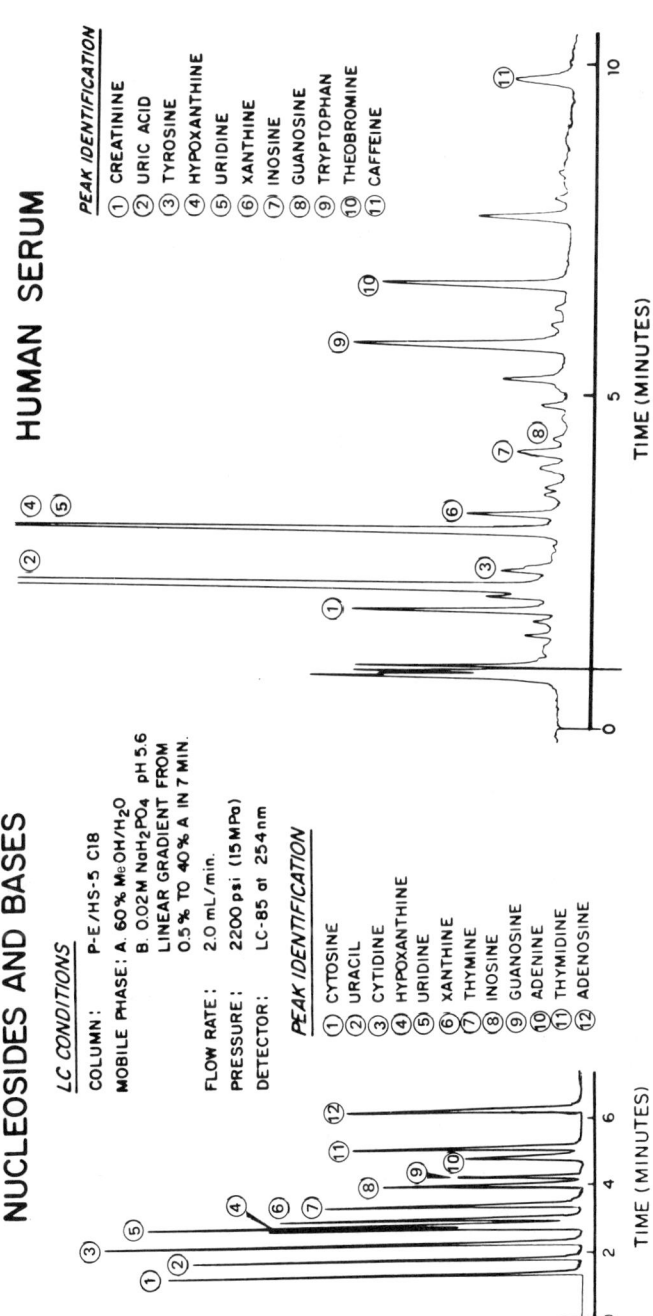

Figure 8. Rapid separation of nucleosides and bases in a standard solution and in serum using a short column. Column: PE HS-5 C18; elution mode: 0.5% to 40%, A in 7 min; eluents: A, 60% MeOH/H_2O, B, 0.02 NaH_2PO_4, pH 5.6; flow rate: 2.0 ml/min; detector: PE LC-8J at 254 nm. (Courtesy of The Perkin-Elmer Company, Norwalk, Conn.)

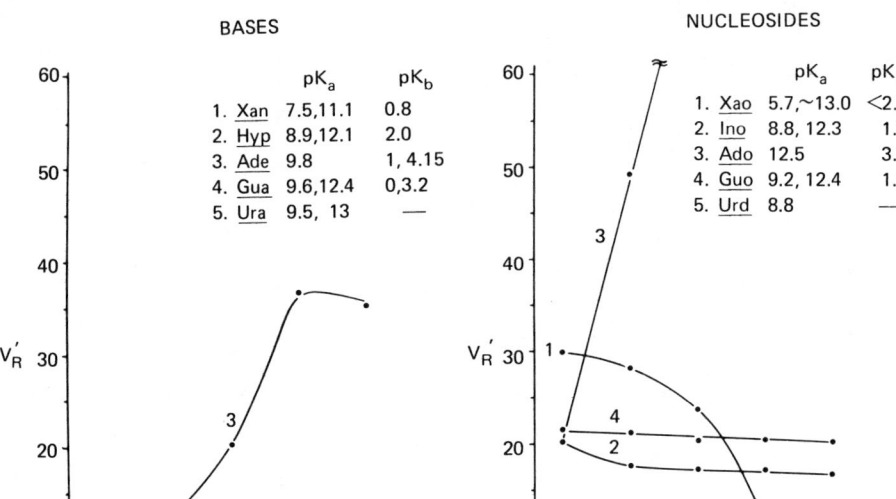

Figure 9. Graph of pH vs. V_R of nucleosides and their bases. The ionic strength of the mobile phase was held constant by making all eluents 0.01 M in KCl. (From Ref. 11.)

For a biological sample which contains compounds of varying molecular weights, structures, and polarities, isocratic elution (i.e., a solvent system which does not change composition during the analysis) may not give the desired resolution; thus a gradient elution system or solvent programming may be required. With a gradient, the solvent composition changes with time. The gradient may be stepwise or continuous. If continuous, the gradient may be a linear change with time, or any number of types of gradients may be generated. The change may be a pH gradient, a change in ionic strength, or gradual addition of an organic modifier or modifiers. For example, in the reversed-phase separation of nucleosides and bases in serum, typically we use 0.01 M KH_2PO_4, pH 5.6, as the initial eluent. The second solvent system used to make the linear gradient is a 60% $CH_3OH\text{-}H_2O$ solution [11]. Good separations have also been obtained using ammonium ions instead of potassium ions [24], a pH of 3.5-5.6, and varying amounts of an

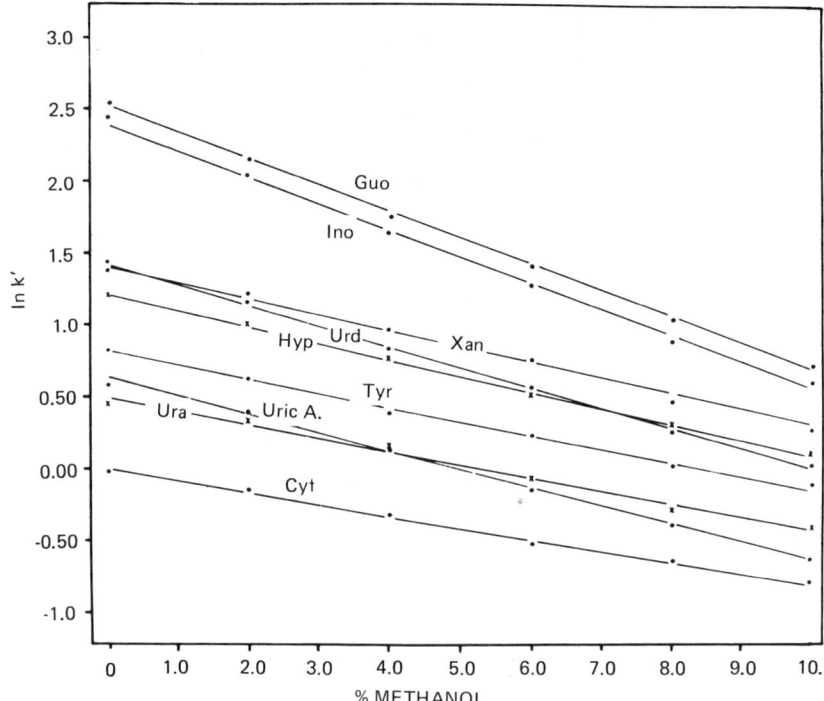

Figure 10. Graph of percentage of methanol in the mobile phase vs. ln k' of eight of the nucleosides and their bases and the amino acid tyrosine in a reversed-phase separation. All data points represent the average of two injections. Column: Whatman Partisil-5, C_8; particle size 5 nm; mobile phase: 0.005 M KH_2PO_4, pH 5.0; flow rate: 1.0 ml/min; temperature: 25°C. (Reprinted with permission from Ref. 26. Copyright 1979 American Chemical Society.)

organic modifier. An example of an improvement in the separation of nucleotides is shown in Fig. 14, where the time of analysis was decreased from 110 to 65 min by adding KCl to the eluent and slightly modifying flow rate and gradient slope [25]. In addition, by varying only the gradient slope, α values as well as k' values may be varied (Fig. 15).

C. Temperature

Temperature programming is rarely used in HPLC because of the time required for thermal equilibrium of liquids. Most analyses are carried out at ambient temperature, especially in reversed-phase separations.

It is important, however, that the temperature remain constant for reproducibility of chromatographic characteristics. As can be seen from the Van Deemter equation (Eq. 8), temperature will affect efficiency because of its influence on the k' factor and on the viscosity of solvents, which affects the D_S term. In the separation of nucleotides by ion exchange using pellicular ion exchangers, elevated temperatures

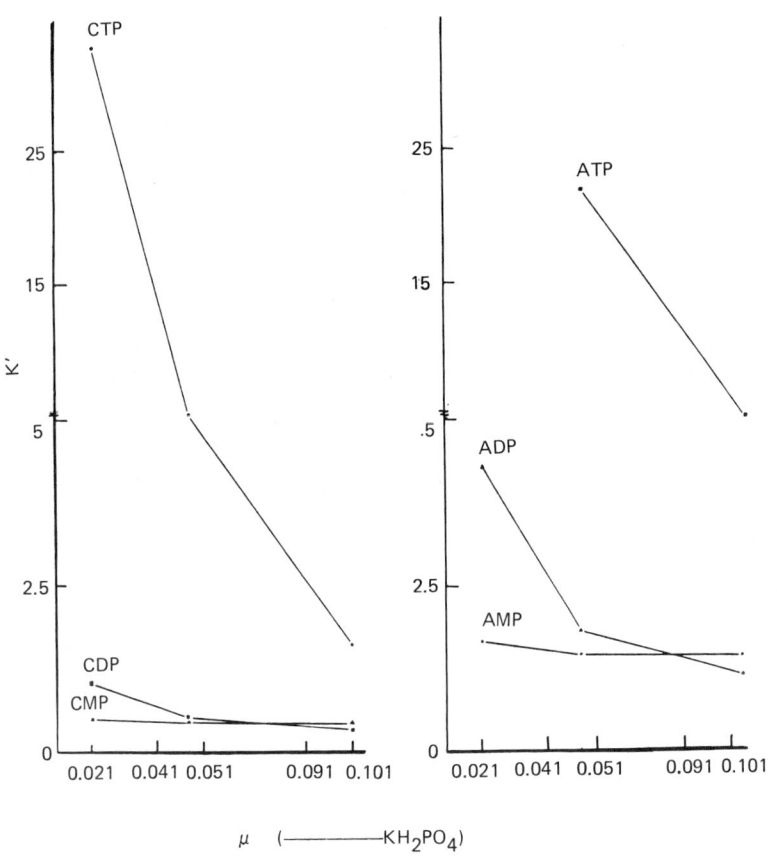

Figure 11. Graph of ionic strength vs. k' of mono-, di-, and triphosphate nucleotides in a reversed-phase separation. Column: Whatman Partisil PXS -10/25 ODS-3; mobile phase: 0.02 M KH_2PO_4 adjusted for ionic strength with KCl; flow rate: 1.5 ml/min; temperature: ambient. (Reprinted with permission from Ref. 13. Copyright 1983 American Chemical Society.)

(~70°C) were required to achieve good separation of the nucleotides [7,14]. However, using ion exchangers chemically bonded to microparticle silica spheres, excellent separations were obtained at ambient temperatures (Fig. 14).

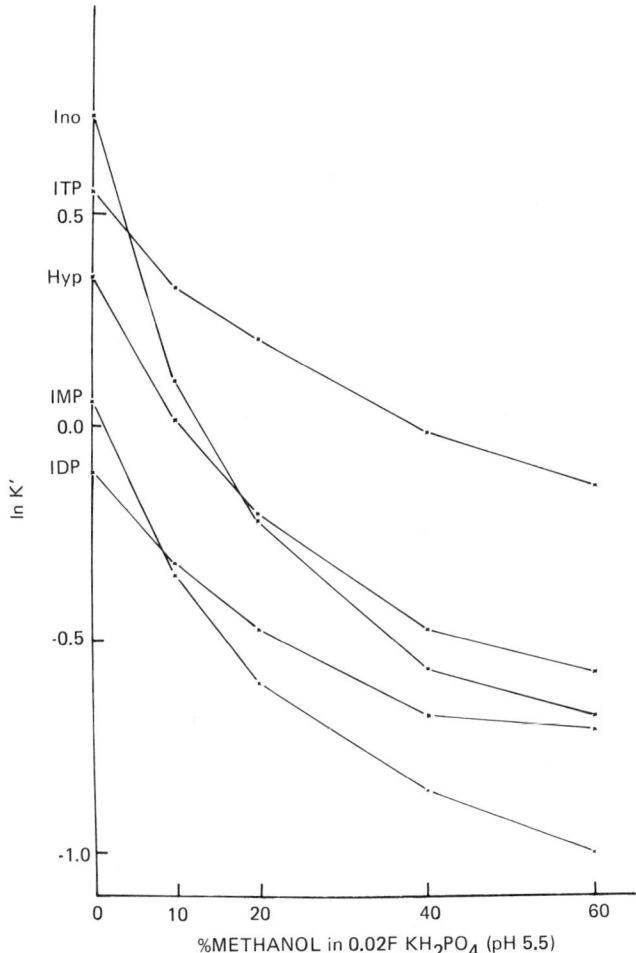

Figure 12. Graph of percentage of methanol vs. ln k' of nucleotides in a reversed-phase separation using an eluent of pH 5.5. Conditions as in Fig. 11, except the mobile phase is adjusted for percentage of methanol. (Reprinted with permission from Ref. 13. Copyright 1983 American Chemical Society.)

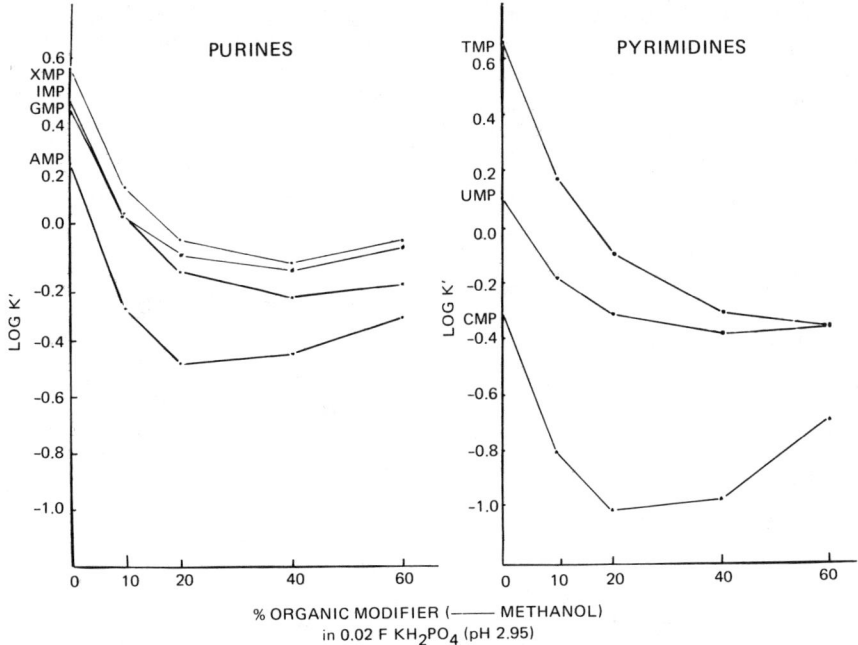

Figure 13. Graph of percentage methanol vs. log k' of nucleotides in reversed-phase separation using an eluent of pH 2.95. Conditions as in Fig. 12. (Reprinted with permission from Ref. 13. Copyright 1983 American Chemical Society.)

IV. OPTIMIZATION AND PROBLEM SOLVING

In chromatography, the perfect separation may not be readily obtainable. However, the problem at hand may be solved with *an adequate* separation. Thus, in tackling any analytical problem, the first step is to ask some questions:

1. Do I need qualitative or quantitative information from the chromatographic separation; i.e., is the chromatography only a first step prior to quantitation by another means (NMR, IR, MS), or is it the complete analysis?
2. Do I need a complete analysis of all components in the mixture, or do I need an analysis of only one or two components? (For example, are all the nucleosides in a serum sample of interest in the study or, as in some cardiac investigations, just adenosine and/or inosine?)

3. If the concentrations of only one or two compounds are to be determined, are these compounds present in the sample in large amounts or approximately equal amounts compared to the other constituents in the sample?
4. If trace analyses are desired, are interfering substituents present in equal amounts or in large amounts? (For example, a dye used in a cardiac study interfers with the relatively small amounts amounts of inosine present, as shown in Fig. 16.)
5. Will this analysis be used routinely, or will it be used once or just a few times? In other words, is the analysis being set up for routine clinical use or for use in a specialized biomedical study?
6. Will this analysis be used for large numbers of samples; thus is throughput time very critical? In the clinical laboratory, an assay used routinely must be rapid or it is not cost-efficient.
7. Are large amounts of sample available or only very small amounts of sample? (For example, only small volumes of blood are available for pediatric and sometimes geriatric samples, but larger volumes of blood are readily available from healthy adults.)
8. Can the sample be analyzed directly by HPLC, or is some sample preparation necessary prior to HPLC analysis?
9. What methods are available for peak identification? Will only nucleic acid constituents be present in the sample, or will other endogeneous compounds be present (e.g., in serum, aromatic amino acids and creatinine are present). Furthermore, will there be just the normally occurring compounds, or will unusual analogs of these compounds also be present? (For example, are just the nucleosides present which are normally in DNA or RNA, or will methylated analogs or psuedouridine also be in the sample?)

After asking these questions, the problem at hand and the purpose of the analysis should be defined. A literature search should then be carried out to determine if the compounds of interest have already been separated by HPLC. Sometimes some of the compounds can be readily separated and conditions for good resolution have been reported. However, if there are other compounds of interest which must be analyzed at the same time, the reported conditions must be modified or optimized. For example, the RPLC separation of nucleosides and their bases in serum is fairly routine [11,28,29]. However, if the concentration of pseudourdine and uric acid must also be determined, the conditions must be changed, since these compounds are poorly retained and elute in a big peak near the void volume (see Fig. 16).

For a good HPLC analysis, the total procedure must be considered. The steps involved in an analysis are (a) collection of sample, (b) preparation of sample, (c) chromatography, (d) identification of peaks,

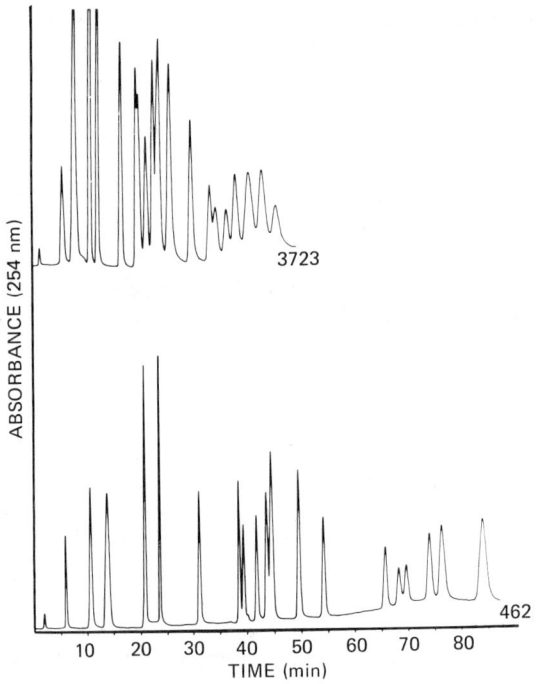

Figure 14. Chromatogram of nucleotides separated by anion exchange under different eluent conditions. Column for both chromatograms: Whatman Partisil 10-SAX.

	Conditions for bottom chromatogram	Conditions for top chromatogram
Low-concentration eluent	0.007 M KH_2PO_4, pH 4.0	0.007 M KH_2PO_4 0.007 M KCl } pH 4.0
High concentration eluent	0.25 M KH_2PO_4 0.25 M KCl } pH 4.5	0.25 M KH_2PO 0.50 M KCl } pH 5.0
Flow rate	1.5 ml/min	2.0 ml/min
Isocratic elution of low-concentration eluent	15 min	5 min
Gradient	Linear 0-100% in 45 min	Linear 0-100% in 35 min
Isocratic elution	30 min	10 min

(e) integration of peaks, (f) conversion of peak areas to concentrations, (g) analysis of data, and (h) interpretation of results. We will discuss steps (a) and (b) in Chap. 3, step (d) in Chap. 5, and steps (e), (f), (g), and (h) in Chap. 6. The specific chromatography of each class of nucleic acid constituents (nucleic acids, oligonucleotides, nucleotides, etc.) will be discussed in the applications chapters. However, in this chapter, we will set up *general* guidelines for optimizing separations of compounds of interest in nucleic acid research.

A. Choice of chromatographic mode

The first step in setting up an analysis is to determine the chromatographic mode to be used: ion exchange, reversed phase, normal phase, or exclusion chromatography. In dealing with nucleic acid constituents, a rule of thumb is that if the compounds are ionic (nucleic acids, oligonucleotides, nucleotides), then ion exchange is usually used. However, the reversed-phase mode can also be used with ionic compounds if techniques such as ion pairing or ion suppression are utilized (Chap. 9). With nonionic compounds such as purines and pyrimidines and their nucleosides, the reversed phase gives the best separation. However, these compounds have also been separated by ion exchange using a technique called ion-exclusion chromatography, in which a basic buffer is used with an anion-exchange column and an acid buffer with a cation-exchange column [30]. Exclusion chromatography is usually used to separate by class compounds which differ greatly in molecular weight (nucleic acids, oligonucleotides, nucleotides) or to separate various types of nucleic acids which may differ in size or molecular weight. Often exclusion chromatography is used as a preliminary step to separate the groups. Each fraction is collected and then the compounds in each "peak" are analyzed by the appropriate mode of chromatography, e.g., IE for nucleic acids or nucleotides and RP for nucleosides or bases (Fig. 17).

Figure 14. (Continued) 1 = Cytidine 5'-monophosphate; 2 = adenosine 5'-monomonophosphate; 6 = guanosine 5'-monophosphate; 7 = xanthosine 5'-monophosphate; 8 = thymidine 5'-diphosphate; 9 = uridine 5'-diphosphate; 10 = cytidine 5'-diphososphate; 11 = iosine 5'-diphosphate; 12 = adenosine 5'-diphosphate; 13 = guanosine 5'-diphosphate; 14 = xanthosine 5'-diphosphate; 15 = uridine 5'-triphosphate; 16 = thymidine 5'-triphosphate; 17 = cytidine 5'-triphosphate; 18 = inosine 5'-triphosphate; 19 = adenosine 5'-triphosphate; 20 = guanosine 5'-triphosphate; 21 = xanthosine 5'-triphosphate. (From Ref. 25.)

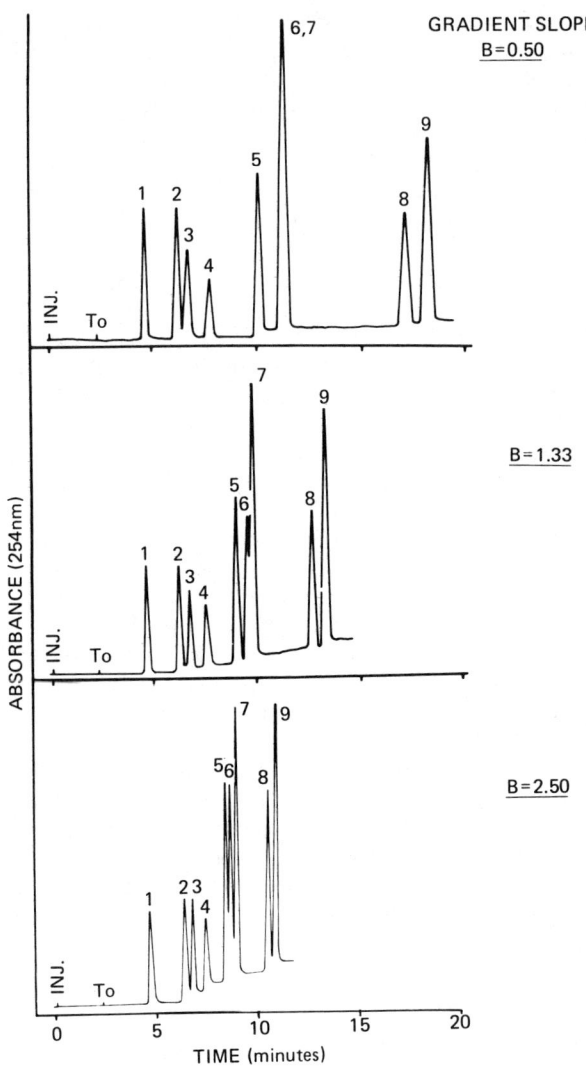

Figure 15. Three chromatograms showing the effect of change in gradient slope in the reversed-phase separation of nucleosides and bases. Peak identities are as follows: (1) Cyt, (2) Ura, (4) Tyr, (5) Hyp, (6) Urd, (7) Xan, (8) Ino, (9) Guo. Injection volume, 5 µl; sample concentration, $\sim 1 \times 10^{-4}$ mol/liter in each compound. (Reprinted with permission from Ref. 26. Copyright 1979 American Chemical Society.)

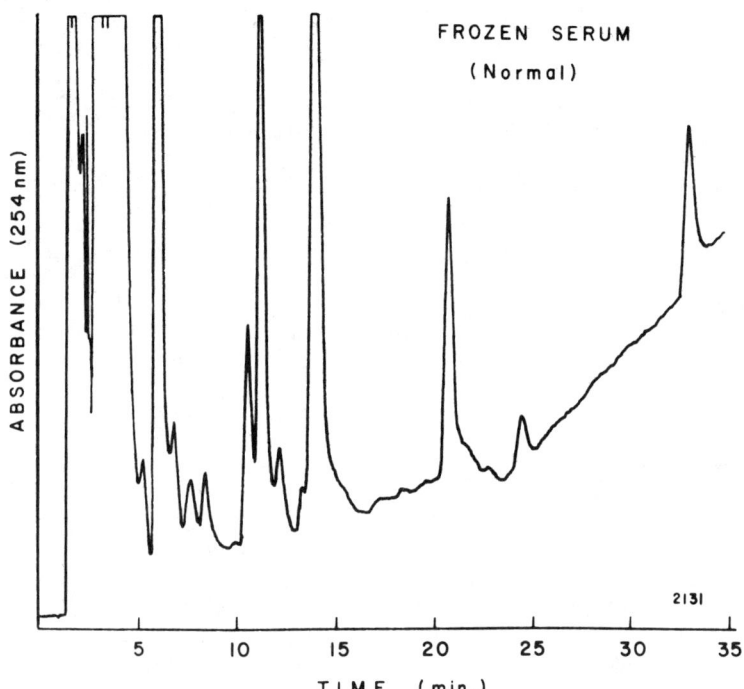

Figure 16. Chromatogram of pig serum using the reversed-phase mode of HPLC. Column: Whatman Partisil ODS-3 (10-μm particles); elution mode: linear gradient, 0-24% in 35 min; eluents: A, 0.002 M KH_2PO_4, pH 5.7; B, CH_3OH; flow rate: 1.5 ml/min; temperature: ambient.

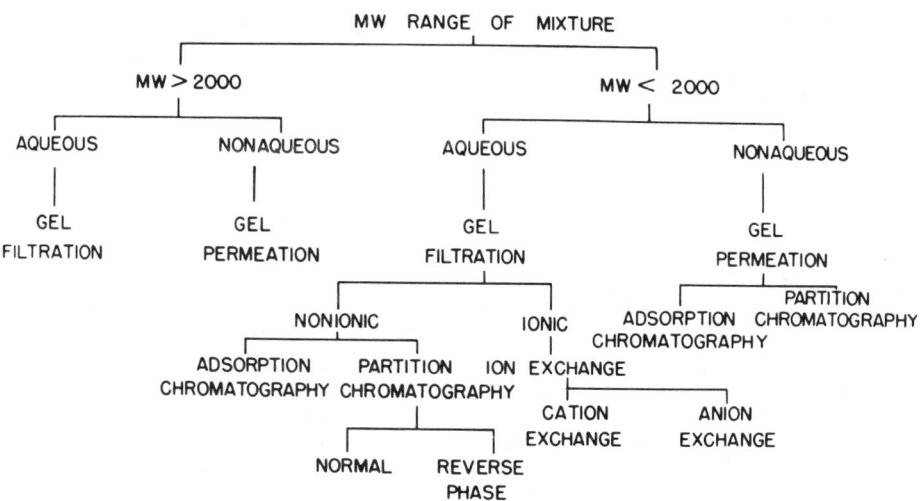

Figure 17. Chromatographic scheme. (From Ref. 7.)

It is possible to separate bases, nucleosides, and nucleotides on one column using anion exchange [31,32]; however, analysis time is relatively long and the potential for band spreading increased. Moreover, in cell extracts, nucleotides are present in 100-fold or greater concentrations than nucleosides and bases; conversely, in blood fluids, the concentrations of nucleosides and bases greatly overshadow concentrations of nucleotides. Therefore, it is difficult to determine quantitatively the concentrations of the compounds in one group in the presence of the other. However, the technique of column switching can be useful when both the phosphorylated and non-phophorylated nucleosides are present.

B. Choice of stationary and mobile phases

Once the mode is determined, the type of column must be chosen. The stationary phase parameters that can be varied are column length and diameter, particle size, shape and porosity, type of bonded phase, bonding linkage and surface coverage. There are a large number of excellent columns commercially available, and the applications department of many of the companies selling columns will be glad to help you choose an appropriate column for your particular problem.

After choosing the column, a suitable mobile phase or phases and mode of elution must be determined. It is the combination of stationary and mobile phases that will produce optimal resolution. With the mobile phase, extremes of eluents should be tried. For example, for the determination of concentrations of adenosine in serum, first water (or a mildly buffered aqueous solution of 0.007 M KH_2PO_4) was tried and then a 1/1 buffer/CH_3OH solution. With the aqueous solution the retention time was much too long, and with the 1/1 solution the adenosine did not separate from the other nucleosides. However, with a 9/1 buffer/CH_3OH solution an excellent separation of adenosine from other bases, nucleotides as well as nucleosides, was achieved (Fig. 18). HPLC has tremendous potential for achieving difficult separations of complex mixtures in a biological sample because of the versatility of its liquid mobile phase. Thus the mobile phase may be organic or aqueous or any miscible combination. If the eluent is organic, it can be anywhere from heptane to methanol in polarity. If the eluent is aqueous, pH, ionic strength, and type of cation or anion may be varied. In addition, the elution mode may be varied to obtain optimal resolution. Isocratic (a single solvent system) elution is usually used to determine the concentration of one or a small number of similar compounds—e.g., hypoxanthine, inosine, and adenosine, which are the compounds of interest in some cardiac studies. However, in separating compounds that are more dissimilar, such as the mono-, di-, and triphosphate nucleotides, which have one, two, and three charges, respectively, gradient elution is required (Fig. 14). Although some separations may be accomplished without gradient, often the peaks with

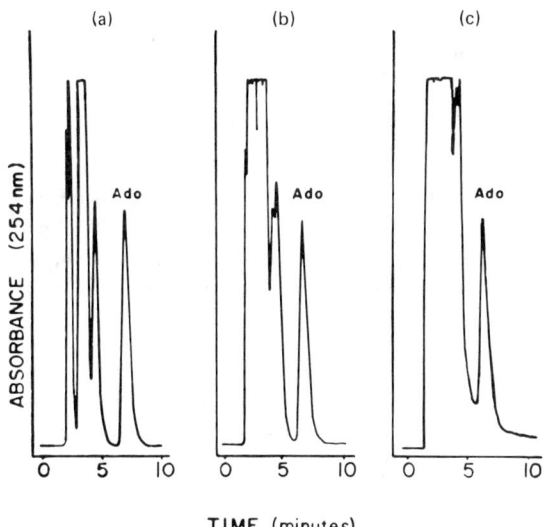

Figure 18. Chromatogram of the separation of adenosine from nucleotides, bases, and other nucleosides. Column, µBondapak C_{18}; mobile phase, 0.007 M KH_2PO_4; anyhdrous CH_3OH (10.90), pH of solution 5.8; flow rate, 2.0 ml/min; temperature: ambient. (From Ref. 12.)

longer retention times are too broad for good resolution; thus a gradient is used to decrease t_R and/or to sharpen the peaks. However, isocratic elution is preferred whenever possible, because retention behavior is more reproducible. With gradient elution, reequilibration of the column must be carefully controlled to obtain reproducibility, and this reequilibration time lengthens the overall time of analysis. In addition, the column must be washed regularly with a high concentration of the eluent used (e.g., high ionic strength for IE or high CH_3OH concentration for RP) to remove regularly constituents of the biological sample that are adsorbed to the stationary phase.

Although equations are now available for choosing mobile phase conditions for optimal resolution [26,33-36], they are not yet useful in setting up new analyses or for unknown samples.

To optimize an analysis, the relative importance of optimum k', α, and t_R values must be considered. Sometimes one parameter must be sacrificed to obtain the maxima in the other parameters. Using the adenosine example again, in serum samples, the separation of all the other nucleosides and bases was sacrificed to obtain a fast determination of the concentration of adenosine alone; thus low k' and t_R values for adenosine were obtained (Fig. 18). In addition, the total analysis time was lowered because no reequilibration time was required.

If k' values are too high (>10), they can be reduced in RPLC by adding an organic modifier to the eluent or in IE by using an ionic modifier to create an eluent of higher ionic strength (Fig. 14). If k' values are too low (<1), then adding an ion pair reagent such as t-butylammonium chloride may help in RPLC (see Chap. 9). Changing the pH or lowering the ionic strength may accomplish the purpose in IE. The selectivity is improved when the mobile phase parameter affects one compound (or group of compounds) and not another. For example, by changing the pH to 3.8, the bases xanthine, hypoxanthine, adenine, guanine, and uracil can be separated by RPLC; whereas at a pH of 6.8, xanthine, hypoxanthine, and guanine are not resolved (Fig. 9). An organic modifier may also be added to the eluent to change both k' and α values in isocratic separations. More than one modifier may be used to create the improved selectivity—for example, both methanol and dioxane or EDTA may improve a desired separation. Other reagents that can affect selectivity of ionic compounds such as nucleotides are solutions containing metal ions [37] or ion-pairing reagents such as tetrabutylammonium ion [38]. Mobile phase control using the techniques of ion pairing, ion suppression, and ion exclusion will be discussed in greater detail in Chap. 9.

Although flow programming will affect t_R, k' values, and peak width, it will not affect selectivity. In HPLC it is not usually used, nor is temperature programming. However, increased temperature may improve peak sharpness and sometimes resolution, especially in IE. It is vitally important for reproducibility that a HPLC analysis be run at constant temperature.

There are other techniques that can be used for complex mixtures. For example, column switching is a technique that can be used to separate solutes with widely differing properties, such as the purine and pyrimidine bases from the nucleotides. It is possible to use for the first column an ion exchanger on which the bases are not retained but the nucleotides are. The nonretained fraction elutes first and the eluent is directed onto a reversed-phase column where the bases are separated [39]. Many variations on this technique are possible, especially with microprocessor-controlled automatic switching valves.

Another technique for the separation of different classes of compounds is the use of mixed beds in one column. Good separation of nucleotides and nucleosides have been obtained using this method [40].

Still another technique used to obtain better resolution of very similar compounds is the recycling technique, in which the effluent is passed repeatedly through the column. This technique results in improved resolution; however, it can be used only with isocratic elution.

REFERENCES

1. H. J. Maggs, Chromatographia, *1*, 43 (1968).
2. R. P. W. Scott and J. G. Laurence, J. Chromatogr. Sci, *8*, 446 (1970).
3. L. R. Snyder, Anal. Chem., *7*, 105 (1967).
4. L. R. Snyder, *Principles of Adsorption Chromatography*, Marcel Dekker, New York, 1968.
5. L. R. Snyder, J. Chromatogr. Sci., *7*, 352 (1969).
6. N. G. Anderson, J. G. Greene, M. L. Barber, and F. C. Ladd, Sr., Anal. Biochem., *6*, 153 (1963).
7. P. R. Brown, J. Chromatogr., *52*, 257 (1970).
8. C. G. Horvath, B. A. Preiss, and S. R. Lipsky, Anal. Chem., *39*, 1422 (1967).
9. F. Regnier, Abstract #6519, 66th Annual Meeting of FASEB, New Orleans, La., April 1982.
10. C. G. Horvath and S. R. Lipsky, Anal. Chem., *41*, 1229 (1969).
11. R. A. Hartwick and P. R. Brown, J. Chromatogr., *126*, 679 (1976).
12. R. A. Hartwick and P. R. Brown, J. Chromatogr. Biomed. Appl., *143*, 383 (1977).
13. M. Zakaria and P. R. Brown, Anal. Chem., *55*(3), 457 (1983).
14. R. A. Hartwick and J. Crowthers, in *Advances in Chromatography* (A. Zlatkis et al., Eds.), University of Houston, Tex., 1981.
15. H. Petermann, *Gel Chromatography*, 2nd ed., Springer-Verlag, New York, 1969.
16. K. J. Bombaugh, W. A. Dark, and R. F. Levaugie, in *Advances in Chromatography* (A. Zlatkis, Ed.), Preston Technical Abstracts, Evanston, Ill., 1969, p. 334.
17. A. M. Krstulovic and P. R. Brown, *Reversed-Phase High Performance Liquid Chromatography: Theory, Practice and Biomedical Applications*, Wiley-Interscience, New York, 1982.
18. J. P. Martin and R. L. M. Synge, Biochem. J., *35*, 1358 (1948).
19. J. C. Giddings, *Dynamics of Chromatography, Part I, Principles and Theory*, Marcel Dekker, New York, 1965.
20. J. J. Van Deemter, F. J. Zuiderweg, and A. Klinenberg, Chem. Eng. Sci., *5*, 271 (1956).
21. AAAS, American Association for the Advancement in Science, *Guide to Scientific Instruments*, New York, 1982-1983.
22. Analytical Chemistry, Annual Lab Guide, *10*, 1982.
23. Whatman, Inc., Bulletin #116, "Analysis of Nucleic Acid Constituents by High Performance Liquid Chromatography, 1976.

24. F. S. Anderson and R. C. Murphy, J. Chromatogr., *121*, 251 (1976).
25. M. McKeag and P. R. Brown, J. Chromatogr., *152*, 253 (1978).
26. R. A. Hartwick, C. Grill, and P. R. Brown, Anal. Chem., *51*, 34 (1979).
27. R. A. Hartwick and P. R. Brown, J. Chromatogr., *112*, 651 (1975).
28. R. A. Hartwick and P. R. Brown, in CRC Rev., *10*, 279-337 (1981).
29. R. A. Hartwick, A. M. Krstulovic, and P. R. Brown, J. Chromatogr., *186* (1979).
30. R. P. Senghal and W. E. Cohn, Biochem. Biophys. Acta, *262*, 565 (1972).
31. B. Bakay, E. Nissinen, and L. Sweetman, Anal. Biochem., *86*, 65 (1978).
32. A. Floridi, C. A. Palmerini, and C. Fini, J. Chromatogr., *138*, 203 (1977).
33. P. Jandera and J. Churacek, J. Chromatogr., *91*, 223 (1974).
34. P. Jandera and J. Churacek, J. Chromatogr., *91*, 207 (1974).
35. P. Jandera and J. Churacek, J. Chromatogr., *93*, 17 (1974).
36. P. J. Schoenmakers, H. A. H. Billiet, R. Tijssent, and L. DeGalan, J. Chromatogr., *149*, 519 (1978).
37. F. K. Chow and E. Grushka, J. Chromatogr., *185*, 361 (1979).
38. N. E. Hoffman and J. C. Liao, Anal. Chem., *49*, 2231 (1977).
39. Waters Bulletin, 1982.
40. J. B. Crowther and R. A. Hartwick, Chromatographia, *16*, 349 (1983).

5
Peak Identification

MARY JO WOJTUSIK University of Rhode Island, Kingston, Rhode Island

I. Introduction 82
II. Specific Techniques Using UV Detectors 82
    A. Retention characteristics 82
    B. Absorbance ratios 83
    C. Co-chromatography 83
    D. Enzymatic peak shifts 83
    E. UV scanning spectra 84
    F. Applications 84
    G. Conclusion 89
III. Using Other Detectors 91
    A. Introduction 91
    B. Fluorescence detectors 92
    C. Electrochemical detectors 93
    D. Reaction detectors 93
    E. Inductively coupled plasma emission spectroscopy 94
    F. Mass spectrometers 94
    G. Nuclear magnetic resonance and infrared spectroscopy 95
IV. Conclusion 95
    References 96

I. INTRODUCTION

A limiting factor in the analysis of biological samples for nucleic acid constituents is often the identification of the peaks in the chromatogram. Because of the sensitivity of most high performance liquid chromatography (HPLC) systems, which is invaluable in many research and clinical assays, it is often difficult to characterize unambiguously the peaks which represent compounds in complex mixtures. In the last few years, there have been substantial advances in interfacing HPLC with various spectroscopic techniques such as mass spectrometry (MS), nuclear magnetic resonance (NMR), and infrared (IR) spectrometry; however, these instruments are still not available for routine use in most chemical and biomedical laboratories.

Multiple identification techniques utilizing the detection of ultraviolet (UV) absorbance have been and are being used to identify eluting compounds of a biological matrix (Table 1). Identification methods which have been used on line with UV absorbance detectors include retention times of eluting constituents, co-chromatography with standards, absorbance ratios, and UV scans of particular peaks. In addition, methods which are done off line, such as the enzymatic peak shift technique, can be monitored by UV absorbance detectors. Data can also be obtained from other types of detection systems, such as a fluorescence detector which can be on line with a UV detector or by an electrochemical detector which is usually interfaced separately with a chromatograph.

II. SPECIFIC TECHNIQUES USING UV DETECTORS

A. Retention characteristics

In the analysis of physiological fluids or tissue extracts, the retention behavior of the eluents is insufficient to provide unambiguous identification of the compounds present. A comparison of the retention times of eluting constituents with those of standard reference compounds that have been chromatographed previously under identical conditions provides only tentative identification. However, once the identity of a peak is established, the retention time can be used to distinguish this particular component from other peaks in the chromatogram.

Table 1 UV Detection: Identification Techniques

1. Retention characteristics
2. Absorbance ratios
3. Co-chromatography
4. UV scanning spectra
5. Enzymatic peak shifts

The retention behavior of the constituents in a biological sample when chromatographed on different columns and/or when chromatographed with different mobile phase composition can indicate the classes of compounds present. In other words, the retention characteristics of an unknown sample component can be used to predict its structural features [1].

B. Absorbance ratios

The determination of absorbance ratios at two different wavelengths is a simple method for establishing both peak identity and peak purity. It is convenient to use ratios at 254 nm/280 nm when studying UV-absorbing biological compounds, since a measurable difference is usually realized in both peak area and peak height at these two wavelengths. Tentative peak identification is achieved by comparing the absorbance ratios of a peak from a standard reference compound with that of the peak under study in the chromatogram. Absorbance ratios can also be used to indicate peak purity, since a peak composed of two co-eluting compounds will have an absorbance ratio that is not characteristic of either compound.

A multiwavelength UV absorption detector has been developed that is capable of monitoring up to four different wavelengths simultaneously [2]. The advantage of such an instrument is that three absorbance ratios can be obtained which provide more definitive identification of eluting compounds. Also, plots of absorbance ratios vs. time can be used to establish peak purity.

C. Co-chromatography

Co-chromatography of a biological sample with a known standard compound also provides only tentative peak identification. To characterize an unknown peak in a chromatogram using this technique, a known amount of a standard compound that has a similar retention time to the unknown is added to the sample. The sample is then rechromatographed using the same conditions. If a quantitative increase is observed in the unknown peak, the compound representing that peak could be identical to the standard compound that was added.

D. Enzymatic peak shifts

The relative specificity of enzymes can be used to identify peaks and establish peak purity in chromatograms of biological samples. The enzymatic peak shift technique is used to identify a peak that has been tentatively characterized previously by retention data, absorbance ratios, and co-chromatography, provided that a specific enzyme is used. Basically, an enzyme is used that converts the substrate, the compound represented by the peak, to a different compound, the product.

The choice of enzyme and the conditions under which the reaction is performed are extremely important, since cross-reactivity with undesired compounds of the same class is possible. If the specificity of the enzyme for the compound under study is reasonably assured, a decrease in the peak in the chromatogram is observed and tentative identification is obtained. One further consideration is that the product(s) and/or the reactant(s) must be UV-absorbing.

Enzymatic peak shifts are also used to qualitatively establish peak purity. The appearance of unusual peak shapes—for example, shoulders—as the original peak is converted to a different compound by the enzyme would indicate the presence of a co-eluting compound.

E. UV scanning spectra

The UV scanning technique, which yields the entire UV spectrum of a compound responsible for a peak, also provides only tentative identification. The UV spectra of reference compounds are compared to the UV spectrum of the peak under study. Close agreement between a reference spectrum and that of the peak in question indicates that there is a possibility that the same compound is responsible for both spectra. Dissimilarities in the two spectra would indicate positively that the compounds are not the same.

Rapid scanning UV detection systems are now commercially available for high resolution liquid chromatography which provide the entire UV spectra of eluting compounds in a very short time, making fast qualitative peak characterization possible. Three-dimensional chromatograms of total absorption vs. elution time are obtained (Fig. 1). Another advantage of a rapid scanning UV detector is that a plot of a selected absorption band vs. elution time can be recorded in post-run time, thus facilitating selective detection at different absorption maxima for each eluting compound.

F. Applications

Multiple identification techniques are used to identify chromatographic peaks that represent nucleic acid constituents. The retention times of eluting compounds are first compared with those of reference compounds for tentative identification. The other techniques, such as co-chromatography with reference standards, absorbance ratios, enzymatic peak shifts, and UV scans, are then used to characterize further the chromatographic peaks in question. The accumulated evidence from all of the multiple identification techniques can yield unambiguous peak identification. If, however, appropriate reference compounds or enzymes are not available, peak identification can be difficult.

The chromatographic analysis of a serum sample from a normal subject with no known diseases is shown in Fig. 2. Many of the chromatographic peaks have been tentatively identified on the basis of retention

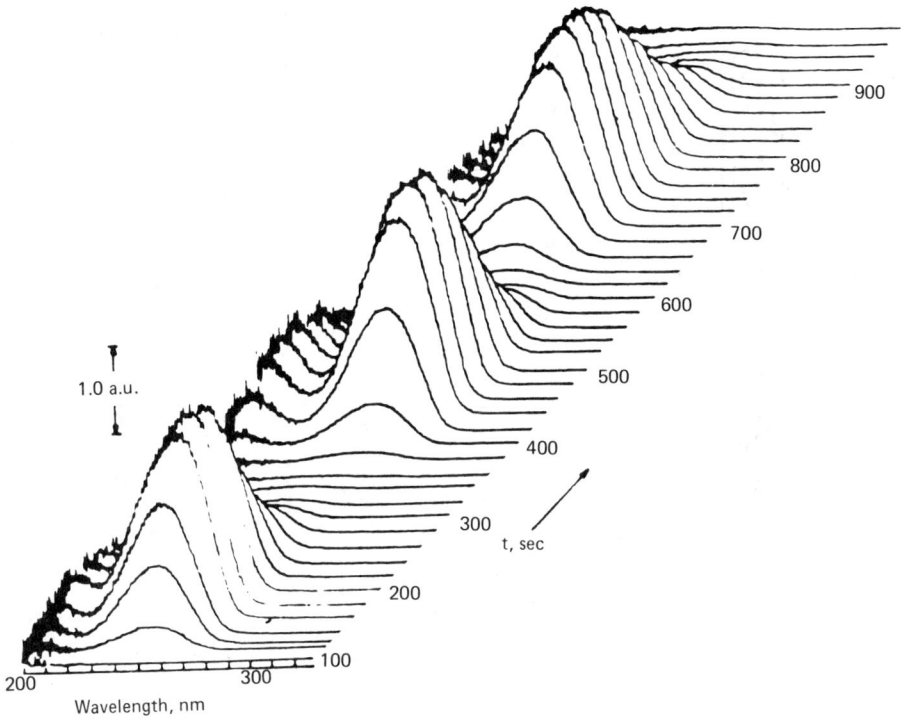

Figure 1. Three-dimensional chromatogram with rapid scanning UV detector for separation of uracil, cytosine, and adenine. (Reprinted with permission from Ref. 3. Copyright 1976 American Chemical Society.)

data. Figure 3 is a chromatogram of the same serum sample to which reference compounds have been added. In the co-injected chromatogram, the appearance of shoulders or double peaks indicates that a peak is comprised of compounds that have similar retention times but the reference compound is not identical to that present in the sample. Also, it should be emphasized that retention data and co-chromatography with standard compounds provides only tentative peak identification.

Absorbance ratios measured at 245 nm/280 nm can also be used to characterize chromatographic peaks. The peak height or peak area ratios of standard reference compounds (Table 2) are compared with the ratios of unknown peaks in a chromatogram (Fig. 4, Table 3). Absorbance ratios are most helpful when distinguishing between two closely eluting compounds, since the absorbance ratios can differ significantly.

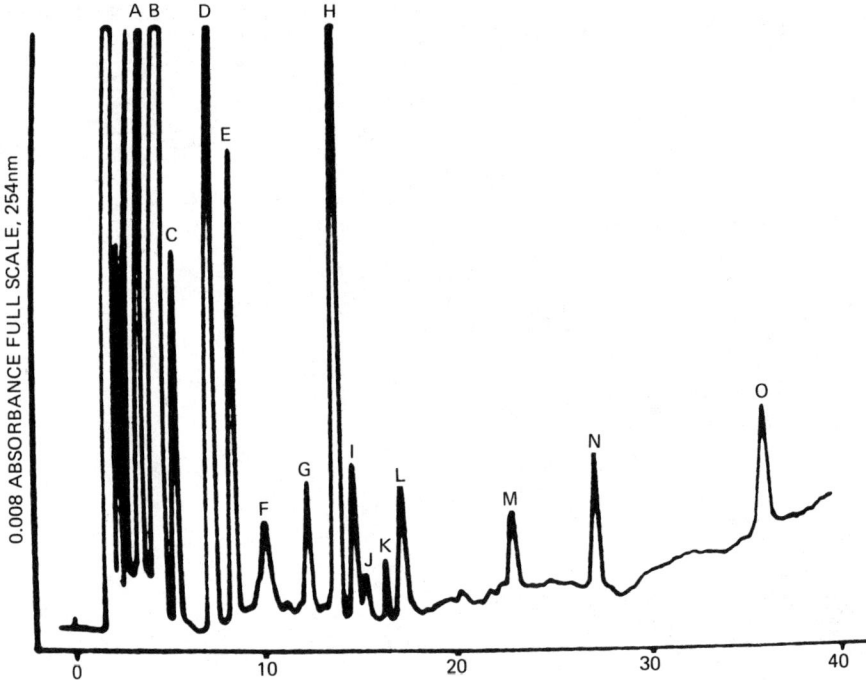

Figure 2. Chromatogram of a control serum ultrafiltrate sample from a normal donor using UV detection at 254 nm. Injection volume: 80 μl. Column: reversed-phase 10-μm particle diameter. Eluents: (A) 0.02 M KH_2PO_4, pH 5.6; (B) 60% methanol-water. Gradient linear, 0-11% of B in 87 min, slope 0.06% methanol/min. Flow rate: 1.5 ml/min. (From Ref. 4. Elsevier Scientific Publishing Company, Amsterdam.)

The enzymatic peak shift technique is used to identify chromatographic peaks representing biological compounds or classes of compounds. Peak purity can also be determined using this technique. A sample that had been chromatographed previously is incubated with an enzyme that catalyzes the reaction of a certain compound(s) under study in the chromatogram. An aliquot of the sample is rechromatographed, and the disappearance of the substrate peak and/or the appearance of the product peak(s) establishes the identity of the compound in question. Table 4 lists some enzymes that have been used to identify some UV-absorbing compounds found in human serum [3].

The stopped-flow UV scanning technique is used to provide the entire UV spectrum of a peak under study. This technique, used in combination with the evidence from the techniques mentioned previously, can provide unambiguous peak identification.

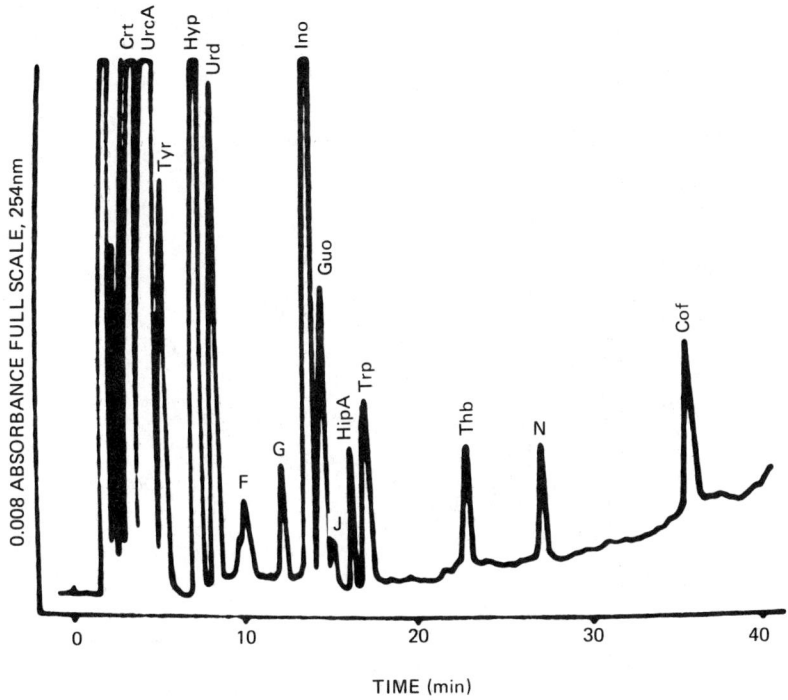

Figure 3. Control sample coinjected with a solution containing creatine (Crt), uric acid (UrcA), tyrosine (Tyr), hypoxanthine (Hyp), uridine (Urd), inosine (Ino), guanosine (Guo), hippuric acid (HipA), tryptophan (Trp), theobromine (Thb), and caffeine (Caf). Conditions same as in Fig. 3. (From Ref. 4. Copyright Elsevier Scientific Publishing Company, Amsterdam.)

Table 2 Peak Height Ratios (254 nm/280 nm) for Some Standard Nucleosides and Bases

Cytosine (CYT)	1.91 ± 0.02
Uracil (URA)	3.20 ± 0.04
Cytidine (CYD)	1.20 ± 0.03
Hypoxanthine (HYP)	19.70 ± 0.02
Xanthine (XAN)	1.73 ± 0.03
Uridine (URD)	3.42 ± 0.02
Thymine (THY)	2.22 ± 0.04
Xanthosine (XAO)	1.57 ± 0.03
Inosine (INO)	8.52 ± 0.01
Guanosine (GUO)	2.18 ± 0.03
Adenine (ADE)	11.00 ± 0.04
Thymidine (dTHD)	1.42 ± 0.04
Adenosine (ADO)	8.03 ± 0.04

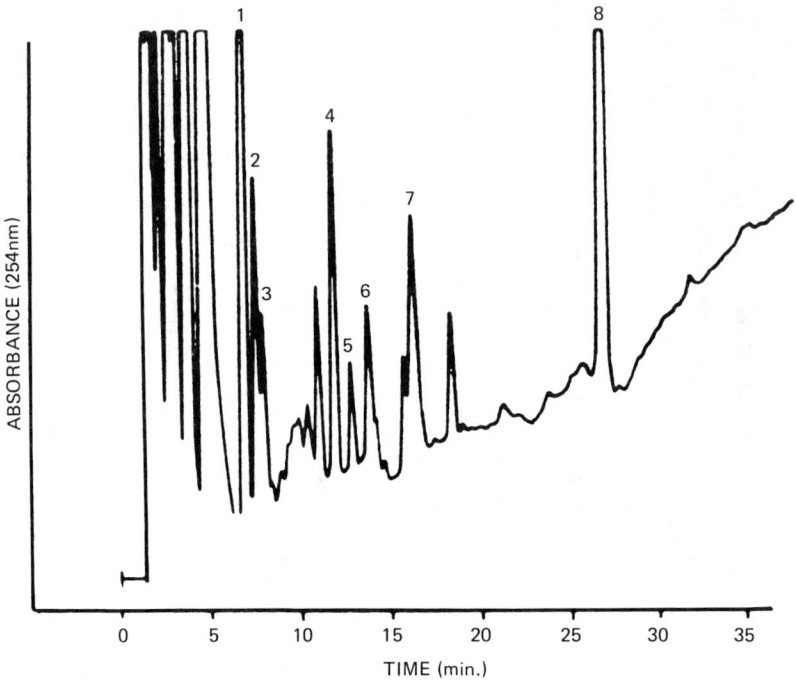

Figure 4. Separation of components in a serum sample from a patient with congestive heart failure. Chromatographic conditions: column, Bondapak C18; gradient, linear in 35 min from 0% of 0.02 F KH_2PO_4; pH 5.5 to 40% of 60/40 (v/v) CH_3OH/H_2O; flow rate, 1.5 ml/min; temperature, ambient. (Reprinted with permission from Ref. 5. Copyright 1977 American Chemical Society.)

Table 3 Peak Height Ratios (254 nm/280 nm) for Some Standards and Peaks with the Same Retention Time Characteristics in Serum Sample

Peak no.		Standard	
1.	19.00 ± 0.03	HYP	19.70 ± 0.02
2.	3.56 ± 0.02	URD	3.42 ± 0.02
3.	1.78 ± 0.03	XAN	1.73 1 0.03
4.	1.62 ± 0.02	XAO	1.57 ± 0.03
5.	8.42 ± 0.04	INO	8.52 ± 0.01
6.	2.25 ± 0.01	GUO	2.18 ± 0.03

Table 4 Enzymes Useful for the Identification of Some UV-Absorbing Compounds Found in Human Serum

Substrate	Enzyme	Product(s)
Adenosine	Adenosine deaminase	Inosine
Guanine	Guanase	Xanthine
Guanosine	Purine nucleoside phosphorylase	Guanine
Hypoxanthine, xanthine	Xanthine oxidase	Xanthine, uric acid
Inosine	Purine nucleoside phosphoyrlase	Hypoxanthine, uric acid
L-Tryptophan	Tryptophanase	Indole
Uric acid	Uricase	Allantoin

Source: Ref. 4.

Figure 5A represents the chromatogram of the serum profile of a patient with severe depression. The peak eluting at approximately 15 min was tentatively identified as inosine on the basis of retention behavior. Further peak characterization was accomplished by co-chromatographing the sample with a standard inosine reference compound (Fig. 5B), which resulted in an increase in the peak representing the compound under study. The UV spectrum of an inosine standard was compared to the UV scan of the chromatographic peak tentatively identified as inosine (Fig. 6); no significant dissimilarities were observed. The enzymatic peak shift technique using purine nucleoside phosphorylase was then performed as final confirmation for the identity of the peak (Fig. 5C). Therefore, by using the accumulated evidence of multiple identification techniques, unambiguous peak identification can be accomplished.

G. Conclusion

The identification of peaks in a chromatogram of a biological sample can be accomplished using a combination of several of the UV detection techniques: retention characteristics, co-chromatography with standards, enzymatic peak shifts, absorbance ratios, and UV spectra. Retention times and absorbance ratios are used for initial, tentative peak identification, since these techniques provide immediate information without addition of reagents to the sample and subsequent injections. Two chromatograms are required to characterize a peak using the co-chromatography technique. Subsequent alterations of a sample prior

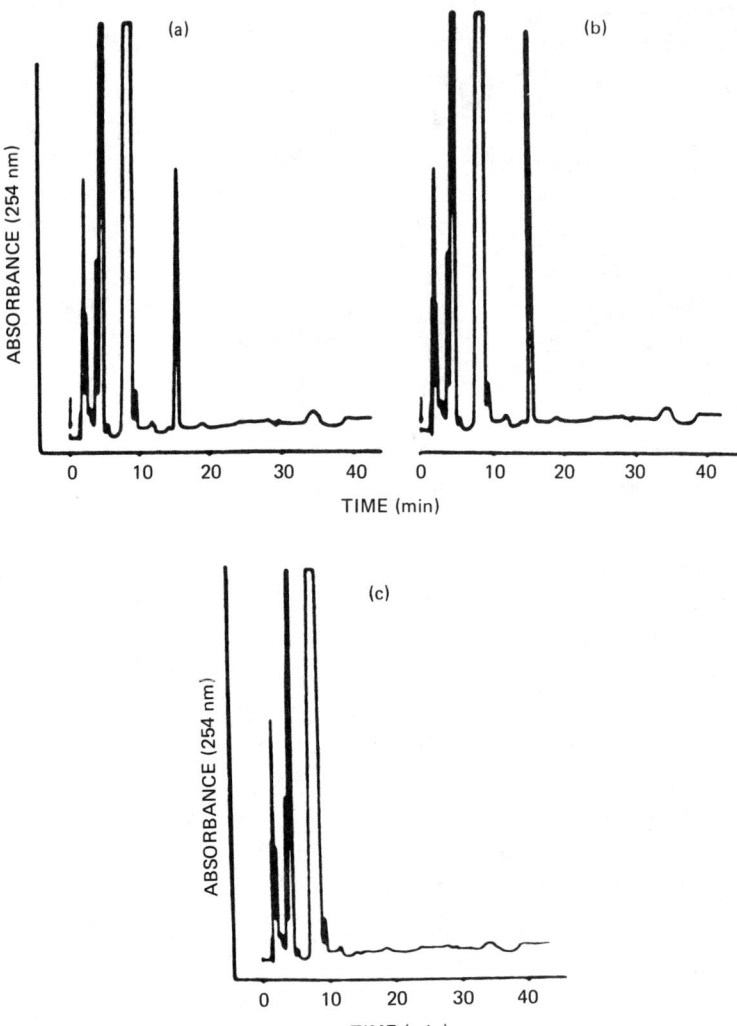

Figure 5. (a) Chromatogram of a serum sample from a patient suffering from severe depression. (b) Chromatogram of the same serum sample co-injected with inosine. (c) Purine nucleoside phosphorylase peak shift of patient serum sample. Conditions are the same as Fig. 2 except as follows: gradient linear from 1 to 40% B in 35 min. (From Ref. 6. Copyright Elsevier Scientific Publishing Company, Amsterdam.)

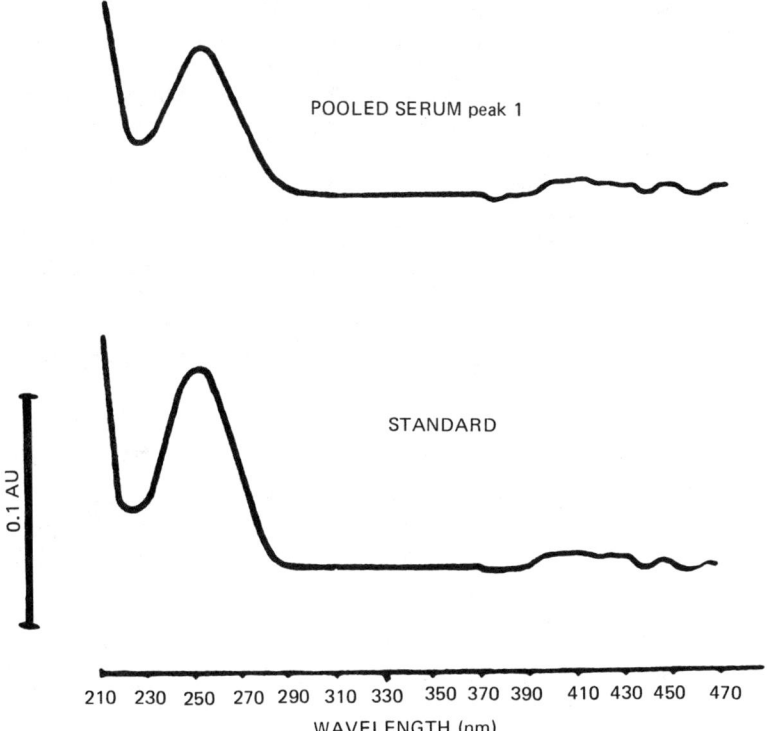

Figure 6. Stopped-flow UV spectra of chromatographic peak in question from Fig. 5 and the inosine standard. (From Ref. 6. Copyright Elsevier Scientific Publishing Company, Amsterdam.)

to injection are needed when using enzymatic peak shifts for peak identification. The scanning UV absorbance technique, which is normally performed by the stopped-flow method, is time-consuming and often provides only negative evidence with regard to the identity of a peak. However, the accumulated evidence from all of the identification techniques mentioned above reasonably establishes peak identity.

III. USING OTHER DETECTORS

A. Introduction

UV detection employing multiple peak identification techniques in many cases provides only tentative peak characterization. In addition it can be very time-consuming and can require relatively large sample volumes, since an aliquot may be needed for each identification technique.

Also, since many of the compounds present in biological samples are UV-absorbing, chromatograms are complex, further complicating peak identification. Therefore, it is often advantageous to use more selective detectors, for example, fluorescence or electrochemical detectors, since only those compounds that are naturally fluorescent or electroactive will produce a response. Reaction detectors, which use specific chemical reactions, are also very selective detectors for use with HPLC. Another extremely selective detector, the inductively coupled plasma (ICP) emission spectrometer, is specific for certain elements and has been successfully interfaced with a liquid chromatograph. Detection systems such as mass spectrometry (MS), infrared spectroscopy (IR), and nuclear magnetic resonance (NMR) are potential universal detectors for liquid chromatography and can provide unambiguous peak identification.

B. Fluorescence detectors

Fluorescence detectors provide extremely sensitive and selective detection of certain biological compounds eluting from a chromatographic system. However, the compound of interest must fluoresce or a fluorescent derivative must be produced upon reaction with a suitable reagent. Fluorescent detection systems provide more information than absorbance detectors, since both the excitation and emission spectra can be obtained. Since particular excitation maxima are required for individual compounds, selective detection is possible, thereby providing information on the identity of a compound. Also, selective detection can eliminate interferences from undesired compounds that either do not fluoresce or cannot be excited at the particular wavelength chosen [4].

Peak identification with a fluorescence detector includes several of the same identification techniques commonly used with UV absorbance detectors. For example, the comparison of retention times and peak height ratios of eluting compounds with those of standard compounds is used to identify tentatively a peak. However, since both excitation and emission spectra are monitored, peak height ratios at two different excitation and emission wavelength combinations can be obtained, providing additional information for peak identification [6].

Peak scanning with a fluorescence detector can be performed by scanning either the excitation and/or the emission spectra. Peak characterization is accomplished by comparing the spectrum of a sample peak with spectra of standard reference compounds. A synchronous scanning technique can also be used in which the excitation and emission wavelengths are varied simultaneously while maintaining a constant difference between the two wavelengths [7]. A simplified spectrum results where the maximum intensities are highly characteristic of particular compounds. Therefore, selective detection of certain constituents in a biological sample can be accomplished. The synchronous scanning technique used in the stopped-flow mode to scan an individual peak provides a powerful method of peak identification [8].

Newer fluorescence detection systems which greatly facilitate peak identification are the rapid scanning fluorometer [9,10] and the video fluorometer [11,12]. The rapid scanning fluorometer is similar to the rapid scanning UV absorbance spectrometer in that three-dimensional chromatograms are produced in real time. Multicomponent fluorescence analysis is performed with the video fluorometer [13].

C. Electrochemical detectors

Peak identification with electrochemical (EC) detectors is based on the specific oxidation potentials and the thin-layer cyclic voltammograms of the compounds under study [14-16]. Obviously, the compounds of interest in a chromatogram must be electroactive in order to be detected. The comparison of the current-potential responses of an unknown compound with the responses of standard reference compounds establishes tentatively the identity of the unknown peak. New electrochemical detectors, which include multiple electrode systems, greatly facilitate peak identification [17,18].

Multiple electrode systems include two electrodes acting either in series or opposed, providing selective detection of the electroactive components of a column eluent. Two electrodes operating in the parallel mode provide two chromatograms that are obtained simultaneously by operating the system at two different potentials. Therefore, information is provided concerning the redox identity of eluting compounds. Electrodes operated in series yield improved selectivity, since the product of the first electrode are monitored downstream by the second electrode. Both oxidative and reductive current-potential information is obtainable with series dual electrodes. The ratio of the downstream peak current to the upstream current, known as the collection efficiency, qualitatively indicates the type of functional groups present in the compounds eluting from the chromatograph. Also, this ratio is an important measure of peak purity, since a pure sample component will have a characteristic collection efficiency.

The application of electrochemical detectors in analyzing for nucleic acid constituents has been limited to such compounds as NADP, NADPH, and uric acid [19-22]. However, with the development of derivatizing agents specifically for electrochemical detection [23], it is expected that EC detectors will be more readily used in the analysis of nucleic acid constituents.

D. Reaction detectors

Reaction detectors, which are designed using specific chemical derivatization techniques, are selective detection systems for liquid chromatograms. Frei and coworkers have published extensively on the use of chemical derivatization for reaction detectors [24-26]. Specific chemical reactions are used to react selectively particular sample components, making peak identification possible. However, since most chemical reactions are specific for certain functional groups, care must be taken when using chemical reactions for peak identification.

Post-column reaction systems are most often used to produce derivatives of a column eluent that can subsequently be detected by other types of detectors. For example, post-column reactors have been used on line with fluorescence detectors [27-30] and electrochemical detectors [14,31]. Krull and Lankmayr have recently published a review of post-column systems [32]. A review of post-column reactions for use with electrochemical detectors has been written by Kissinger et al. [33].

Continuous-flow enzyme detectors, which use immobilized enzymes, are another type of reaction detector for liquid chromatography. The relative specificity of enzymes is used to react selectively certain sample components eluting from a chromatograph. Also, enzymatic methods are sufficiently specific so that enzymatic reactors are often used on-line with other nonspecific detection systems, providing two different detection systems on which to base peak identification. For example, Kito et al. performed an immobilized enzyme reaction on eluted purines in which the UV absorbance of the products was detected [34]. Several enzyme reactors used in series can selectively react several different compounds in a column eluent, thereby providing additional information for peak characterization [35]. The use of immobilized enzyme reactors for peak identification is, however, limited by the irreproducibility of enzymatic reactions if reaction conditions are not carefully controlled from sample to sample. Frei and Lawrence have reviewed the use of immobilized enzymes in post-column reactors [26].

E. Inductively coupled plasma emission spectroscopy

Inductively coupled plasma (ICP) emission spectrometers are extremely specific in that they respond only to certain inorganic elements [36-38]. ICP detection systems also have multielement capability [39], thus enhancing their use for the analysis of complex samples. A liquid chromatograph interfaced with a phosphorus-sensitive ICP detector has been used for the determination of nucleotides in biological samples [40-41]. Since the detector responds only to phosphorus-containing compounds, extremely selective detection results in which peak identification is facilitated by a simplified chromatogram as opposed to the complex chromatograms that result from a UV absorbance detector.

F. Mass spectrometers

Mass spectrometry (MS), which has been recognized as a universal detector for gas chromatography, is now being interfaced with liquid chromatographic (LC) systems [42-47]. Since mass spectra can provide not only the molecular weight but also the fragmentation patterns of chemical structures depending on the ionization techniques used, unambiguous peak identification is possible. Also, the identity of unknown compounds can be established by searching existing spectral data. LC-MS has been used for the analysis of purine and pyrimidine bases, nucleosides, and nucleotides with subsequent characterization of the sample components in the column effluent [48,49].

G. Nuclear magnetic resonance and infrared spectroscopy

The quest for a universal detector that provides unambiguous peak identification has lead to the investigation of both nuclear magnetic resonance spectroscopy (NMR) and infrared spectroscopy (IR). Both techniques provide sufficient structural information to obtain the identity of sample components under study.

NMR spectra provide chemical shift values, coupling constants, and integration ratios that if evaluated together establish the structure of a compound. Bayer et al. have demonstrated a liquid chromatograph nuclear magnetic resonance system in which sufficient structural information is obtained to provide classification of unknown compounds [50].

The interfacing of IR spectroscopy with HPLC has in the past been limited to gel chromatography [51]. However, IR spectroscopy has been used in two different ways to monitor column eluents: by comparing the spectra of a chromatographic peak with a standard reference spectrum or by monitoring a frequency characteristic of a certain functional group [52,53]. Also, the combination of a liquid chromatograph with a Fourier transform IR spectrometer has recently been achieved which will greatly aid chromatographic peak identification [54,55].

Further work concerning the interfacing of both NMR and IR spectroscopy with HPLC must be done before these detection systems are readily available commercially and can be used routinely in the laboratory. However, when LC-NMR and LC-IR are available for routine use, both techniques will provide powerful methods for absolute sample component identification.

IV. CONCLUSION

At present the identification of nucleic acid constituent peaks in HPLC chromatograms of biological samples is routinely accomplished using an UV detector and multiple identification techniques: retention times, absorbance ratios, co-chromatography with standards, and enzymatic peak shifts. However, many other biological compounds also absorb in the same region of the UV spectrum; thus chromatograms are complex and peak characterization difficult. Therefore, it is often advantageous to use a selective detector, which responds only to particular compounds, depending on the molecules of interest and the type of detector used.

Recently, substantial advances have been made in the area of interfacing HPLC with such spectroscopic techniques as MS, NMR, and IR. These spectroscopic methods, when used on-line with a liquid chromatograph, can provide unambiguous peak identification of the biological compounds eluting from a column. LC-MS, LC-NMR, and LC-IR systems are as yet not readily available on the commercial market. However, the successful development and routine use of these chromatographic systems is inevitable, and these systems will and have great potential in the HPLC analysis of nucleic acid constituents.

REFERENCES

1. P. R. Brown and E. Grushka, Anal. Chem., 52, 1210, 1980.
2. T. Catterick, J. Chromatogr., 259, 59 (1983).
3. M. S. Denton, T. P. DeAngelis, A. M. Yacynych, W. R. Heineman, and T. W. Gilbert, Anal. Chem., 48, 20 (1976).
4. R. A. Hartwick, A. M. Krstulovic, and P. R. Brown, J. Chromatogr., 186, 659 (1979).
5. A. M. Krstulovic, P. R. Brown, and D. M. Rosie, Anal. Chem., 49, 2237 (1977).
6. A. M. Krstulovic, R. A. Hartwick, P. R. Brown, and K. Lohse, J. Chromatogr., 158, 365 (1978).
7. J. B. F. Lloyd, Nature (London), 231, 64 (1971).
8. J. L. DiCeasare and L. S. Ettre, J. Chromatogr., 251, 1 (1982).
9. R. P. Cooney, T. Vo-Dinh, G. Walden, and J. D. Winefordner, Anal. Chem., 49, 939 (1977).
10. J. R. Jadamec, W. A. Saner, and Y. Talmi, Anal. Chem., 49, 1316 (1977).
11. D. W. Johnson, J. B. Callis, and G. D. Christian, Anal. Chem., 49, 747A (1977).
12. L. W. Hershberger, J. B. Callis, and G. D. Christian, Anal. Chem., 53, 971 (1981).
13. C. J. Appellof and E. R. Davidson, Anal. Chem., 53, 2053 (1981).
14. P. T. Kissinger, Anal. Chem., 49, 447A (1977).
15. T. C. Pinkerton, K. Hajizadeh, E. Deutsh, and W. R. Heineman, Anal. Chem., 52, 1542 (1980).
16. R. Samuelsson, J. O'Dea, and J. Osteryoung, Anal. Chem., 52, 2215 (1980).
17. G. S. Mayer and R. E. Shoup, J. Chromatogr., 255, 533 (1983).
18. D. A. Roston and P. T. Kissinger, Anal. Chem., 54, 429 (1982).
19. G. C. Davis, K. L. Holland, and P. T. Kissinger, J. Liq. Chromatogr., 2, 663 (1979).

20. J. Moiroux and P. J. Elring, J. Am. Chem. Soc., 103, 6533 (1980).
21. R. E. Shoup, C. S. Bruntlett, W. A. Jacobs, and P. T. Kissinger, Am. Lab., 13(10), 144 (1981).
22. K. Vohra, Am. Lab., 13(5), 66 (1981).
23. K. Shimada, M. Tanaka, and T. Nambara, Anal. Lett., 3, 1129 (1980).
24. R. W. Frei and A. H. M. T. Scholten, J. Chromatogr. Sci., 17, 152 (1979).
25. R. W. Frei, L. Michel, and W. Santi, J. Chromatogr., 126, 665 (1976).
26. R. W. Frei and J. F. Lawrence (Eds.), *Chemical Derivatization in Analytical Chemistry. Volume 1: Chromatography*, Plenum Press, New York, 1981.
27. J. C. Gfeller, G. Frey, and R. W. Frei, J. Chromatogr., 142, 271 (1977).
28. G. Schwedt, J. Chromatogr., 143, 463 (1977).
29. M. A. Moye, S. J. Scherer, and P. A. St. John, Anal. Lett., 10, 1049 (1977).
30. G. Schwedt, Chromatographia, 10, 92 (1977).
31. Y. Takata and G. Muto, Anal. Chem., 45, 1864 (1973).
32. I. S. Krull and E. P. Lankmayr, Am. Lab., 14(5), 18 (1982).
33. P. T. Kissinger, K. Bratin, G. C. Davis, and L. A. Pachla, J. Chromatogr. Sci., 17, 137 (1979).
34. M. Kito, R. Tawa, S. Takeshima, and S. Hirase, J. Chromatogr., 231, 183 (1982).
35. R. S. Schifreen, D. A. Hanna, L. D. Bowers, and P. W. Carr, Anal. Chem., 49, 1929 (1977).
36. C. H. Gast, J. C. Kraak, H. Poppe, and F. J. M. J. Maessen, J. Chromatogr., 185, 549 (1979).
37. D. M. Fraley, D. Yates, and S. E. Manahan, Anal. Chem., 51, 2225 (1979).
38. M. Morita, T. Uehiro, and K. Fuwa, Anal. Chem., 52, 349 (1980).
39. I. S. Krull and S. Jordon, Am. Lab. 12(10), 21 (1980).
40. D. R. Heine, M. B. Denton, and T. D. Schlabach, Anal. Chem., 54, 81 (1982).
41. K. Yoshida, H. Haraguchl, and K. Fuwa, Anal. Chem., 55, 1009 (1983).
42. M. A. Baldwin and F. W. McLafferty, Org. Mass Spectrum., 7, 1111 (1973).
43. R. P. W. Scott, D. G. Scott, M. Munroe, and J. Hess, J. Chromatogr., 99, 395 (1974).
44. E. C. Horning, D. I. Carroll, I. Dzidic, K. D. Haegele, M. S. Horning, and R. N. Stillwell, J. Chromatogr., 99, 13 (1974).
45. W. H. McFadden, H. L. Schwartz and S. Evans, J. Chromatogr., 122, 389 (1976).

46. P. J. Arpino and G. Guiochon, Anal. Chem., *51*, 683A (1979).
47. B. L. Karger, D. P. Kirby, P. Vouros, R. L. Foltz, and B. Hidy, Anal. Chem., *51*, 2324 (1979).
48. C. R. Blakley, J. J. Carmody, and M. L. Vestal, Anal. Chem., *52*, 1636 (1980).
49. C. R. Blakely, J. J. Carmody, and M. L. Vestal, Clin. Chem., *26*, 1467 (1980).
50. E. Bayer, K. Albert, M. Nieder, E. Grom, and T. Keller, J. Chromatogr., *186*, 497 (1979).
51. D. W. Vidrine and D. R. Mattson, Appl. Spectrosc., *32*, 502 (1978).
52. F. M. Mirabella, Jr., J. F. Johnson, and E. M. Barrall, II, Am. Lab., 7(10), 65 (1975).
53. E. G. Bartick, J. Chromatogr. Sci., *17*, 336 (1979).
54. D. Kuehl and P. R. Griffiths, J. Chromatogr. Sci., *17*, 471 (1979).
55. K. H. Shafer, S. V. Lucas, and R. J. Jakobsen, J. Chromatogr. Sci., *17*, 464 (1979).

6
Quantitative Analysis in Liquid Chromatography

HUBERT A. SCOBLE Massachusetts Institute of Technology, Cambridge, Massachusetts

    I. Peak Measurements 99
        A. Peak height 100
        B. Peak area 100
   II. Calibration Methods 104
        A. Response factors 104
        B. Normalization 105
        C. Internal standardization 105
        D. External standardization 106
        E. Standard addition 106
 III. Evaluation of Results 108
        A. Accuracy and precision 108
        B. Analytical errors 108
        C. Detection limits and sensitivity 109
        D. Statistics 109
        References 111

I. PEAK MEASUREMENTS

Quantitative analysis in liquid chromatographic systems requires the precise and accurate measurement of peak heights or peak areas. While many suitable methods are available for relating detector response to concentration, the chromatographic detector response must be linear

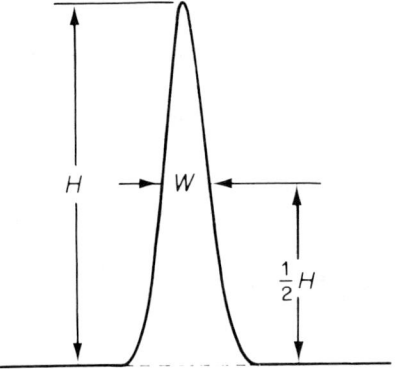

Figure 1. Peak height measurements.

with concentration and constant flow-rate. At present, quantitation is usually carried out by electronic integration of peak areas. This method of quantitation is rapid and precise. Electronic integrators are commercially available, relatively inexpensive, and easy to operate. However, if an electronic integrator is not available, measurement of peak heights is preferable because the peaks resulting using current instrumentation are too narrow for accurate area measurement. Thus, there can be great deviations in peak areas because of the variation in peak width.

A. Peak height

Peak height measurements represent a simple, satisfactory method for calculating detector response in the absence of mechanical or electronic signal integrators. In practice, a baseline is drawn from the leading edge of the chromatographic peak to the trailing edge. The vertical distance from the peak apex to this predetermined baseline represents the peak height. Peak height measurements for quantitative purposes require that the peak of interest be symmetrical (Fig. 1). In cases where peaks are represented as asymmetrical distributions, other measurement techniques must be employed.

B. Peak area

1. *Methods of measurement*

Quantitative measurements based on peak areas can be performed by several methods such as triangulation, peak height times width at one-half height, trapezoid construction, or cut-and-weigh method. Refer to other references for a discussion of these methods, since the use of electronic integration is so widespread [1,2].

2. Electronic integrators

Computing electronic integrators are the simplest and most popular method for the determination of chromatographic peak areas. Most integration systems offer such features as correction for baseline drift, various algorithms for integrating asymmetrical and incompletely resolved peaks, and methods storage for routine quantitative procedures. The following is a general description of the principle of operation and common features of typical computing integrators.

There are generally seven steps to convert the input analog signals to a usable report indicating identities and concentrations of the compounds detected in the chromatographic run:

1. Conversion of the analog signal to digitial form representing the area under the signal.
2. Recognition of chromatographic peaks in the stream of digital data. This requires discrimination of peaks from one another, from noise, and from baseline irregularities.
3. Determination of peak initiation, peak termination, peak apex, and retention time.
4. Definition of a baseline under each peak to determine final peak area and height.
5. Storing of peak identification information for post-run calculations.
6. Using retention times, performing cross-referencing with known components for identification. Using peak areas or heights, performing calculations to find the concentration of each component detected in the sample.
7. Printing the report.

The analog signal is converted to a series of digital values, each of which represents a slice of discrete area under the signal. Each of these discrete slices may be considered as a rectangle of constant width in time, the height of which is the voltage between the absolute baseline and the signal voltage. This series of discrete data points may next pass through some type of digital filter to remove transient power line or other noise. Some integrators may perform some type of summation, perhaps averaging four discrete data points to form one new data point.

A very important feature of computing integrators is that of peak recognition. The integrator must be capable of determining the start of a peak, peak apex, retention time, end of peak, peak area and height, and width (Fig. 2).

Determination of these factors is not a trivial task; some change in the analog-to-digital (A/D) stream must be recognized as indicating the start of a chromatographic peak as opposed to noise and/or simply the changing of the chromatographic baseline. If a peak is recognized, its apex must be determined accurately to define both height and

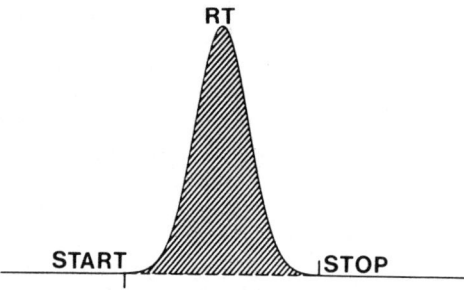

Figure 2. Various peak parameters.

retention time. Finally, a suitable chromatographic baseline must be constructed under the peak, and its final area and height determined.

In order for the above processes to occur, conditions must be defined for signal filtering and for the sampling rate of the raw data. The sampling rate matches the integrator to the rate at which the data vary due to their information content. When the signal varies rapidly, the sampling rate must be rapid. Conversely, when the signal varies slowly, the sampling rate can be slow. Effectively, we choose the number of discrete data points needed to approximate the original signal. Each point is separated from the next by a constant time interval set by the sampling rate.

Most integrators include provisions for setting thresholds. This threshold value sets the level of significance at which a signal change should be considered meaningful, as opposed to noise and/or components in concentrations small enough to be ignored.

A detailed description of how peak initiation, peak termination, and apex calculations are conducted is beyond the scope of this chapter. However, since baseline construction markedly effects peak area, a brief description of baselines and baseline construction has been included.

In electronic integration three baselines are normally considered. The first is an absolute electrical baseline imposed by the A/D converter. A signal voltage below this is ignored. The second is the integrator baseline, which is always horizontal and whose level above electrical baseline is established at the time the chromatographic run is initiated. Raw peak areas and heights are calculated from this baseline (Fig. 3). Finally, the chromatographic baseline follows changes in signal voltage. Peak starts and stops as well as final peak area and height are determined from this baseline. This baseline is horizontal only if forced to be so. Normally, the baseline drifts up and down according to the chromatographic signal.

Peak initiation and termination information is used to construct vertical droplines to the integration baseline to give raw peak areas and heights. If the peak of interest is a shoulder of a much larger peak, vertical droplines may give artificially large areas. In such situations, it is best to construct baselines using a tangent skimming technique. In this procedure, the baseline is forced by tangentially skimming the shoulder of the larger peak until the end of the smaller peak is encountered.

Since the data are in the form of discrete area slices measured above some fixed horizontal baseline, and since the beginning and end of each peak is defined and its baseline constructed, we need only subtract the height and area of the trapezoid under the constructed chromatographic baseline from the total raw area (or raw height) to arrive at the actual peak area.

Most integrators incorporate some type of method reporting for routine analyses. In practice a reference standard mixture is chromatographed, with the resulting component retention times and concentrations entered into a calibration table. Samples of unknown concentration can next be chromatographed and each peak of the chromatogram matched with its response factor, which is stored in the calibration table. Reference peaks act as landmarks to identify nonreference peaks by correcting the time at which the integrator expects to find each nonreference peak in the sample run (Fig. 4).

Although this process appears to be simple, there are complications. Individual retention times may vary slightly due to random factors, all retention times may shift with a small change in chromatographic conditions or technique, or some peaks may not appear in some samples while unexpected peaks might appear in others.

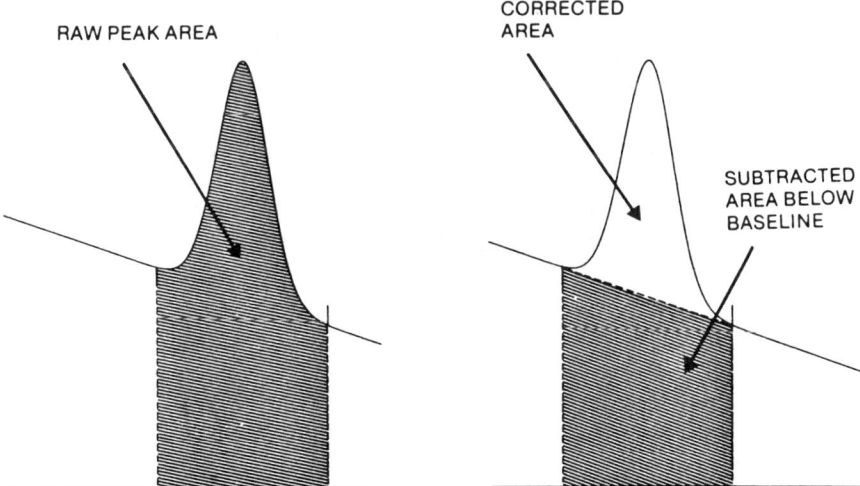

Figure 3. Peak baselines.

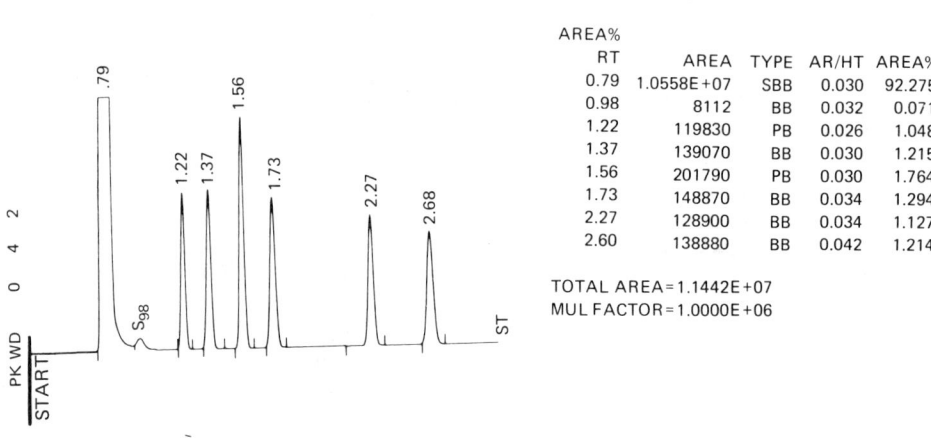

Figure 4. Integrator report.

The integrator first searches peak retention times from a chromatogram for matches to reference peaks, working forward in retention time order. A peak from the chromatographic run is identified as a reference peak in the calibration table if the run peak's retention time is within the "windows" expected from the calibration table. If two or more peaks are within the reference peak's window, the peak with the largest area is chosen. Upon finding reference peak matches, the integrator will construct a calibration curve of retention times for a particular peak in the table vs. observed retention time for the corresponding peak identified in the chromatographic run. Since it is based on identified reference peak retention time, the calibration curve corrects for shifts in retention time caused by minor variation in chromatographic conditions or technique.

Identification works best when there are multiple reference peaks in the calibration table, chosen to be present in all samples, and distributed evenly across the range of retention times for the run.

II. CALIBRATION METHODS

A. Response factors

Several calibration methods are traditionally used for quantification in liquid chromatography: normalization, internal standardization, external standardization, and standard addition. The method generation functions of most computing integrators include provisions for performing quantitative analysis using the first three methods.

A response factor must be calculated for each component by dividing the concentration of a component C_i by the area generated by the detector response to this concentration A_i in the calibration standard, i.e.,

$$RF_i = \frac{C_i}{A_i}$$

This gives a response factor for each compound, which represents a concentration per unit area, to correct for different detector response to each component of the reference mixture. These response factors are affected neither by changes in concentration nor by the absence of a component from a chromatographic run. This assumes, of course, that detector response is linear with concentration.

B. Normalization

In the normalization procedure the relationship for the calculation is

$$\% \text{ of compound "i"} = \frac{RF_i \times A_i \times 100\% \times M}{RF \times A}$$

where M is a multiplication factor that can correct for dilution or express the result in certain concentration units.

Normalization is relatively insensitive to the amount of sample injected from one run to the next. Because measured areas are corrected for variations in detector response, better quantitative results are possible than simply using the relative area percentages.

With respect to disadvantages, response factors must be determined for all components of a mixture, and all components must be eluted and detected.

C. Internal standardization

Precision and accuracy in quantitative analysis can be greatly improved by the use of internal standards. In the internal standard procedure, an additional component, not present in the sample to be analyzed, is added in known concentration to the standard reference mixture. Samples are prepared by adding a known concentration of the internal standard to the sample. The concentration of the unknown analyte is calculated according to the following relationship:

$$\text{concentration of "i"} = \frac{RF_i \times A_i \times IS(\text{concentration}) \times M}{RF(IS) \times A(IS)}$$

Desirable qualities of an internal standard are as follows: (a) It must be chemically similar to the substance to be quantitatively determined; (b) it must be completely resolved from neighboring peaks; (c) it must have a k' value similar to the peak to be quantified; (d) it must be at a concentration similar to the peak of interest; (e) it must be stable, available in high purity, and must not be present in the sample.

The use of internal standards are extremely beneficial when sample pretreatment involves extraction or recovery of the analyte of interest from a complex matrix, or in other cases where sample recovery may not be complete. If several dissimilar compounds are to be quantified in a sample, multiple internal standards should be used.

In the analysis of tRNA hydrolysis products, nucleosides, and bases in physiological fluids, Gehrke et al. [3,4] have used 8-bromoguanosine as an internal standard. In addition to having the desirable qualities required of an internal standard, it elutes in a clear portion of the chromatogram, thus eliminating any separation problems which would occur as a result of small changes in chromatographic conditions or selectivity characteristics of the column (Fig. 5).

D. External standardization

The relationship for the external standard procedure is

$$\text{concentration of "i"} = A_i \times RF_i \times M$$

Since each response factor is independent of the other components, they need be calculated only for the compounds of interest. The presence of other compounds does not affect quantitation. In addition, it is unnecessary that all components be eluted or detected.

With respect to disadvantages, composition of the calibration mixture must be close to that anticipated for samples, chromatographic conditions must be carefully maintained, and injected sample size must be reproducible to a high degree of precision. Also, calibration should be repeated frequently to detect drift in the chromatographic system response.

To eliminate possible background interferences or sample matrix effects, reference standards should be added to the same matrix as the sample to be determined.

E. Standard addition

Accuracy in quantitative chromatographic analyses can also be improved with the use of standard addition methods. In this procedure, it is necessary to have a known concentration of the material for which quantitative measurements are required. The sample of

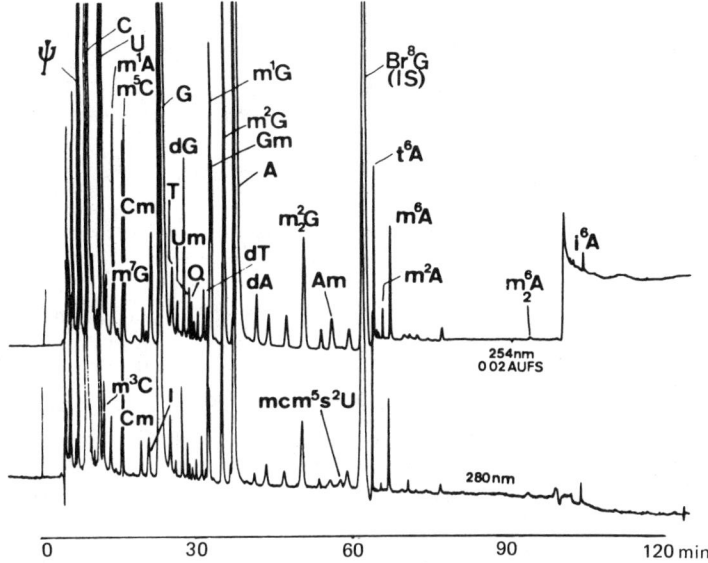

Figure 5. Reversed-phase liquid chromatogram of major and modified nucleosides. Sample: rat liver tRNA, unfractionated. Column: Supelcosil LC-18-DB. Flow rate: 1.0 ml/min. Temperature: 40°C. (From Ref. 3.)

unknown concentration, C_u, is initially chromatographed to give a peak of area A_u. Next, a known concentration of the compound to be determined, C_k, is added to the sample of unknown concentration. Injection of the same volume of sample spiked with the known concentration will result in a new peak of area A_{total}. The original concentration of C_u can be determined from the following equation:

$$C_u = \frac{C_k \times A_u}{A_{total} - A_u}$$

The above relationship holds only if the volume change upon addition of C_k is negligible to the total volume.

An alternative method is to add a constant amount of the sample of unknown concentration to a series of standards of the pure substance, and to plot the peak areas as a function of the known concentrations of the original standard. Extrapolation of the straight-line graph back to the abscissa gives the unknown concentration.

III. EVALUATION OF RESULTS

A. Accuracy and precision

There is considerable confusion as to the meaning of and difference between the terms accuracy and precision. The accuracy indicates the nearness of a measurement to its accepted value and is expressed in terms of error. Accuracy involves a comparison with respect to a true value. Precision describes the reproducibility of results and is defined as the agreement between the numerical values of two or more measurements that have been made in an identical manner.

For example, in determining the concentration of adenosine in a sample of serum, we cannot determine the accuracy of the value since no true value exists. However, we can determine the precision of our measurement if replicate analyses of the same sample are made.

Two additional descriptions of precision are reproducibility and repeatability. Reproducibility indicates the closeness of agreement between individual results obtained with the same method but under different conditions. Repeatability indicates the closeness of agreement between successive results obtained with the same method under the same conditions.

In summary, the accuracy of a measurement can be ascertained only if the true or accepted value of that measurement is available. Given a set of replicate measurements, the precision can always be expressed.

B. Analytical errors

There are generally two broad categories of error, depending on their origin. Determinate errors are those that have a definite value which can be accounted for and measured. Indeterminate errors are brought about by the effects of uncontrolled variables. This error does not have a definite measurable value but fluctuates in a random manner. Depending on the specific type of error being considered, the two categories tend to merge.

Determinate errors can further be classified as personal errors, instrumental errors, or method errors. Personal errors arise from the use of improper technique or the introduction of a personal bias in making a measurement. For example, in the measurement of peak areas via manual triangulation, personal bias may be introduced in the placement of the tangential lines. Instrumental errors arise from the improper calibration of the analytical instruments or equipment involved in the analyses, or imperfections in this equipment. Method errors could be caused from interferents or contaminants in an analysis. Blank determinations are often used to expose errors due to the introduction of interfering contaminants.

Indeterminate errors, which are usually small, occur in every analysis. The source of the errors is the uncertain nature of the measurement process which is intrinsic to the system.

C. Detection limits and sensitivity

Whenever a sample containing a compound in a very low concentration has to be measured by an analytical procedure, the response signal from the detector is usually small. It is difficult to decide whether the response signal emerges from the inherent noise produced by the procedure or from the instrument. This uncertainty gives rise to the term limit of detection. In defining the limit of detection, a criterion must be selected that is applicable to the decision whether a signal can be used for stating that a component is present or not present. If the component is present, it must be determined with what reliability it can be quantified. There are many definitions in the literature. Qualitatively, the detection limit is defined as the concentration which produces an output signal which is two times the square root of the background noise, i.e.,

$$X = 2(N)^{1/2}$$

where X is the lowest limit of detection and N is the background noise of the system. For a more elaborate explanation of sensitivity and limits of detection, the reader is referred to Ref. 5.

A procedure such as an analytical method with an input variable and one or two output variables can be characterized by the sensitivity S, which is the ratio of the quantitative output and the input, for a given qualitative range. Thus sensitivity can be defined as

$$S(X,Y) = \frac{dY}{dX}$$

for the range of X and Y values in which the relationship holds. The range for which X exists and has an unambiguous value is referred to as the dynamic range of the procedure. For reasons of convenience, methods are developed in which S has a constant value in a range as large as possible. This range is called the linear dynamic range, and is expressed in an order of magnitude for which S is constant. The dynamic range is limited at the lower level by the value of X where Y cannot be distinguished from the noise in X. The upper limit of the dynamic range will be set by saturation of the detector.

D. Statistics

Each measurement in an analysis is subject to uncertainty. These uncertainties are responsible for the variations observed among measurements of the same quantity. Hence replicate measurements are often made, the reasoning being that the confidence in which results can be viewed is increased by demonstrating this reproducibility. Having

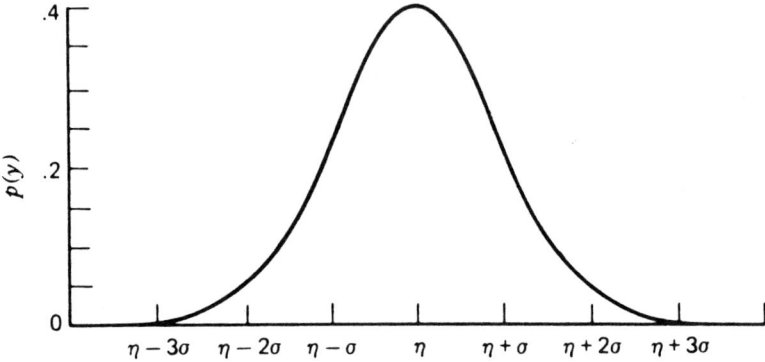

Figure 6. The Gaussian distribution.

obtained several values for the same quantity, the analyst is then confronted with defining the best value for that measurement. The arithmetic mean is used for this purpose. It is obtained by dividing the sum of a set of replicate measurements by the number of individual results in the set, i.e.,

$$X = \frac{1}{N} \sum_{i=1}^{n} X_i$$

The standard deviation σ provides a means of relating the magnitude of a given indeterminate error to its probability of occurrance, assuming that the error involved is randomly and normally distributed. Thus it may be stated that in a normally distributed population, 68.3% of the results will differ from the mean by less than one standard deviation ($\pm 1\sigma$), 95.5% by less than two standard deviations ($\pm 2\sigma$), 99.7% by less than three standard deviations ($\pm 3\sigma$), and so forth (Fig. 6).

The mean and standard deviation form the basis of many statistical tests. While further discussions of these tests is beyond the scope of this chapter, the reader should be aware of such tests for estimating the confidence interval of measurements, rejection of data, and the testing of data for agreement between two methods, among others.

REFERENCES

1. L. Condal-Bosch, J. Chem. Educ., *41*, A235 (1964).
2. John A. Dean, *Chemical Separation Methods*, Van Nostrand Reinhold, New York, 1969, pp. 254-259.
3. C. W. Gehrke, K. C. Kuo, R. A. McCune, K. D. Gerhardt, and P. F. Agris, J. Chromatogr. *230*, 297 (1982).
4. C. W. Gehrke, K. C. Kuo, and R. W Zumwalt, J. Chromatogr., *188*, 129 (1980).
5. G. Kateman and F. W. Pijpers, *Quality Control in Analytical Chemistry*, John Wiley & Sons, New York, 1981, pp. 72-82.

7
Microbore Columns

DEREK DEZARO and RICHARD A. HARTWICK Rutgers, The State University of New Jersey, New Brunswick, New Jersey

I. Introduction 113
   A. Historical overview 114
   B. Rationale for microbore HPLC 115
II. Instrumentation 120
   A. Design requirements 120
   B. Available instrumentation 121
III. Applications 127
   A. Moderate to high efficiency separations 127
   B. Recycle separations 129
   C. High-speed separations 130
IV. Summary and Future Trends 132
   References 135

I. INTRODUCTION

There has been a trend in recent years toward the miniaturization of many analytical instruments and sensors [1]. Most notably, a gas chromatograph has been developed along the lines of an integrated circuit, which can be worn as a personal minotor for environmental gases [2]. There has also been a general trend in high performance liquid chromatography (HPLC) to miniaturize instrumentation. This represents a continued maturation of liquid chromatography, begun in the late 1960s [3-5]. The continuing development of HPLC was made possible by the work of theoreticians such as Van Deemter [6], Sternberg [7], and Giddings [8]. Giddings in particular achieved a needed unification of the dynamics of chromatography and gave researchers a firm direction in the development of modern, high performance systems. The trend toward miniaturization while maintaining high performance is a natural one, since HPLC has found great applicability in the separation and quantitation of trace constituents in complex biochemical matrices.

A. Historical overview

Interestingly enough, the earliest developments in high performance liquid chromatograhy produced what is today referred to as microbore HPLC. For example, Horvath and Lipsky [9] and Burtis et al. [10] utilized 1- and 2-mm-i.d. packed pellicular columns, as did Kirkland [11,12], Snyder [13,14], and Brown [15,17]. The columns were packed with 30-50 μm packings, and were often a meter or more in length. However, with the advent of microparticulate silica materials [18-20], columns of 4.6 mm i.d. and between 10 and 30 cm in length became the standard configuration. With the reduced length, extra-column effect became more pronounced; thus column diameters and dead volumes were increased to match the detectors/injectors available. In addition, studies by Knox et al. [21-24] indicated that columns of "infinite diameter" would be more efficient due to the elimination of wall effects.

The term "microbore" HPLC is generally used to refer to packed stainless steel columns of between 0.5 and 2.0 mm i.d. bore. Packed microbore columns represent only one of three general miniaturized column types currently in use. In a recent paper, Knox [24] has classified these types as (a) small-bore packed columns (PC) (viz.), as developed primarily by Scott and Kucera [25-29,43,44]; (b) packed capillary columns, such as those developed by Novotny et al. [33-35], produced by drawing, through a drawing machine, glass tubing packed with the desired material, producing a type of hybrid packed capillary column with diameters on the order of 50-200 μm; (c) true open tubular columns (OTC), as pioneered by Ishii, Tsuda et al. [32,36-39], Nota [40], and Knox and Gilbert [41]. Knox and Gilbert have shown theoretically that such columns must be on the order of 10-30 μm in internal diameter if they are to compete with present packed conventional column performance.

More recently, Yang [47,48] has pioneered the slurry packing of fused silica gas chromatography (GC) columns of 50-500 μm. Such columns seem to exhibit properties bridging those of categories (a) and (b). The main advantage of such solumns seems to be the ease with which long lengths can be efficiently packed. A problem, however, is that the dead volume requirements are nearly as severe as with capillary HPLC, while the theoretical performance is only as great as with other packed bed columns.

The demands placed on the injection and detection volumes increase drastically in going from packed microbore to the open tubular type columns. In the latter case, injection/detection volumes of less than 10 nl must be achieved [24,42], with flow rates of only nanoliters per minute being required. Such miniscule volumes will one day revolutionize methods of analysis, but for the present time these methods remain in the research laboratories.

Packed microbore columns, on the other hand, require injection/detection volumes on the order of 0.2-0.5 µl to maintain their intrinsic efficiency, with flow rates of 10-80 µl/min [24,30,31,42,45,46]. It is feasible to attain such volumes with modified commercial equipment. Due largely to the stringent dead volume requirements of OTC systems, both Knox [24] and Guiochon [42] have shown that packed microbore systems have superior performance and speed at efficiencies up to about 200,000 plates.

Miniaturized HPLC has great advantages for biomedical analyses, since efficient separations can be achieved using only microliters per minute of solvent, and submicroliters of sample. Therefore it is useful to examine the theoretical and practical requirements for microbore HPLC, and the general rationale behind its use.

B. Rationale for microbore HPLC

1. *Performance potential of microbore*

Narrow-bore packed columns are not intrinsically more efficient than larger-bore packed columns, nor are they faster. In addition, for the same efficiency and sample loading, microbore separations tend to be less sensitive than larger-bore columns, since the detector flow cell must either use shorter path lengths or smaller apertures to keep the volume low. Despite these limitations, there are tangible reasons for using microbore HPLC for separations of nucleic acid constituents.

When conventional 4.6-mm-i.d. columns are coupled, a linear increase in efficiency is not usually observed, mainly due to fitting design. However, well-designed microbore columns may be coupled to any desired length with nearly linear increases in efficiency [43]. This is a very real practical advantage, since the separation problems of a laboratory range from several hundred plates for simple separations to as many as several hundred thousand plates for very complex mixtures. Thus, either the desired efficiency for a separation can be achieved simply by coupling columns, or higher speed of separation with less efficiency can be obtained by deleting or cutting column sections of the same material. Figure 1 shows the separation of an essential oil

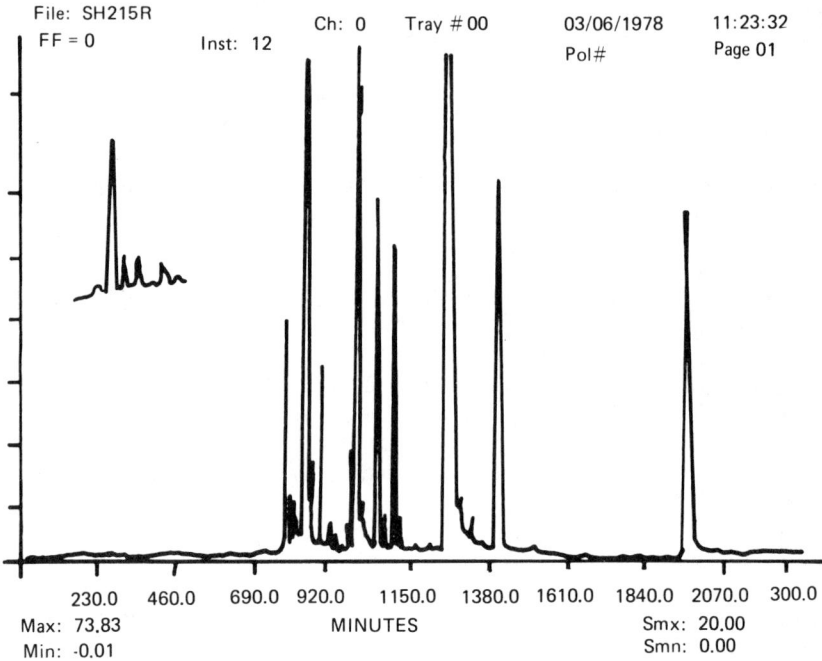

Figure 1. Chromatogram of an essential oil. 14-m × 1-mm-i.d. column. (5:95) ethyl acetate:n-heptane mobile phase, 510,000 theoretical plates. (From Ref. 27.)

[27] using a 14-m column with an efficiency of 510,000 theoretical plates and a separation time of more than 35 hr. In spite of the long analysis time, the fact that such separations can be performed at all is one of the assets of microbore HPLC. Such high efficiency separations are especially important for complex biochemical mixtures, since the number of solutes separable by an elution method can be approximated as [42]

$$N_{req} \simeq 2n^2 \tag{1}$$

Thus with the 5,000-10,000 plate separations common today, 50 peaks could be baseline-resolved within a k' of 10, whereas at 500,000 plates more than 350 solutes could be resolved.

At the other end of the performance sepctrum, microbore can be used to advantage for high-speed, simple separations. Scott et al. [28] were able to separate a seven-component mixture of common organic reagents within 30 sec (Fig. 2). In more recent work, Dezaro et al. [55] separated the clinically significant purines, theobromine and theophylline within 5 sec total time, using 3.6-cm × 1-mm-i.d. columns, packed with 5 μm reversed-phase c18 (Fig. 3). The speed of separation is not related to bore diameter, but rather to the mass transfer properties of the stationary phase, the particle size, the pressure limitation of the instrumentation, solvent viscosity, and the number of theoretical plates required. However, for the separation of theobromine and theophylline using a 4.6-cm-i.d. column, a flow rate of more than 30 ml/min would be required, a flow rate expensive to maintain and beyond the capacity of most HPLC pumps. Such simple high-speed separations will be of great utility for routine, low-efficiency separations in quality and process control, simple laboratory separations such as for DNA hydrolysate mononucleotide determination. The ability to switch readily from separations of low efficiency and high speed to those of moderate efficiency, to those of extremely high efficiency (several hundred thousand plates), without modification of the microbore instrument, is one of the most important features of narrow-bore column separations.

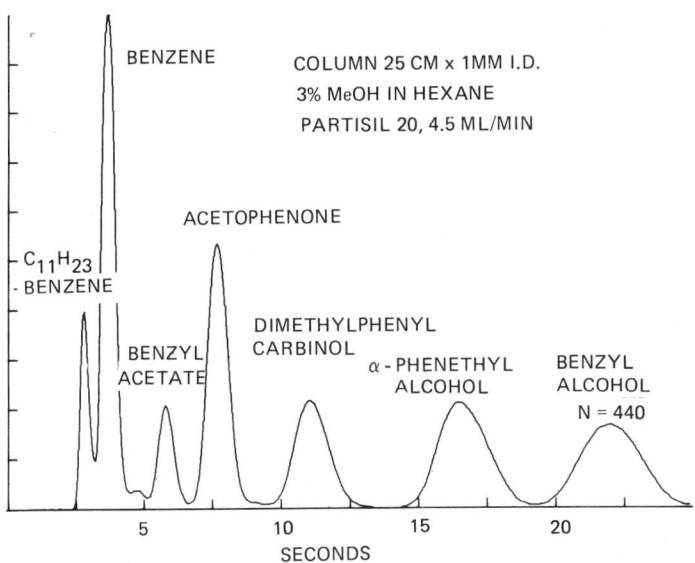

Figure 2. Thirty-second separation of a seven-component mixture. 25-cm × 1-mm-i.d. column, Partisil 20. (50:50) 3% methanol in pentane:hexane mobile phase. Peaks: 1 = phenylundecane; 2 = benzene; 3 = benzyl acetate; 4 = acetophenone; 5 = dimethylphenylcarbinol; 6 = a=phenylethyl alcohol; 7 = benzyl alcohol. (From Ref. 28.)

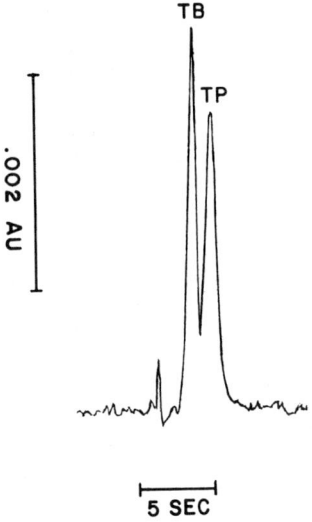

Figure 3. Five-second separation of theobromine and theophylline. 3.6-cm × 1-mm-i.d. column, 5-μm C_{18} packing. 10% acetonitrile with pH 4.5 phosphate buffer mobile phase. 1.5 ml/min flow rate.

2. Cost

Microbore (packed column) HPLC is simply HPLC on a scale smaller than that used routinely today. The entire system must be reduced proportionately. In order to maintain the same linear velocity through the packed bed at the same pressure, the HPLC is scaled down by the ratio of column diameters (d_1^2/d_2^2), as shown in Table 1. Thus, with a 4.6-mm-i.d. column and a flow rate of 1.0 ml/min, then for a microbore column with the same particle size packing, a flow rate of 0.048 ml/min would be required, with a scaling factor of 1/21. A 0.25-mm-i.d. column would require only 0.29 μl/min, or 18 ml per 10-hr day.

The cost savings of microbore HPLC for a hypothetical instrument year are calculated in Table 2 for a variety of column diameters. For the 1-mm-i.d. columns, a cost savings of $5392 would be accumulated per instrument year (with solvent at $43.00/gal). With solvents, such as acetonitrile and ion-pair reagents, the cost savings in solvents can be appreciable. With such low solvent consumption, the use of expensive solvents such as deuterium oxide becomes feasible, allowing interfacing with instruments such as nuclear magnetic resonance (NMR) spectrometers.

Table 1 Scaling Factors for Microbore and Larger Bore HPLC

d_c(mm)[a]	Scale factor d_1^2/d_2^2[b]	Equivalent f_v(cm$^3 \cdot$ min^{-1})[c]
4.6	1.0	1.0
2.0	5.3	0.19
1.0	21	0.048
0.50	85	0.012
0.25	339	0.0029

[a]Column diameter (mm).
[b]Ratio of the square of column diameters (d_1 = 4.6 mm i.d.).
[c]Flow rate (cm^3/min).

Of even greater significance is the reduced cost of column packing materials. However, it is doubtful that great savings would be realized in practice, since much of the cost of a column lies in the quality control of the packing procedure. Nevertheless, for those who pack their own columns, the reduction in materials cost from $70 to only $3 is highly significant. In addition, the 1/16-in. tubing and hardware are less expensive than standard 1/4-in. blanks. The most significant potential value of such scale reductions will be in the area of novel but expensive stationary phases, such as diamond dust [57] and noble metal packings.

Table 2 Comparative Expenses for Conventional and Microbore HPLC

Column i.d. (mm)	Required flow rate (mm^3/min^{-1})	Solvent expense ($43.00/gal)	Total dollars saved/year	Packing cost ($)
0.25	12.9	$ 16.64	$5642.59	$ 0.21
0.50	51.8	66.64	5592.41	0.83
1.0	270.0	267.03	5392.20	3.35
2.0	830.0	1070.70	4588.53	13.28
4.6	4387.0	5659.23		70.40

II. INSTRUMENTATION

A. Design requirements

1. Injection volumes

Just as the solvent flow and packing mass must be reduced by a factor of 21 in going from conventional to 1-mm packed microbore columns, so must the sample injection volume be reduced. The maximum injection volume for any eluted peak can be approximated as

$$V_{inj} = \frac{0.33 V_p}{\sqrt{N}} \qquad (2)$$

where V_p is the elution volume of the first peak of interest and N is the number of theoretical plates [58]. Calculations will show that injection volumes of between 0.2 and 2.0 µl will be required for 1-mm-i.d. columns producing separations of 5,000-10,000 plates. As the plate count increases with constant column length (e.g., using smaller, more efficient particles), the injection volume must be reduced. When using long columns (in microbore of up to several meters in length) with a constant particle size, the injection volume may be increased. Just as in standard HPLC, if some resolution can be sacrificed, the detector response can be increased by injecting larger-than-optimal sample sizes. If several microliters are injected into a microbore column, it becomes in effect a microprep column, which can be very useful for concentrating microgram quantities of oligonucleotides and other rare compounds in small volumes of solvent.

2. Detection volumes

The maximum detector cell volume is similar in magnitude to the injection volumes, and can be approximated as [24]

$$V_{cell} \simeq \sigma_v^0 \qquad (3)$$

where σ_v^0 is the standard deviation of the volume of an unretained peak. For typical microbore separations, detector volumes should thus be kept between 0.5 and 1.0 µl for maximum system performance. Also, virtually no connecting tubing can be added to the system after the injector or before the detector, since severe band spreading can result. At best, less than 10 cm of 0.007 in. tubing can be tolerated before severe efficiency losses occur when using a 25-cm × 1-mm column packed with 10-µm reversed-phase material. In practice, this is not a problem, since the 1/16-in. tubing used is usually fitted directly into the flow cell and injector.

3. Sensitivity

The 0.5-µl flow cell volume calculated as necessary for successful microbore operation imposes some real limitations on the sensitivity of microbore HPLC. In order to reduce the cell volume, both the cell path length and the cell diameter must be reduced, which in turn decreases the sensitivity. Many commercially available flow cells utilize 1-mm path lengths, a reduction of 10-fold over cells used in conventional HPLC. Therefore there is a reduction of at least 10 in absolute detector sensitivity (for the same relative mass injected and the same column efficiency) without taking into account the increased noise from the reduced aperture. If the cell length is increased and the diameter is reduced, the absorbance will increase at the expense of greater noise. This 10-fold loss in absolute sensitivity can be partially overcome by increasing the injected sample size. In the above case, a net reduction of about two-fold over conventional HPLC in terms of absolute mass injected would result. It is apparent that with absorbance detectors, microbore HPLC can never achieve the full 21-fold reduction in sample mass injected and will be at best perhaps several times more "mass sensitive" than the conventional systems. The real power of the microbore instruments will become more apparent as novel detectors are introduced to take full advantage of the reduced flow rates. Thus, utilizing HPLC with mass spectrometry [51,52], flame-based detectors [59], and other novel detectors will greatly strengthen the case for utilizing microcolumns.

B. Available instrumentation

Microbore HPLC using 1-mm-i.d. columns requires flow rates of approximately 50 µl/min, injection volumes of 0.2-5.0 µl, and detector cell volumes of 0.5 µl. Many standard instruments can be converted to achieve these volumes, and the biomedical researcher will be pleasantly surprised at the ease of use of microbore HPLC. The modifications necessary for each component will be discussed briefly, but the chromatographer considering adaptation for using microbore should consult the instrument manufacturer and the general literature for specifications on each particular instrument.

1. Pumping systems

While not every HPLC is suitable for microbore use, new pumps are becoming available with microliter-per-minute flow rates, and many existing pumps can be readily modified for microbore use. In general, the standard HPLC pump can be operated at slower than normal flow rates with good results. Check valves and pistons must be in good condition, since insignificantly small leakages at 1 ml/min can become major ones at 0.05 ml/min. The only modification that must be made is in the driving circuits, since most pumps have a lower limit on the dial setting of 0.1 ml/min.

There are two major types of motors used in HPLC pumps; stepper and analog motors. Application of analog motors is usually more difficult, since piston and internal drive friction can cause erratic operation at microbore flow rates below the manufacturers design specifications.

Stepper motor pumps, which are probably the most common, operate by the application of pulses or voltage transitions to the drive sequencing circuitry. Each pulse produces a fixed fractional motor rotation with maximum torque delivered at the lowest speed. Lowering the flow rate becomes a matter of driving the stepper motor with a lower-frequency signal of suitable form. The frequency from the dial setting can usually be modified by changing an RC element or by inserting a frequency divider in the internal pulse-generation circuitry. Some researchers have their pumps modified so that a toggle switch on the front cover can be switched to either range of flow rates. In all cases, the pump manufacturer should be consulted for details.

For external control, the easiest controller is a simple signal generator, such as HP 3311 (Hewlett Packard). If one has some expertise in digital electronics, it will be found that a simple 5-V digital signal will drive many pumps satisfactorily, so that a small microcomputer can be used. Such a microcomputer can also form the basis for a sophisticated control and data acquisition system.

The Waters M6000 pump has proven to be well suited for microbore use, as shown in Fig. 4. This calibration curve was obtained using a computer-driven TTL signal, and exhibits excellent flow linearity down into the microliter-per-minute range, at the lower end of the calibration curve. A point to keep in mind is that large volumes are contained in the internal plumbing of the pump, so that solvent changeover times can be long unless the pump is flushed at high flow rates. In general, the pump modifications require very little investment and do not disable the pump for normal use.

2. Injectors

Fortunately, several excellent commercial microbore injectors are now available at reasonable cost, including the Rheodyne models 7413 and 7411 and the Valco microinjector with 0.2- or 0.5-µl injection volume. The Rheodyne is probably a more versatile valve for the biomedical researcher, since a choice of injection volumes ranging from 0.5 to 5.0 µl is available using internal loops. The microbore column, usually 1/16 in. stainless steel or 1/8 in. glass-lined steel, is connected directly into the injector port, reducing extracolumn volumes to an absolute minimum. For some microbore column designs, as in the Whatman microbore columns, very short lengths of 0.004 in. tubing are used to

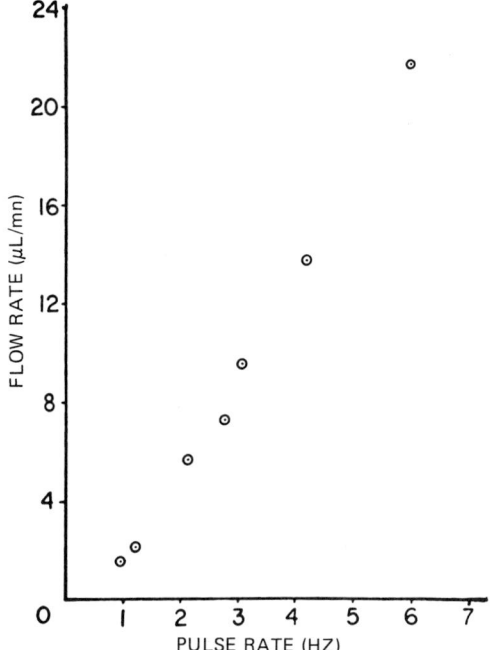

Figure 4. Low flow rate calibration curve for M 6000 pump (Waters, Inc., Milford, Mass.) Computer-generated, TTL-compatible driving signal input through external gradient control cable.

connect the column to the injector/detector. Only moderate losses in efficiency will occur as long as these are kept as short as possible.

An injection volume of 5 µl on a 1-mm-i.d. column is equivalent in terms of sample loading to a 100-µl injection on a 4.6-mm-i.d. column. These injection sizes are usually adequate for many biomedical separations. If greater injection masses are desired, sample preconcentration must be performed.

3. Detectors

Absorbance: A number of detector companies now sell microflow cells for their instruments, either by special order or as stock items. Figure 5 shows a diagram of the Kratos 0.5-µl flow cell, which can be used in any of their line of absorbance detectors with no modification. As noted above, a loss in sensitivity for an absorbance detector must invariably occur with decreasing path length, at least as much as the path length reduction itself (Beer's law), and in addition, to the

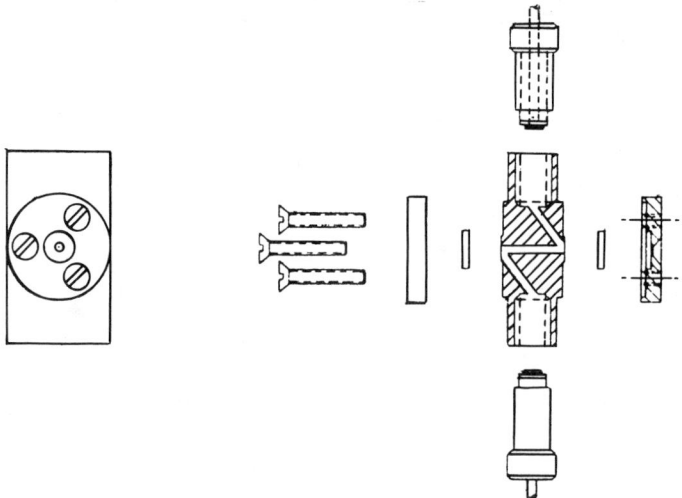

Figure 5. Micro flow cell (0.5 µl) for UV detection. (From Kratos, Inc., Ramsey, N.J.)

increased noise levels. While cell designs have been improving, a loss of about 5- to 10-fold in sensitivity will usually be encountered, thus conteracting the increased mass sensitivity of narrow-bore columns. Thus, the absolute mass of sample injected can be reduced only by about two- to fourfold in practice, rather than the 21-fold in theory. With the current state of the art of absorbance detectors, microbore will not always reduce the absolute sample mass required as compared to conventional HPLC. One solution to this problem is to enhance the solute sensitivity using post-column reaction systems. Kucera and Manius [56] have shown that miniature post-column systems can be constructed for microbore use.

Fluorescence: Unlike absorbance detection, with proper design, a small-volume fluorescence cell can be as sensitive as a larger cell. This is illustrated in Fig. 6, where a 1.0-µl flow cell is fitted into a Farrand fluorescence detector, for the separation and detection of quinine and quinidine, using a 25-cm × 1-mm-i.d. Whatman glass-lined microbore column, at a flow rate of 100 µl/min. The absolute lower limit of detection was found to be 35 fmol for the microbore flow cell, which was a fewfold more sensitive than the same instrument operated with a 10-µl flow cell. If precolumn derivatization is used, fluorescence detection in microbore HPLC can offer significant sensitivity increases and reduce the absolute sample mass requirements for the separation of nucleic acid constituents.

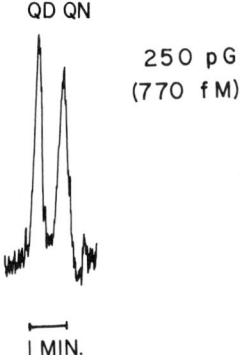

Figure 6. Fluorescense detection of quinidine (QD) and quinine (QN) in microbore HPLC. 25-cm × 1-mm-i.d. column, Partisil 10 ODS-3. 33% methanol with pH 3.25 phosphate buffer mobile phase. 0.5-µl injection of 500 ppb (250 p, 770 fmol) of each.

Flame-based ionization: Perhaps the greatest potential for microbore HPLC lies in the use of novel detectors which take advantage of the reduced flow rates of microbore separations. These include both flame-based (specific element, such as P, N, and S) and mass spectral applications. Figure 7 shows a separation of organophosphorus pesticides using a P/N sensitive flame ionization detector [59]. This separation was performed using a hybrid packed capillary column of 10 m length. Microbore flow rates allow for the direct introduction of selected organic solvents into the flame without significant quenching. Although such flame-based detectors are not widely employed with 1-mm-i.d. packed columns, there is a tremendous potential for such detection modes, especially for nucleic acid constituent separations, which invariably contain either nitrogen or phosphorus.

LC-MS: One of the more exciting uses of micro-LC is that of interfacing liquid chromatography with mass sepctrometry [51,52]. Liquid chromatography is inherently not well suited for mass spectral interfacing, since LC by definition is best used with liquid-soluble, nonvolatile solutes. However, modern mass spectrometry has advanced to the state that the ionization and analysis of nonvolatile solutes is routine using such ionization techniques as chemical ionization, laser methods, and fast atom bombardment (FAB).

Figure 8 shows a diagram of an interface designed by Henion et al. [51,52], whereby the effluent from a 1-mm-i.d. microbore column is introduced directly into the vacuum chamber of a chemical ionization quadrapole mass spectrometer. Operation is routine, and buffer salts must be limited to formate and similar volatile salts. Figure 9 shows

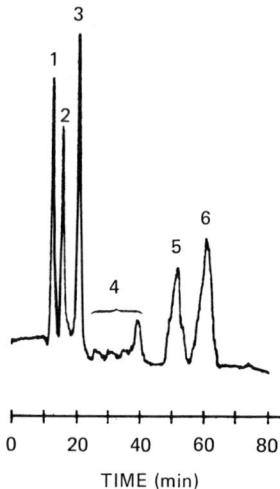

Figure 7. Chromatogram of organophosphorus pesticides. 10-m C8 microcolumn. 42% methanol mobile phase. Peaks: 1 = solvent and phosphorus-containing impurity; 2 = cygon; 3 = DDVP; 4 = phosphorus-containing impurities; 5 = malathion; 6 = guthion. (From Ref. 59.)

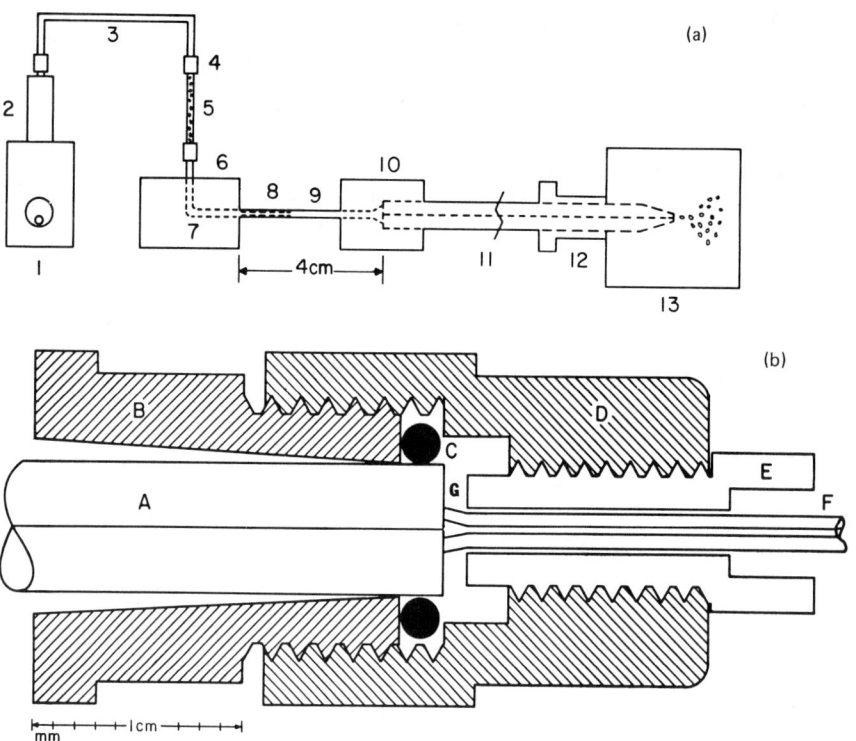

the total ion current profile and chemical ionization mass spectra for microbore cortisone and dexamethasone drugs in equestrian urine. The routine use of LC-MS, while not without problems, will at least make available the absolute identification of nonvolatile solutes in complex mixtures. Such identification, which can also exploit the inherent mass sensitivity of micro-HPLC techniques, will truly offer a quantum improvement in HPLC separations of nucleic acid constituents.

4. Column designs

Numerous manufacturers are making available microbore columns of 1-2 mm i.d. so that the biochemical researcher can construct and use microbore HPLC routinely in the laboratory. Column configurations are generally 1/16-in stainless steel tubing, although at least one manufacturer (Whatman) has opted for a glass-lined 1/8-in. column which is connected to the injector-detector via short lengths of very-small-bore tubing. Most commercial microbore columns are packed with 10-µm particles, but it should not be long before 5- and perhaps even 3-µm columns are available.

III. APPLICATIONS

A. Moderate to high efficiency separations

One of the advantages of microbore HPLC is that columns can be coupled without serious losses in efficiency. For example, Kucera et al. [27,43] were able to couple meter-long lengths of microbore columns together to generate over several hundred thousand theoretical plates. In practice, then, one is able to adjust the column length to generate only the number of theoretical plates required, thus minimizing separation times.

For normal separations of several thousand theoretical plates, microbore offers performance equivalent to conventional HPLC, but with reduced operating costs, and with good detectors, some gain in mass

---

Figure 8. (a) Schematic drawing of micro-LC/MS system. Components: 1 = pump; 2 = gas-tight syringe; 3 = PTFE tubing; 4 = sample inlet; 5 = microbore column; 6 = UV detector; 7 = 0.3-µl micro flow cell; 8 = stainless steel capillary; 9 = PTFE tubing; 10 = glass-to-Teflon connector; 11 = glass capillary micro-LC/MS probe; 12 = direct insertion probe inlet of MS; 13 = CI mass spectrometer ion source. (b) Schematic drawing of the polypropylene glass-to-Teflon connector (10 above) between PTFE tubing and the glass capillary probe. (From Ref. 52.)

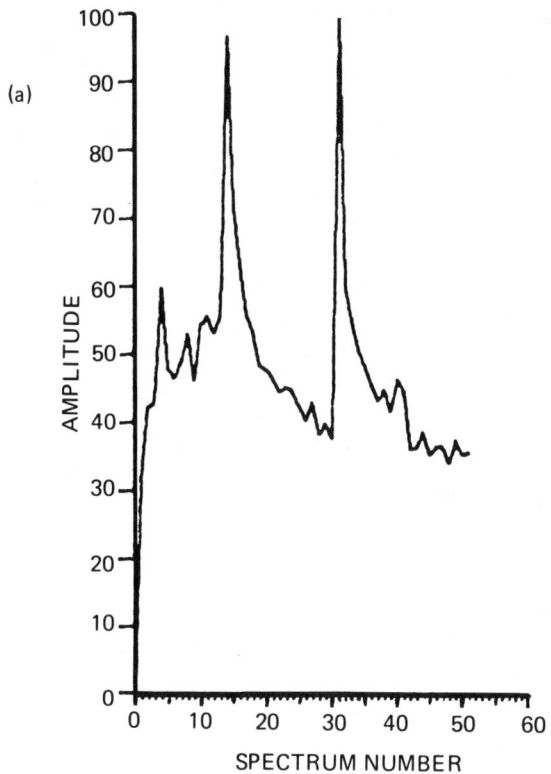
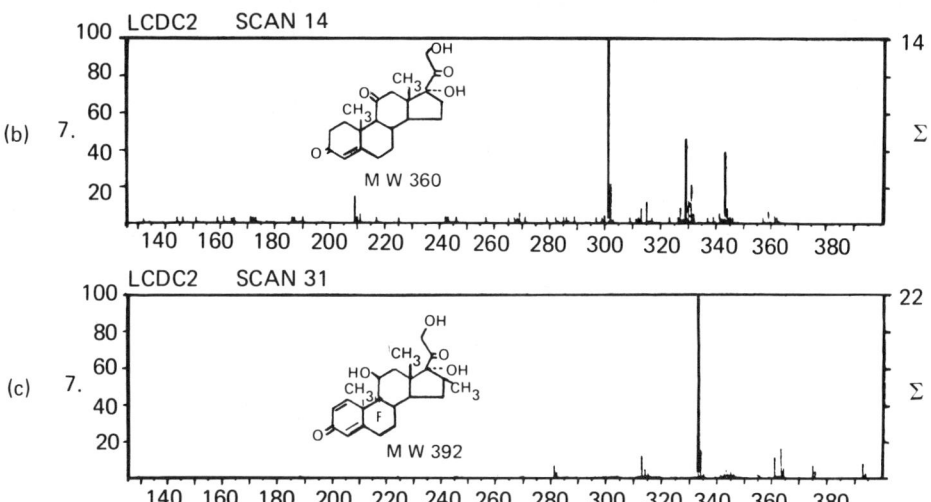

sensitivity. Figure 10 shows the separation of the four major deoxyribonucleosides using a 1-mm-i.d. column of 50 cm length, packed with 10-μm reversed-phase material. The most noticeable aspect of this separation is its equivalence to conventional HPLC separations. Excellent efficiencies and peak symmetry are observed, while the flow rate is only 50 μl/min, rather than the 1.0 ml/min which would be necessary if a 4.6-mm-i.d. column had been used.

The utility of microbore HPLC as a separation technique for rare solutes can be seen in Fig. 11, where several dimers of the ribonucleotides are separated on a 1-mm-i.d. column. Since the purine and pyrimidine rings are strong ultraviolet absorbers, detector sensitivity is good, and only nanomoles of solute need be used. In addition, if microbore is used as a preperative system, micromoles of oligonucleotides can be collected in only a few microliters of solvent, thus greatly simplifying the collection and purification of synthetic or natural oligonucleotide chains.

B. Recycle separations

One way to generate large numbers of theoretical plates without using very long columns is by recycle chromatography. This technique that has been successfully adopted to microbore HPLC by Kucera et al. [44]. In this method, the effluent is continually recycled through the column until separation is achieved. The only constraints are that the earliest eluting peaks do not overlap with the last emerging ones, and that the dispersion introduced by the connecting fittings, etc., does not introduce unacceptable dispersion. Figure 12 shows the recycle separation of a deuterated from a nondeuterated benzene derivative using recycle LC with a 1-mm-i.d. column of 1 m length. This separation which uses only submicroliter quantities of solute and microliters of solvent, demonstrates the potential of narrow-bore systems to achieve easily very high plate counts.

Figure 9. LC/MS of dexamethasone and cortisone. Separation and micro-LC/MS using a 7-cm × 0.5-mm-i.d. SC-01 column at a flow rate of 8 μl/min. 40% acetonitrile used as the LC mobile phase/CI reagent gas. (a) Total ion current profile for 20 ng of cortisone and 20 ng of dexamethasone. (b) Micro-LC/MS CI mass spectra of 20 ng of cortisone. (c) Micro-LC/MS CI mass spectra of 20 ng of dexamethasone. (From Ref. 52.)

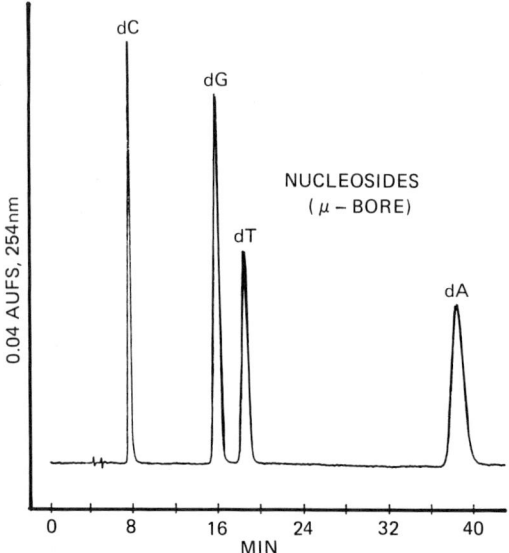

Figure 10. Chromatogram of the separation of the deoxy nucleosides. 50-cm × 1-mm-i.d. C8 column. 10% methanol with pH 4.8 phosphate buffer mobile phase.

C. High-speed separations

In the development of high-speed separations, it is important to remember that there is no minimum time or particle size for a given separation, except within the context of certain practical limitations. Thus, the number of plates, as determined necessary for a separation, can be achieved in a certain time using a given particle size and pressure (flow velocity). If the particle size is reduced, then the time of analysis will be reduced as well, all other parameters being held constant. Thus, high-speed separations, via microbore or conventional HPLC, are really only carefully optimized separations. If several hundred thousand plates are required, then several hours will be needed to achieve this, using even small particles at maximum (6000 psi) pressures. On the other hand, if only several hundred plates are necessary, then in theory it should be possible to achieve these in only several seconds, or even milliseconds, using small particles at maximum pressures. Microbore has no influence on the speed of separation. Its advantages are practical ones, since inordinate flow velocities would be needed to achieve rapid, low-efficiency separations on a large-bore column.

An example of practical high-speed automated microbore HPLC is shown in Fig. 13, where the bronchodilator theophylline is separated in four samples of human serum in 50 sec. The retention time for theophylline is 10.5 sec, while the cycle time between injections was 12.5 sec. The cycle time was chosen so that caffeine would be eluted before theophylline in the next injection without interference. Figure 14 shows a series of injections of theophylline, theobromine, and caffeine. Details on the laboratory-constructed microcomputer controller and on the theory of high-speed optimization can be found in the literature [54,55].

Figure 15 shows the separation of the four major ribonucleosides in under 24 sec, using a short, 1-mm-i.d. column packed with C_{18} stationary phase. A flow rate of 0.50 ml/min was used, which is equivalent to 10.5 ml/min in a 4.6-mm-i.d. column. Such high-speed separations, when integrated into a specially designed small instrument, should have a great impact on clinical and routine biomedical separations or on repetitive assays where low to moderate efficiency, high-volume separations are required.

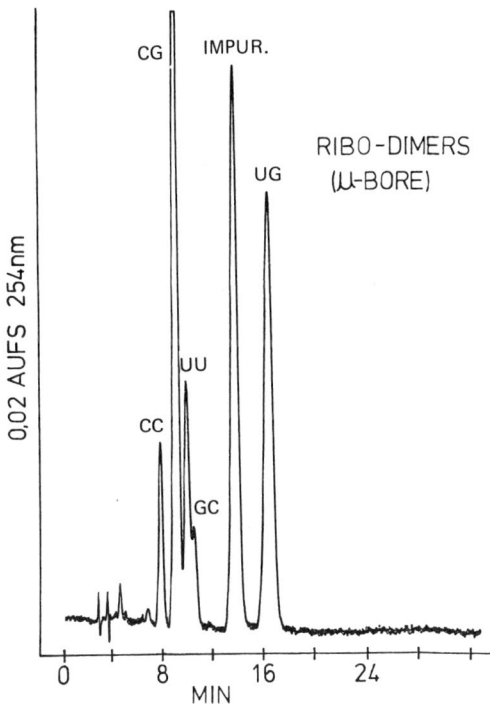

Figure 11. Chromatogram of the separation of various 3'-5' dimers. 50-cm × 1-mm-i.d. C8 column. 10% methanol with pH 4.8 phosphate buffer mobile phase.

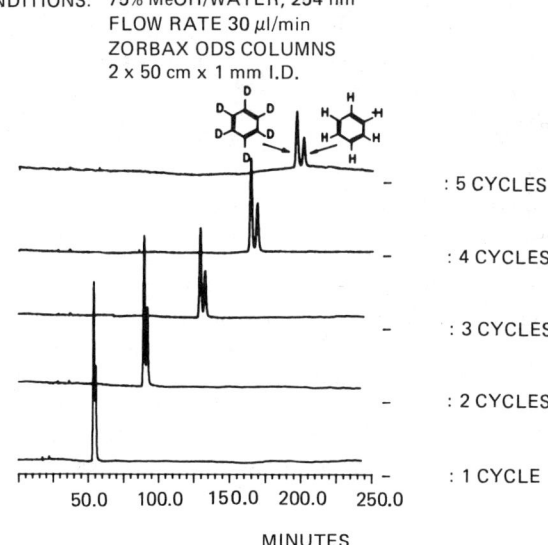

Figure 12. Recycle microbore separation of benzene isotopes. (From Ref. 44.)

IV. SUMMARY AND FUTURE TRENDS

In summary, packed microbore and other emerging small-column techniques promise to improve the performance of biological separations. Currently, small-bore packed columns of 1-2 mm i.d. seem to offer the most practical compromise between reduced solvent consumption and increased demands on small-volume detectors and injectors. As has been shown, packed microbore HPLC is today a practical technique, requiring surprisingly little in new instrumentation. The researcher should be aware that microbore is not a panacea for biological separation problems, since detection problems still exist. Indeed, for some problems, microbore is actually more demanding in terms of absolute solute mass required than is conventional HPLC. The great promise of microbore lies in the development of novel detectors and interfaces, a promise which is only beginning to be achieved.

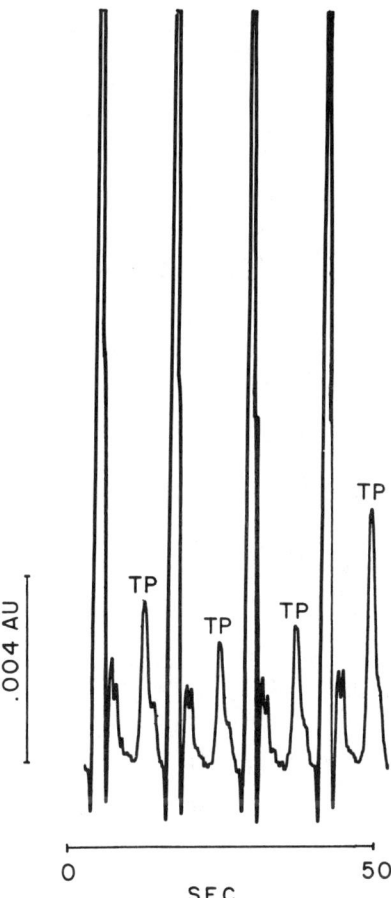

Figure 13. Separation of theophylline (TP) in four consecutively injected (0.5-μl) samples of human serum (protein precipitation with trichloroacetic acid). 6-cm × 1-mm-i.d. reversed-phase column. 15% acetonitrile with pH 4.5 phosphate buffer mobile phase. UV detection at 274 nm.

In practice, microbore can be compared to other useful techniques, such as gradient elution, which extend the versatility and scope of chromatographic separations. It assumes a special role in producing either very efficient (and long) separations, or very fast ones, two areas where conventional HPLC has difficulty competing. In between these extremes, it has the attractive features of economy with equivalent efficiency. Such advantages are not inconsequential in larger laboratories keeping several instruments running. The problems of detector sensitivity remain, and limit the attainment of the real sample reductions of which narrow-bore columns, capillary or otherwise, are capable. These problems will undoubtedly be solved, however, and as they are, microseparation methods will become increasingly common in the biomedical research laboratory. It seems unlikely that

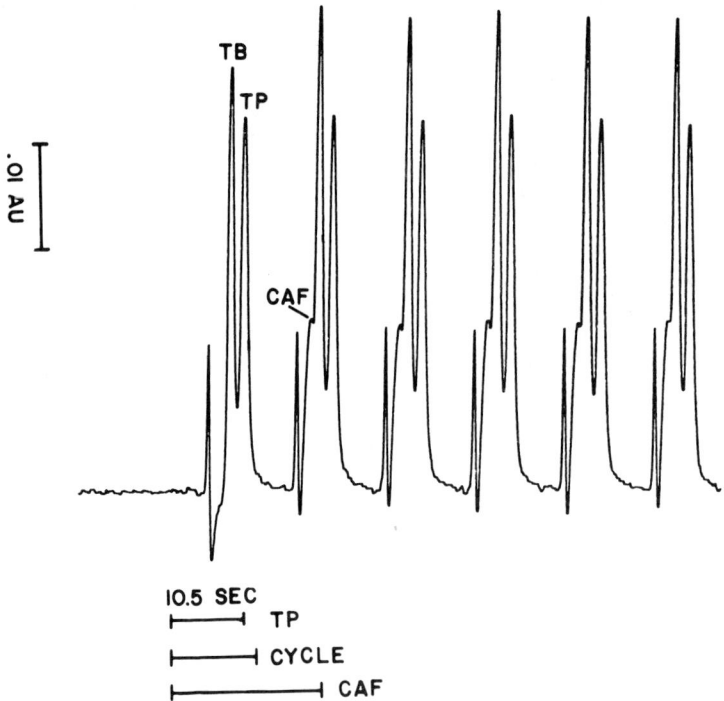

Figure 14. Multiple injections (0.5 µl) of xanthine standards (30 µg/ml), theobromine (TB), theophylline (TP), and caffeine (CAF).

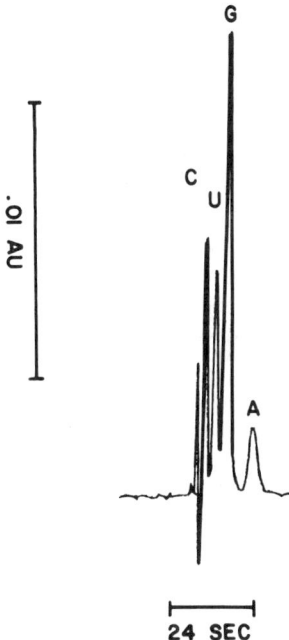

Figure 15. Chromatogram of the separation of the ribonucleosides. 4.5-cm × 1-mm-i.d., 5-μm C_{18} column. (5:10:85) acetonitrile:methanol:water with pH 4.5 phosphate buffer mobile phase. 0.3 ml/min flow rate at 4400 psi. UV detection at 254 nm.

microbore instruments will replace HPLC as it is currently practiced any time in the near future. Micro-HPLC will instead supplement the field and further broaden the range of tools available to the biomedical researcher.

REFERENCES

1. P. A. Peaden, J. C. Feldsted, M. L. Lee, S. R. Springston, and M. Novotny, Anal. Chem., 54, 1090 (1982).
2. P. W. Barth, Chemtech, 12(11), 666 (1982).
3. B. L. Karger, L. R. Snyder, and C. Horvath, An Introduction to Separation Science, Wiley-Interscience, New York, 1973.
4. L. R. Snyder, J. Chromatogr., 63, 15 (1971).
5. J. J. Kirkland, J. Chromatogr. Sci., 8, 72 (1970).

6. J. J. van Deemter, F. J. Zuiderweg, and A. Klinkenberg, Chem. Eng. Sci., 5, 271 (1956).
7. J. C. Sternberg, J. Chromatogr. Sci., 8, 692 (1969).
8. J. C. Giddings, *Dynamics of Chromatography*, Marcell Dekker, New York, 1965.
9. C. Horvath and S. R. Lipsky, Anal. Chem., 41, 1227 (1966).
10. C. A. Brutis, M. N. Munk, and F. R. MacDonald, Clin. Chem., 16, 667 (1970).
11. J. J. Kirkland, J. Chromatogr. Sci., 7, 7 (1969).
12. J. J. Kirkland, J. Chromatogr. Sci., 8, 72 (1970).
13. L. R. Snyder, J. Chromatogr. Sci., 8, 692 (1970).
14. L. R. Snyder and D. L. Saunders, J. Chromatogr. Sci., 7, 195 (1969).
15. P. R. Brown, J. Chromatogr., 52, 257 (1970).
16. P. R. Brown, Anal. Biochem., 43, 305 (1971).
17. P. R. Brown, Anal. Chem., 47, 784 (1975).
18. K. Unger, *Porous Silica: Its Properties and Uses as Support in Liquid Chromatography*, Elsevier, New York, 1979.
19. I. Halasz, H. Schmidt, and P. Vogtel, J. Chromatogr., 126, 19 (1975).
20. L. R. Snyder and J. J. Kirkland, *Introduction to Modern Liquid Chromatography*, 2nd ed., John Wiley & Sons, New York, 1979.
21. D. S. Horne, J. H. Knox, and L. McLaren, Sep. Sci., 1, 531 (1966).
22. J. H. Knox, G. R. Laird, and P. A. Raven, J. Chromatogr., 122, 129 (1976).
23. J. H. Knox, Ed., *High-Performance Liquid Chromatography*, Edinburgh University Press, Edinburgh, 1980.
24. J. H. Knox, J. Chromatogr. Sci., 18, 453 (1980).
25. R. P. W. Scott, and P. Kucera, J. Chromatogr., 125, 251 (1976).
26. R. P. W. Scott, and P. Kucera, J. Chromatogr., 185, 27 (1979).
27. R. P. W. Scott and P. Kucera, J. Chromatogr., 169, 51 (1979).
28. R. P. W. Scott, P. Kucera, and M. Munroe, J. Chromatogr., 186, 475 (1979).
29. P. Kucera, J. Chromatogr., 198, 93 (1980).
30. Z. Yukuei, B. Mianaheng, L. Xiouzhen, and L. Peichang, J. Chromatogr., 197, 97 (1980).
31. M. Goto, Y. Koyanagi, and D. Ishii, J. Chromatogr., 208, 261 (1981).
32. T. Tsuda and M. Novotny, Anal. Chem., 50, 271 (1978).
33. M. Novotny, Clin. Chem., 26(10), 1474 (1980).
34. Y. Hirata, M. Novotny, T. Tsuda, and D. Ishii, Anal. Chem., 51, 1807 (1979).
35. Y. Hirata and M. Novotny, J. Chromatogr., 186, 521 (1979).
36. T. Tsuda, K, Hibi, T. Nakanishi, T. Takeuchi, and D. Ishii, J. Chromatogr., 158, 227 (1978).

37. D. Ishii, K. Hibi, T. Yoshimura, T. Nakanishi, and I. Fugi, 26th IUPAL Conference, Tokyo, Japan, 1977.
38. K. Hibi, D. Ishii, I. Fugishima, T. Takeucci, T. Nakanishi, J. High Res. Chromatogr., 1, 21 (1978).
39. K. Hibi and D. Ishii, J. Chromatogr., 189, 179 (1980).
40. G. Nota, G. Marino, V. Buonocore, and A. Ballio, J. Chromatogr., 46, 103 (1970).
41. J. H. Knox and M. T. Gilbert, J. Chromatogr., 186, 405 (1979).
42. G. Guiochon, Anal. Chem., 53, 1318 (1981).
43. P. Kucera and G. Manius, J. Chromatogr., 216, 9 (1981).
44. P. Kucera and G. Manius, J. Chromatogr., 219, 1 (1981).
45. R. P. W. Scott, J. Chromatogr. Sci., 18, 49 (1980).
46. C. E. Reese and R. P. W. Scott, J. Chromatogr. Sci., 18, 479 (1980).
47. F. J. Yang, J. High Res. Chromatogr. and Chromatogr. Commun., 4(2), 83 (1981).
48. F. J. Yang, J. Chromatogr., 236, 265 (1982).
49. J. Hermensson, Chromatographia, 13(12), 741 (1980).
50. K. Lohse, Kratos Analytical Instruments, Inc. (personal communication).
51. J. D. Henion, B. A. Thompson, and P. H. Dawson, Anal. Chem., 54, 451 (1982).
52. J. D. Henion, J. Chromatogr. Sci., 19, 57 (1981).
53. G. Guiochon, J. Chromatogr., 185, 3 (1979).
54. R. A. Hartwick and D. Dezaro in Microcolumn High-Performance Liquid Chromatography (P. Kucera, Ed.), Elsevier, Amsterdam, 1984.
55. D. Dezaro and R. A. Hartwick, manuscript in preparation.
56. P. Kucera and H. Umagat, J. Chromatogr., 255, 563 (1983).
57. M. J. Telepchak, Chromatographia, 6, 234 (1973).
58. K. Karch, I. Sebastian, and I. Halasz, J. Chromatogr., 122, 3 (1976).
59. V. L. McGuffin and M. Novotny, Anal. Chem., 53, 946 (1981).

8
Detection Systems

SEBASTIAN P. ASSENZA* University of Rhode Island, Kingston, Rhode Island

I. Introduction 140
   A. General information 140
   B. Classification of detectors 140
   C. Detector characteristics 140
II. Detectors for Purine and Pyrimidine Assay 142
   A. Absorbance detectors 142
   B. Refractive index detectors 148
   C. Electrochemical detectors 149
   D. Fluorescence detectors 150
   E. Radioactivity detectors 152
   F. Combined detectors 154
III. Future Trends 154
   A. Mass spectrometry 154
   B. Atomic absorption 155
   C. Inductively coupled plasma emission 155
   D. Post-column detection systems 156
   E. Enzyme reaction detectors 157
   References 157

Current affiliation: Stuart Pharmaceuticals, Division of ICI Americas, Wilmington, Delaware.

I. INTRODUCTION

A. General information

The weakest link in the liquid chromatographic system is the detector. Unlike some of those available for gas chromatography, the truly ideal liquid chromatographic detector (LCD) has yet to be developed. Moreover, in comparison to gas chromatography, fewer types of LCDs are commercially available [1-4]. This is somewhat surprising, since the LCD has long been the subject of extensive research. During the explosive growth of liquid chromatography, many different types of detectors were developed and evaluated. While some of the early LCDs are still widely used, others have been discarded. At present, most research has focused on extending the capabilities of the readily available LCDs. However, the use of other promising techniques with liquid chromatography is receiving much interest [1].

The major function of the LCD is to detect and measure the individual components of a mixture after separation. Therefore, the LCD should have a predictable response for the compounds of interest. In addition, the LCD must be designed in such a way as to prevent the recombination of the separated compounds. The separation of closely eluting compounds can be obscured with the use of a LCD that has too slow a response or too large a cell volume. Furthermore, the size and configuration of the LCD should allow close placement of the detector to the liquid chromatograph in order to minimize the dead volume between the column and detector; too much tubing can lead to excessive dilution and band broadening.

B. Classification of detectors

The LCDs fall into two basic categories: bulk property (universal) or solute property (selective) detectors. A bulk property detector functions on the differential measurement of properties common to both the solute and mobile phase; examples include the dielectric constant and refractive index detectors. The solute property detector responds to a feature characteristic of the solute; the ultraviolet (UV) absorption and radioactivity detectors are solute property detectors. Table 1 lists some of the different types of LCDs.

C. Detector characteristics

Choosing a detector for a particular application can be difficult. Many factors contribute to the choice of a detector; however, the type of detector necessary for a particular application will be determined primarily by the compounds of interest and the sample matrix. The ideal detector for both routine and specialized work should have the following characteristics:

Table 1 Liquid Chromatographic Detectors

Nondestructive		Destructive	
Selective	Universal	Selective	Universal
Ultraviolet	Refractive index	Amperometric	Flame aerosal
Infrared	Dielectric constant	Polarographic	Flame ionization
Raman	Light scattering	Mass spectrometer[b]	Thermal conductivity
Fluorescence	Nuclear magnetic resonance	Atomic absorption	
Radioactivity	Piezoelectric crystal	Flame emission	
Electron capture[a]	Potentiometric[a]	Plasma emission	
Electron spin resonance	Streaming current	Thermal energy analyzer	
Chemiluminescence[a]	Heat of adsorption[a]	Photoconductivity	
Optical activity	Thermal lens calorimetry		
Photoacoustic[b]			
Thermomechanical analyzer[b]			

[a] Could also be destructive.
[b] Could also be universal.

1. Low drift and noise
2. High sensitivity
3. Universal response to all solutes
4. Predictable specificity or tunability
5. Fast response
6. Wide linear dynamic range
7. Low dead volume and no unswept areas
8. Efficient cell design which eliminates refractive index effects and remixing
9. Insensitivity to changes in solvent, temperature, and flow
10. Reliability and continuous operation

In addition, the detector should be simple to operate and inexpensive, and the compounds should not be destroyed in the detection process. It is also valuable if structural data can be obtained for component identification. Unfortunately, no single LCD can meet all these criteria. Thus, the use of several detectors and/or combined detection techniques may be required to meet the demands of the analytical system.

II. DETECTORS FOR PURINE AND PYRIMIDINE ASSAY

Since the late 1960s, absorbance detectors have been the most widely used for the liquid chromatographic separations of nucleic acid constituents. However, other devices and techniques have emerged which show great promise. Table 2 lists the detectors suitable for purine and pyrimidine separations.

A. Absorbance detectors

The purines and pyrimidines have very high absorbance in the ultraviolet range. The high absorbance of these compounds, combined with the low noise and large linear dynamic range of the absorbance detector (Table 3), make possible in a single separation the measurement of both trace and major components in a mixture. In addition, the relative insensitivity of the absorbance detectors to change with temperature, flow rate, and mobile phase composition contributes to their widespread use. Thus they can be used with different chromatographic modes (reversed-phase, ion-exchange, gel permeation, etc.) in the separation of nucleic acid components.

The absorbance LCD is a conventional spectrophotometer which is adapted for high performance liquid chromatography (HPLC) by use of a flow cell and modified optics. A beam of light irradiates the effluent and a photomultiplier or photodiode monitors the change in transmittance due to the absorption of the solutes as they elute. There are two basic configurations of absorbance LCDs: single-beam (no

Table 2 Detectors for Purines and Pyrimidines

Type	Selective	Gradient capability, reversed phase	Minimum detected	Developed for routine use
Ultraviolet absorption	Yes (tunable)	Yes (all reversed phase)	10^{-9} g	Yes
Fluorescence	Yes (tunable)	Yes (most reversed phase)	10^{-9}-10^{-12}g	Yes
Refractive index	No	No	10^{-6} g	Yes
Radioactivity	Yes (tunable)	Yes (corrections possible)	10^{-3} μCi	Yes
Electrochemical	Yes (tunable)	Yes (no steep gradients)	10^{-9}-10^{-12} g	Few applications
Inductively coupled plasma	Yes	Yes (no phosphate buffers)	10^{-6} g	No
Atomic absorption	Yes	Yes (no phosphate buffers)	10^{-6} g	No

Table 3 Common Specifications for Absorbance Detectors

Parameter	Specification
Noise	$5(10^{-5})-1(10^{-4})$ aufs
Drift	$5(10^{-4})$ aufs/hr
Flow sensitivity	$2(10^{-4})$ aufs/ml
Minimum detection[a,b]	5-10 pmol
Operating range	0.001-2.0 aufs
Temperature sensitivity	$1(10^{-3})$ aufs/°C
Gradient sensitivity	$1(10^{-4})$ aufs @ 254 nm w/MeOH
Linear dynamic range	$1(10^4)$
Response time	$\geqslant 0.05$ sec
Wavelength accuracy	0.02-1.0 nm (fixed and variable)
Flow-cell pressure	1000-2500 psi
Flow-cell volume	4-8 µl
Flow-cell path length	4-10 mm

[a] Ref. 82.
[b] Ref. 83.

reference) and dual-beam (with reference) designs. In addition, there are four general types of absorbance LCDs commercially available: (a) fixed- and dual fixed-wavelength detectors employing a discrete source such as a low-pressure, 254-nm Hg vapor lamp; (b) multiwavelength photometers employing a continuous source, where the wavelength of interest is selected through the use of appropriate filters; (c) variable-wavelength spectrophotometers using a monochromator covering the range of 190 to 400 nm (deuterium lamp) or the range of 350 to 700 nm (quartz-iodine or tungsten halogen lamps) with or without automatic scanning capability; and (d) rapid-scan, variable-wavelength detectors capable of "on-the-fly" spectra or absorbance ratios.

1. Fixed-wavelength and filter photometers

The optical layouts of a single-beam and a dual-beam LCD are shown in Fig. 1. Since the dual-beam design compensates for irregular performance in the source or photodetector, as well as for short-term electrical fluctuations, it is generally preferred for more exacting work.

While microprocessors can correct for constant irregularities in the single-beam design, they cannot correct for inconsistent noise. To operate the dual-beam detector in a differential mode, the inlet of the reference cell can be connected to the outlet of the solvent pump so that the eluent is continuously monitored (Fig. 2). The differential mode of operation is particularly useful in gradient elution and flow programming. However, to operate the detector in this manner, the cells must be able to withstand the column back-pressure; unfortunately, most cells leak at pressures in excess of 2500 psi, which can limit their use to static reference systems. In addition to withstanding high pressure, the cell design should eliminate refractive index effects, have small volumes, and no unswept areas. Figure 3 shows some cell designs commonly used.

The fixed-wavelength LCD has the advantage of high sensitivity at 254 nm because Hg emits strongly at that wavelength. In addition, a phosphor (280 nm) can be used to give efficient dual fixed-wavelength

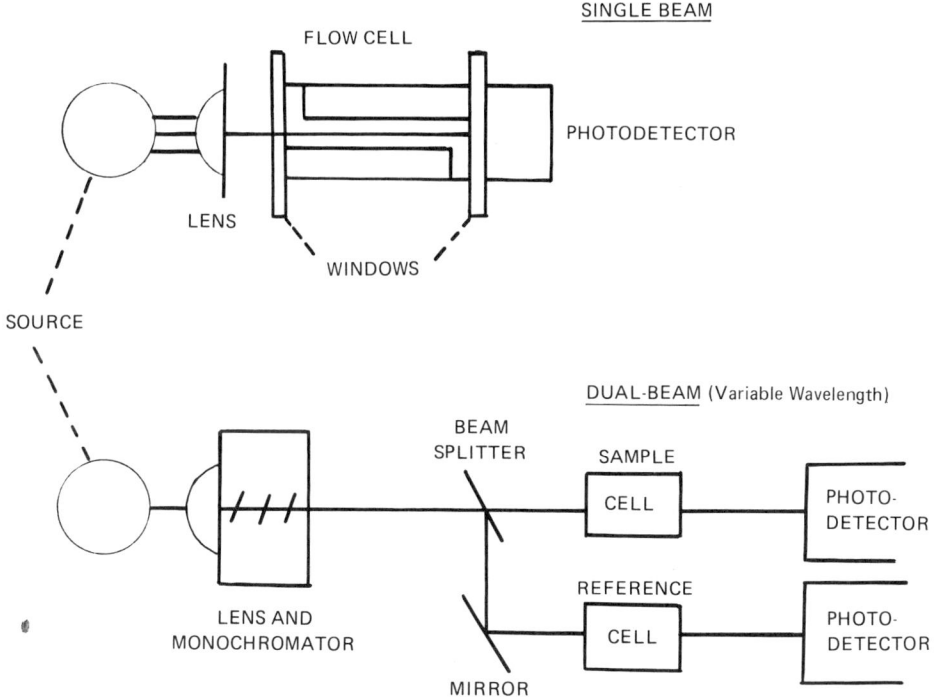

Figure 1. Absorbance detectors for liquid chromatography. Upper: single-beam design. Lower: dual-beam design (shown as a variable-wavelength type).

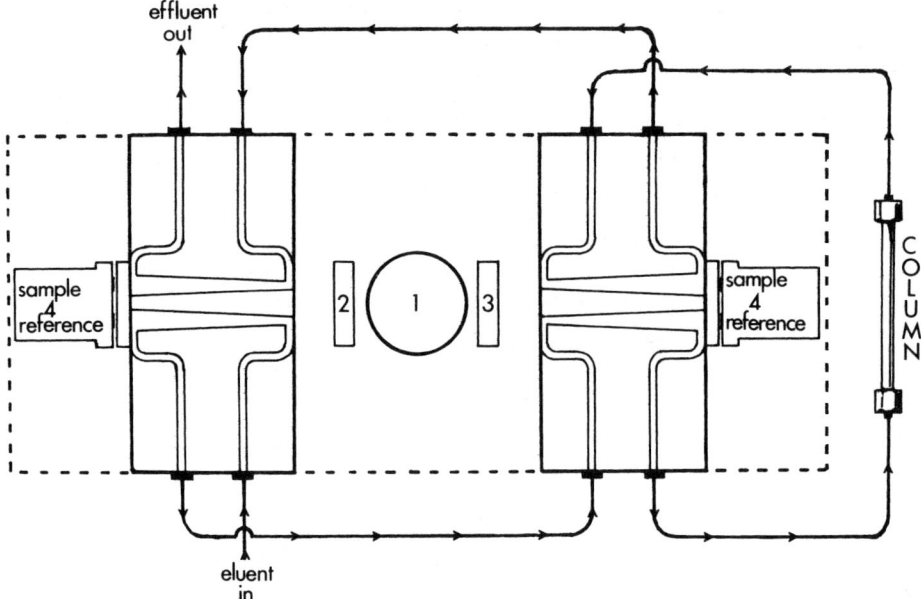

Figure 2. Dual-beam detector with dynamic (flow-through) reference. 1. Source, 2. 254-nm aperture, 3. 280-nm aperture, 4. Photodetector. (Modeled after Waters Associates design.)

detectors (Fig. 2). However, the other wavelengths emitted by Hg are very weak and cannot be used efficiently. Thus, these detectors are more suited for routine applications.

Often, wavelengths other than 254 or 280 nm are required for some purine and pyrimidine separations. The most cost-effective absorbance LCD in which wavelength selection is possible is the filter photometer. The filter photometer uses the same deuterium lamp source as in the more expensive monochromator systems; thus with the use of appropriate optical interference filters any wavelength in the range of 190 to 400 nm can be selected. This type of detector is reviewed by Klotter [5].

In order to evaluate the performance of absorbance LCDs, the ASTM E 685-79 standard for testing linear range, drift, total noise, response, and other parameters was published [6]; following this procedure a number of commercial detectors were evaluated and their specifications listed [7]. In addition, detector stability [8] and linearity [9,10] have been studied in detail.

2. Variable-wavelength and scanning detectors

The variable-wavelength and scanning detectors have several advantages over the fixed-wavelength devices. Although the filter photometer also has these advantages, it is not as convenient to use. Sensitivity can be maximized by tuning in the wavelength of highest absorptivity while minimizing the absorbance of the mobile phase. By choosing a wavelength where only the solutes of interest absorb, the selectivity can be optimized. Moreover, these detectors can be applied to aid in the identification of the separated purine and pyrimidine compounds or to confirm their purity after separation [11,12]. One method for identification uses several variable-wavelength detectors, set at appropriate values, in series to obtain absorbance or peak height ratios which can be compared to known standards [11]. Another method uses the stopped-flow scanning technique to obtain the UV spectra of the individual chromatographic peaks [12].

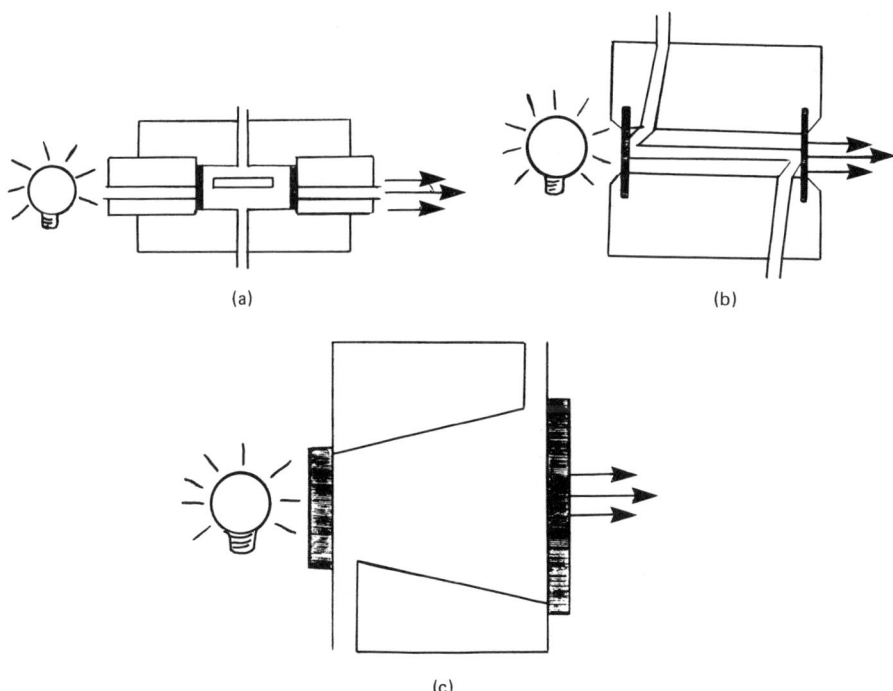

Figure 3. Common flow-cell designs. (a) H-type cell. (b) Z-type cell. (c) Z-taper cell (Waters).

Both the variable-wavelength and scanning detectors usually employ a grating monochromator to select wavelengths. The scanning and some variable-wavelength detectors have additional devices to change wavelength automatically and to correct for background effects. Background corrections for spectra can be achieved with a memory module, usually part of the detector, which stores the prerecorded background and then makes the necessary correction by subtraction from the solute spectrum [12]. Scanning in the stop-flow system is typically performed at a rate of 100 nm/min; thus, a scan requires several minutes. Since the integrity of the separation can be lost in a stop-flow system, microprocessors are used to stop and restart the chromatograph and integrator together. With some variable-wavelength detectors, microprocessors are used for other functions, such as changing wavelength and attenuation during the course of the separation [13].

The rapid-scanning spectrophotometers are gaining popularity as LCDs. The fastest of the rapid-scanning detectors use photodiode arrays or charge-coupled devices. Reviews of the operating principles [14] and the commercially available rapid-scan LCDs were published recently [15-17]. The notable feature of these rapid-scan detectors is that either continuous, "on-the-fly" spectra or absorbance ratios at two or more preprogrammed wavelengths are readily obtained. Since these devices use a continuous source (190-400 nm) and diffraction gratings to split the absorbed wavelengths (Fig. 4), the scan rate is dependent on the sweep of the diode array; usually, the time required for one scan is 50 msec. However, a rapid-scan detector can also employ a fast-moving, servo-controlled grating monochromator to disperse the light onto the flow cell. Using conventional photodiodes, the servo-controlled system can be used to measure absorbance rapidly at several wavelengths in sequence while the sample flows through the cell. The time required for a discrete scan is approximately 1 sec [18].

B. Refractive index detectors

Although the refractive index detector was one of the first LCDs, at present it has few applications in purine and pyrimidine separations. The refractometer responds to changes in the refractive index of the effluent; the eluent is used as the reference. It is considered a universal detector; it responds to all solutes whose refractive index differs from the mobile phase. While these detectors meet many of the criteria of good LCDs, they have poor sensitivity (10^{-6}g), no tunability, and lack of specificity. In addition, they are sensitive to small changes in temperature and flow rate and are best used in isocratic systems, since gradients can cause constant refractive index changes. However, they are valuable to test the purity of standard solutions or collected fractions prior to off-line characterization tests. In addition, they aid greatly in determinations of system dead-volume and gradient lag time.

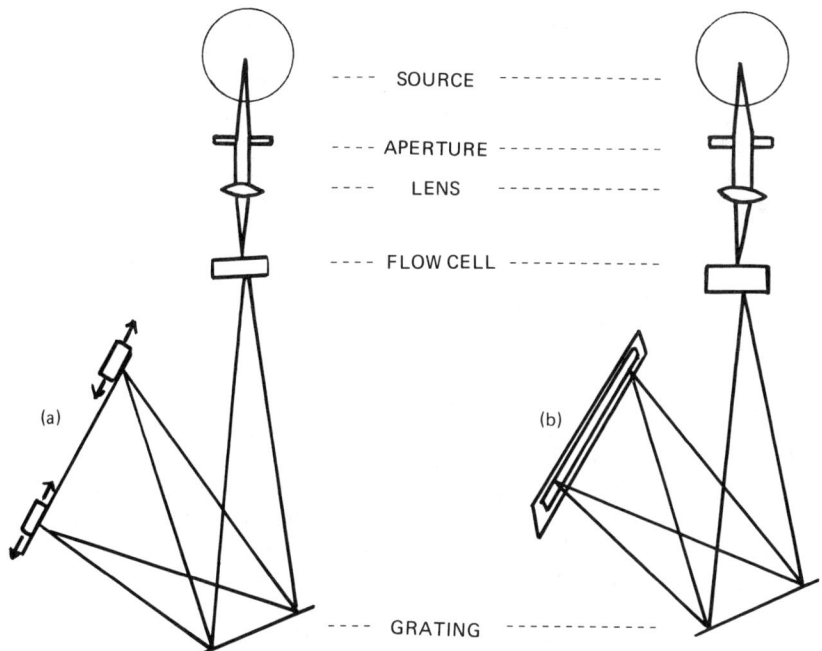

Figure 4. Scanning ultraviolet detectors. (a) Moving sample and reference photodiodes, scan rate ca. 100 nm/min. (b) Photodiode array, scan rate ca. 250 nm/msec. (From Ref. 16.)

C. Electrochemical detectors

Electrochemical detectors are extremely sensitive (10^{-12} to 10^{-15} g) for certain classes of compounds that are easily oxidized or reduced. In these detectors, current is measured which is associated with electrolytic reactions under constant or variable applied potentials. The current varies as compounds, which flow past the electrodes, are oxidized or reduced, depending on the applied potential. The current that results from the electron exchange is monitored as a function of time. Since the rate of material conversion by the electrochemical reaction is proportional to concentration, the current change is directly related to the amount of compound eluted.

The detection limit depends on the rate of the electrochemical reaction and the level of residual current. Chromatographic conditions, such as flow rate, mobile phase composition, and temperature, must be carefully controlled to obtain trace determinations; gradients should not be used because they cause severe baseline drift at a high detector gain.

Two types of electrochemical detectors are available; amperometric and voltametric. Reviews of these detectors are readily available [19, 20]. At present, the amperometric detectors [19] offer better sensitivity and are more applicable for routine use. Although the detector cells and electrodes are still under development, thin-layer cells of 1 µl or less have been used successfully. Electrodes usually employed are made from silver, mercury, platinum, glassy carbon, and gold; early detectors using carbon paste electrodes were not found to be as suitable for routine work. Electrochemical detectors have their greatest applications in separations where polar mobile phases are used. For certain applications, ionic strength, pH, buffer, and organic modifier may have to be modified to maximize the sensitivity of the electrochemical detector. Although the use of electrochemical detector in nucleic acid research has been limited, some applications include NADH, NADPH, and uric acid [19-23]. This list is expected to grow dramatically, since the aminopurines and xanthines readily undergo reduction polargraphically. In the future, it is expected that this detector will find more uses in the separation of other purines and pyrimidines with the recently developed derivatization reagents specifically designed for electrochemical detection [24,25]. Moreover, these detectors could be used in structural elucidations of the separated compounds, since each compound has a specific oxidation potential and thin-layer cyclic voltametric and/or spectroelectrochemical characterizations have been demonstrated [26,27].

D. Fluorescence detectors

The high sensitivity and specificity of fluorometric methods make fluorescence detectors extremely attractive LCDs. Since few compounds of biological importance fluoresce, this detector can be used for very selective applications and, with the detectors which permit selection of both excitation and emission wavelengths, selectivity can be further enhanced. Furthermore, methods based on light emission are inherently more sensitive than absorption methods. With the absorption LCDs, the detection limit is related to the minimum detectable change in intensity between the incident and transmitted radiation. Thus, at small sample concentrations, measurements are based on only small differences between two large numbers. On the other hand, fluorescence detectors involve the measurement of light against a black background; detection limit is set mainly by the noise level of the detector and its components. This can lead to 10-1000 times more sensitive measurements than with the absorbance detectors.

The single most important advantage of the fluorescence detectors is their ability to aid in the identification of the separated compounds. Adsorption and emission spectra of any particular compound reflect its unique energy level diagram. Thus, each substance is essentially fingerprinted by its electronic transitions. Fluorescence yields more structural data than absorption spectra because fluorescence involves up to at least four different parameters related to the energy level fingerprint. The absorption spectrum correlates two variables, absorption intensity and wavelength; fluorescence correlates more variables, such as absorption wavelength, emission intensity, emission wavelength, and quantum yields or decay times. For the purines and pyrimidines, emission intensity and other fluorescence parameters, as a function of pH and retention time, can be used to obtain unambiguous identifications [28].

In its simplest form, the fluorescence detector consists of a source flow cell, and photodetector usually 90° to the source. In the absence of light scattering, the light which strikes the photodetector is due only to fluorescence emission. In most cases, only specific wavelengths will cause absorption leading to fluorescence emission at longer wavelengths. Thus, most fluorescence detectors use a system of excitation and emission filters, excitation monochromator and emission filter, or excitation and emission monochromators to maximize selectivity and specificity of the method. Typically, these detectors use a Xe light source because it provides more intense light over a wider range of the UV region. Some of the newer fluorescence detectors use lasers to greatly enhance the sensitivity of the detector [29,30]. In addition, rapid-scanning fluorometers are gaining widespread popularity to obtain three-dimensional excitation-emission matrix data [31-33].

The typical fluorescence detectors can be operated in any of three ways:

1. Excitation and emission wavelengths are set and kept constant as emission intensity is monitored.
2. Excitation wavelength is kept constant and emission wavelengths are scanned to obtain intensity vs. emission wavelength spectra.
3. The emission wavelength is kept constant and excitation wavelengths are scanned to obtain emission intensity vs. excitation wavelength spectra (an indirectly obtained absorption spectra).

The latter two methods are useful in optimizing the selectivity of the detector, as well as aiding in the identification of compounds. It is important to note that excitation spectra cannot be compared with absorption spectra without making the appropriate corrections for grating efficiency, reflection, etc. However, proof for identifications can be obtained from differences in absorption and corrected excitation spectra; the fluorescence of some purines is attributed to specific tautomers whose existence with the nonfluorescent tautomers give rise to the observed absorption spectra [34].

The use of derivatization agents also adds to the advantageous application of fluorescence detectors. In an ultraselective and sensitive method [35,36], $1,N^6$-etheno derivatives of adenine, adenosine, and its nucleotides are prepared in a precolumn derivatization with chloroacetaldehyde. With this highly specific method, as little as 1 pmol of derivative may be determined [35,36]. However, derivatization reactions comparable to the above which can be used efficiently in a postcolumn technique have yet to be developed.

E. Radioactivity detectors

Radioactivity tracers are perhaps the most efficient means for the systematic study of purine and pyrimidine metabolism. While concentrations of purine and pyrimidine metabolites in a sample can be valuable, these data alone do not reflect the actual metabolic pathway. For example, a defect in one pathway may be associated with compensatory changes in another such that the overall levels of metabolites are not affected. By addition of isotopically labeled compounds to a system, it is possible to determine the distribution between intermediate pathways, assess metabolic fate, and measure rates of production or interconversion.

Most methods using the radiolabeled purine or pyrimidine compounds involve the collection of fractions and off-line analysis with conventional liquid scintillation counters. The off-line method is slow and requires extensive time for sample collection, drying, adding the scintillation fluid, and counting. In addition, the off-line method is expensive and subject to handling errors. Therefore, there is an obvious need for on-line radioactivity detectors which can rapidly and efficiently count the radiolabeled compounds as they elute from the column.

Radioactivity LCDs require several features for rapid and efficient detection of radiolabeled compounds in the effluent. The detectors should be flexible to distinguish between isotopes of different energies (^3H, ^{14}C, and ^{32}P); usually this is accomplished with the use of two channels (photomultipliers) that are set independently from each other. In addition, the detector should be able to detect and count low-energy emitters at a high counting efficiency. Furthermore, the detector should have some method to correct for quenching by the components in the mobile phase, especially when gradient elution or high buffer concentrations are used.

The basic design of a radioactivity LCD is similar to that found in the conventional off-line scintillation counters. A review of a radioactivity LCD is available [37]. The important feature is the flow-cell design [37-40]. Resolution and sensitivity of the detector are dependent on both flow rate and cell volume. Sensitivity can be maximized by decreasing the flow rate (increasing residence time) and/or increasing the cell volume. As the compounds elute from the column, they should be separated by small time intervals. Thus, too slow a flow rate or too large a cell volume will give the appearance of poor separation

Table 4 Types of Flow Cells for the Flow-Through Radioactivity Detector

---

Type 1: solid scintillator (heterogeneous) cell

Basic types

    Plastic scintillator cell—200 μm (20-25% efficiency for ^{14}C)—aqueous solvent system

    Glass scintillator cell—70-90 μm (20-25% efficiency for ^{14}C, 1% for ^{3}H

    XE scintillator cell—100-200 μm europium-activated calcium fluoride (75-80% efficiency for ^{14}C, 5% for ^{3}H)

Advantages of solid scintillator cell

    High salt or buffer concentration affects counting efficiency only slightly

    Nondestructive to sample—entire sample recovered

Disadvantages of solid scintillator cell

    Samples may bind irreversibly to solid support

    Low counting efficiency of low-energy β-emitter (^{3}H)

    CaF cell may become soluble in water and poison biological molecules

Type 2: liquid scintillator (homogeneous) cell

Basic type—PTFE tubing sizes of 10-2000 μl

Advantage of liquid scintillator cell

    High counting efficiency (75-80%) for high-energy β-emitters (^{14}C, ^{35}S, etc.)

    High counting efficiency (20-60%) for low-energy β-emitters (^{3}H)

    No adsorption of PTFE lines, thus no contamination

Disadvantages of liquid scintillator cell

    The sample must be mixed with a nongelling scintillation fluid, and thus no further chemical or biological studies of the material are possible.

Source: Ref. 37.

efficiency. The choice of flow rate and cell volume must be a compromise between the sensitivity and resolution required. Typically, cell volumes of 20-1000 μl are available; those used most are 100-500 μl.

Another factor which affects sensitivity is the type of scintillation counting cell. Table 4 lists the two types of cells, heterogeneous (solid scintillator) and homogeneous (liquid scintillator) cells. The major advantage of the solid cell is that the entire sample can be recovered after analysis. The major advantage of the liquid cell is its high counting efficiencies.

F. Combined detectors

An important function of the LCD is to aid in the identification of the separated compounds. The scanning ultraviolet detectors are valuable in assessing purity and characterizing the compound. As mentioned earlier, however, UV spectra are not as powerful as fluorescence spectra in identifying compounds. Often, UV spectra of the purines and pyrimidines are very similar. However, by combining the scanning fluorescence and UV detectors, much more information on compound structure can be obtained. In addition, unambiguous identifications are possible with the use of these detectors in series with a post-column technique to alter effluent pH [41,42]. Furthermore, characterizations can be based on ratios of the response of detectors in series. For example, two or more detectors of the same type (absorbance ratios or fluorescence ratios) or different types (absorbance/fluorescence, electrochemical/absorbance, or fluorescence/electrochemical) can greatly aid identifications. Any detector that has functional group specificity could be used to the analyst's advantage. Webster and Whaun [43], Bakay and Nissinen [44,45] used an on-line radioactivity detector in series with an UV detector to provide for both quantitative and qualitative analysis of the nucleic acid constituents.

III. FUTURE TRENDS

A. Mass spectrometry

Just as mass spectrometry (MS) revolutionized the on-line identification and detection of separated compounds in gas chromatography, the MS detector is believed to be the ultimate LCD. Unfortunately, the development of the ideal interface has been difficult. Comprehensive reviews of the progress in the coupling of the mass spectrometer with liquid chromatography were published by McFadden [46] and by Willoughby and Browner [84]. Blakely et al. [47,48] developed a liquid chromatograph-mass spectrometer (LC-MS) system which was successfully applied to the analysis of nucleotides and nucleosides; while nucleotides have long been considered difficult, the authors note that further improvements are necessary before the system can be used routinely. Another possibility in this area is the use of microbore LC-MS as demonstrated by Henion [49,50]. Moreover, fast atom bombardment (FAB) and other powerful "soft" ionization techniques [51-55] may ultimately lead to universal detection and identification of the polar, thermally labile nucleic acid components separated with liquid chromatography.

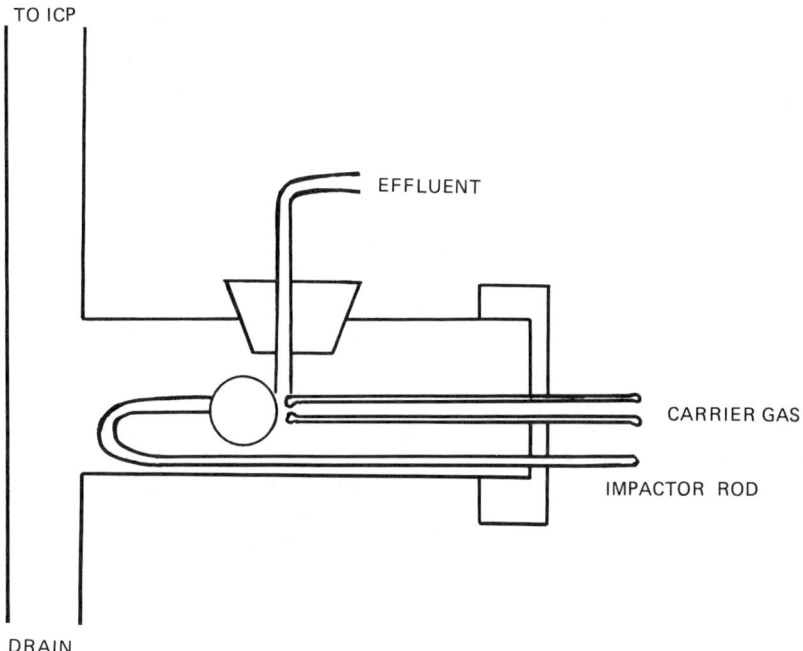

Figure 5. Babington-style nebulizer used for HPLC-ICP of nucleotides. (From Ref. 60.)

B. Atomic absorption

Numerous applications and design aspects of the on-line use of atomic absorption in liquid chromatography were reviewed by Van Loon [56, 57]. At present, the phosphorus atoms of the nucleotides are detected and used in quantitation. In the future, however, applications may include "tagging" the purines and pyrimidines with metallic reagents to enhance and adjust the selectivity as well as the specificity of the system.

C. Inductively coupled plasma emission

Similar to atomic absorption, an inductively coupled plasma emission spectrometer (ICP) can be interfaced (Fig. 5) to the liquid chromatograph [58]. By monitoring the phosphorus emission line, an ICP atomic emission system was demonstrated as a selective detector to determine phosphates [59]. Heine et al. [60] presented the results of studies undertaken to assess the capabilities of the ICP phosphorus-sensitive detector for the quantitative analysis of nucleotides; 750 ng was noted as the minimum amount of nucleotide detected. In addition, it is possible to monitor the carbon emission lines [61] with the ICP, thus extending its capabilities to a universal detector.

D. Post-column detection systems

There has been an increasingly popular trend to use post-column derivatizations to enhance the selectivity and sensitivity of the LCDs. The research of Frei and coworkers [62-65] has greatly facilitated the widespread application of the post-column reaction systems. An excellent review of the post-column systems was published by Krull and Lankmayr [66]. However, there appears to be little need to enhance the detection of the majority of the purines and pyrimidines with the absorbance LCDs. The post-column technique would be best used for identification purposes; however, few compounds could be unequivocally identified since these reactions are specific to functional groups which occur in a large variety of biological compounds. In addition, an elaborate series of post-column reactions and extractions may be necessary to isolate the derivatized purines and pyrimidines from a complex matrix.

The greatest potential in this area is in post-column reactions to make possible sensitive determinations with the electrochemical or fluorescence detectors. Unfortunately, while there are reports which illustrate the possible use of these post-column reaction schemes for purine and pyrimidine assays, none has been applied directly to detect these compounds in complex matrices. Post-column reaction with fluorescamine could be used to determine fluorometrically adenine, adenosine, cytosine, cytidine, and other nucleic acid components with primary amine groups. Frei and Lawrence [64,65] describe other methods which could be applied to the enhanced fluorescence detection of purines and pyrimidines. In addition, they also review post-column reactions for electrochemical detection. Kissinger et al. [67] also presented an overview of possible post-column reactions for the electrochemical detectors. One method which appears applicable to some of the purines and pyrimidines is post-column bromination [68,69]. Bromination was shown in a two-stage electrochemical reaction detector for unsaturated compounds. It may be possible to form many other electroactive compounds with the amine-selective post-column reaction schemes [25,26]. However, to take full advantage of the post-column approach, a derivatization procedure which is specific to the nucleic components must be developed.

A simple approach to post-column reaction is to introduce ion-pairing or micellar probes to enhance either fluorescence or ultraviolet detection [70-72]. These molecules do not react to form a derivative, but interact with the eluted compounds either to increase or decrease the background signal. Similarily, metals such as Tb^{+3} can also be used to enhance the fluorescence of some guanines [73]. In addition, fluorescence or ultraviolet detectability can be enhanced by altering the effluent pH [42,74,75]. This method was applied to enhance the fluorescence detection of some purines and pyrimidines [28].

E. Enzyme reaction detectors

Immobilized enzymes can be employed in a post-column system either to identify or to enhance the detection of some purines and pyrimidines selectively. Frei and Lawrence [64] have reviewed the use of these systems.

In its simplest form, immobilized enzymes can be used to aid in identification. A system was described [76] where the eluted purines were mixed on-line with appropriate buffers followed by the immobilized enzymic reaction. The products were detected with ultraviolet detectors.

Electrochemical and fluorescence detectors can also be employed to detect the products of the enzymic reactions. A possible system would use immobilized xanthine oxidase and electrochemical detection of the by-products formed from uric acid, xanthine, hypoxanthine, or allopurinol. Furthermore, where conditions are similar, a series of immobilized enzymes could be employed for the indirect detection of many of the purines and pyrimidines. The potential of this approach was presented in detail by Schifreen et al. [77] and by Adams and Carr [78].

Another possibility involves immobilized enzymes (e.g., xanthine oxidase) to produce H_2O_2, which then reacts with dichlorofluorescin through another immobilized enzyme, peroxidase, to produce the fluorescent dichlorofluoroscein; this is an on-line liquid chromatographic approach to the method developed by Slowiaczek and Tattersall [79]. Furthermore, methods such as these could lead to a viable chemiluminescence system [80,81].

REFERENCES

1. R. E. Majors, H. G. Barth, and C. H. Lochmüller, Anal. Chem., 54, 323R (1982).
2. W. A. McKinely, D. J. Popovich, and T. Layne, Am. Lab., 12(8), 37 (1980).
3. R. P. W. Scott, *Liquid Chromatography Detectors*, Elsevier Scientific, Amsterdam, 1977.
4. T. M. Vickrey, Ed., *Detectors in Liquid Chromatography*, Marcel Dekker, New York, 1982.
5. K. A. Klotter, Am. Lab., 13(6), 126 (1981).
6. American Society for Testing and Materials, Testing Fixed-Wavelength Photometric Detectors Used in Liquid Chromatography, ASTM E 685-79, 1979.

7. T. Wolf, G. T. Fritz, and L. R. Palmer, J. Chromatogr. Sci., 19, 387 (1981).
8. J. N. Brown, M. Hewins, J. H. M. Van der Linden, and R. J. Lynch, J. Chromatogr., 204, 115 (1981).
9. P. W. Carr, Anal. Chem., 52, 1746 (1980).
10. L. M. McDowell, W. E. Barber, and P. W. Carr, Anal. Chem., 53, 1373 (1981).
11. A. M. Krstulovic, P. R. Brown, and D. M. Rosie, Anal. Chem., 49, 2237 (1977).
12. A. M. Krstulovic, R. A. Hartwick, P. R. Brown, and K. Lohse, J. Chromatogr., 158, 365 (1978).
13. G. L. Bruce and K. Klotter, Am. Lab., 14(3), 74 (1982).
14. J. A. Hall, in *Applied Optics and Optical Engineering* (R. R. Shannon and J. C. Wyant, Eds.), Academic Press, New York, 1980, pp. 349-400.
15. A. F. Fell, Anal. Proc., 17, 512 (1980).
16. S. A. George and A. Maute, Chromatographia, 15, 419 (1982).
17. G. E. James, Cancer Res., 13, 39 (1981).
18. M. Greenbaum, J. Nicholas, and R. Moeller, Pittsburgh Conference on Analytical Chemistry and Applied Spectroscopy, 1981, Paper No. 166.
19. R. E. Shoup, C. S. Bruntlett, W. A. Jacobs, and P. T. Kissinger, Am. Lab., 13(10), 144 (1981).
20. S. K. Vohra, Am. Lab., 13(5), 66 (1981).
21. G. C. Davis, K. L. Holland, and P. T. Kissinger, J. Liq. Chromatogr. 2, 21 (1979).
22. J. Moiroux and P. J. Elring, Anal. Chem., 51, 346 (1979).
23. J. Moiroux and P. J. Elring, J. Am. Chem. Soc., 103, 6533 (1980).
24. K. Shimada, M. Tanaka, and T. Nambara, Chem. Pharm. Bull., 27, 2259 (1979).
25. K. Shimada, M. Tanaka, and T. Nambara, Anal. Lett., 13, 1129 (1980).
26. T. C. Pinkerton, K. Hajizadeh, E. Deutsch, and W. R. Heineman, Anal. Chem., 52, 1542 (1980).
27. R. Samuelsson, J. O'Dea, and J. Osteryoung, Anal. Chem., 52, 2215 (1980).
28. S. P. Assenza and P. R. Brown, J. Chromatogr., 289, 355 (1984).
29. M. J. Sepaniak and E. S. Yeung, J. Chromatogr., 211, 95 (1981).
30. E. S. Yeung and M. J. Sepaniak, Anal. Chem., 52, 1465A (1980).
31. C. J. Appellof and E. R. Davidson, Anal. Chem., 53, 2053 (1981).
32. L. W. Hershberger, J. B. Callis, and G. D. Christian, Anal. Chem., 53, 971 (1981).
33. D. W. Johnson, J. B. Callis, and G. D. Christian, Anal. Chem., 49, 747A (1977).

34. H. C. Borresen, Acta Chem. Scand., *21*, 2463 (1967).
35. J. F. Kuttesch, F. C. Schmalstieg, and J. A. Nelson, J. Liq. Chromatogr., *1*, 97 (1977).
36. M. Yoshioka and Z. Tamura, J. Chromatogr., *123*, 220 (1976).
37. M. J. Kessler, Am. Lab., *14*(8), 52 (1982).
38. L. J. Everett, Chromatographia, *15*, 445 (1982).
39. N. G. L. Harding, Y. Farid, M. J. Stewart, J. Shepard, and D. Nicoll, Chromatographia, *15*, 468 (1982).
40. R. J. Lloyd, J. Chromatogr., *216*, 127 (1981).
41. S. P. Assenza and P. R. Brown, in *Separations and Purifications*, *12*, 177 (1983).
42. S. H. Lee, L. R. Field, W. N. Howald, and W. F. Trager, Anal. Chem., *53*, 467 (1981).
43. H. K. Webster and J. M. Whaun, J. Chromatogr., *209*, 283 (1981).
44. A. B. Bakay, E. Nissinen, and L. Sweetman, Anal. Biochem., *86*, 65 (1978).
45. H. E. Nissinen, Anal. Biochem., *106*, 497 (1980).
46. W. H. McFadden, J. Chromatogr. Sci., *206*, 245 (1981).
47. C. R. Blakley, J. J. Carmody, and M. L. Vestal, Anal. Chem., *52*, 1636 (1980).
48. C. R. Blakley, J. J. Carmody, and M. L. Vestal, Clin. Chem., *26*, 1467 (1980).
49. C. Eckers, D. S. Skrabalak, and J. Henion, Clin. Chem., *228*, 1882 (1982).
50. J. D. Henion and T. Wacks, Anal. Chem., *53*, 1963 (1981).
51. I. Jardine and M. M. Weidner, J. Chromatogr., *182*, 395 (1980).
52. R. D. Macfarlane, C. J. McNeal, and J. E. Hunt, Adv. Mass. Spectrom., *8*, 349 (1980).
53. M. Barker, R. S. Bordoli, G. J. Elliot, R. D. Sedgwick, and A. N. Tyler, Anal. Chem., *54*, 645A (1982).
54. T. Marunka and Y. Vmeno, J. Chromatogr., *221*, 382 (1980).
55. V. T. Vu, C. C. Fenselau, and O. M. Colvin, J. Am. Chem. Soc., *103*, 7362 (1981).
56. J. C. Van Loon, Anal. Chem., *51*, 1139A (1979).
57. J. C. Van Loon, Am. Lab., *13*(5), 47 (1981).
58. I. S. Krull and S. Jordan, Am. Lab., *12*(10), 21 (1980).
59. M. Morita and T. Vehiro, Anal. Chem., *53*, 1997 (1981).
60. Dr. R. Heine, M. D. Denton, and T. D. Schlabach, Anal. Chem., *54*, 81 (1982).
61. M. Morita and T. Vehio, Anal. Chem., *52*, 349 (1980).
62. R. W. Frei and A. H. M. T. Scholten, J. Chromatogr. Sci., *17*, 152 (1979).
63. R. W. Frei, L. Michel, and W. Santi, J. Chromatogr., *126*, 665 (1976).
64. R. W. Frei and J. F. Lawrence, Eds. Chemical Derivatization in Analytical Chemistry, Volume 1: Chromatography, Plenum Press, New York, 1981.

65. J. F. Lawrence and R. W. Frei, *Chemical Derivatization in Liquid Chromatography*, Elsevier Scientific, Amsterdam, 1976.
66. I. S. Krull and E. P. Lankmayr, Am. Lab., *14*(5), 18 (1982).
67. P. T. Kissinger, K. Bratin, G. C. Davis, and L. A. Pachla, J. Chromatogr. Sci., *17*, 137 (1979).
68. W. P. King and P. T. Kissinger, Clin. Chem., *26*, 1484 (1980).
69. P. T. Kissinger, Anal. Chem., *49*, 447A (1977).
70. P. A. Asmus, J. W. Jorgenson, and M. Novotny, J. Chromatogr., *126*, 317 (1976).
71. L. Hackzell and G. Schill, Chromatographia, *15*, 437 (1982).
72. S. Y. Su, A. Jurgensen, D. Bolton, and J. D. Winefordner, Anal. Lett., *14*, 1 (1981).
73. D. P. Ringer, S. Burchett, and D. E. Kizer, Biochemistry, *17*, 4818 (1978).
74. W. F. Bayne, T. East, and D. Dye, J. Pharm. Sci., *70*, 458 (1981).
75. C. R. Clark and J.-L. Chan, Anal. Chem., *50*, 635 (1978).
76. M. Kito, R. Tawa, S. Takeshima, and S. Hirose, J. Chromatogr., *231*, 183 (1982).
77. R. S. Schifreen, D. A. Hanua, L. D. Bowers, and P. W. Carr, Anal. Chem., *49*, 1929 (1979).
78. R. E. Adams and P. W. Carr, Anal. Chem., *50*, 944 (1978).
79. P. Slowiaczek and M. H. N. Tattersall, Anal. Biochem., *125*, 6 (1982).
80. S. Kobayashi and K. Imai, Anal. Chem., *52*, 424 (1980).
81. S. Kobayashi, J. Sekino, K. Honda, and K. Imai, Anal. Biochem., *112*, 99 (1981).
82. C. W. Gehrke, K. C. Kuo, and R. W. Zumwalt, J. Chromatogr., *188*, 129 (1980).
83. R. H. Hartwick and P. R. Brown, J. Chromatogr., *126*, 679 (1976).
84. R. C. Willoughby and R. F. Browner, *Trace Analysis*, Vol. 2 (J. F. Lawrence, Ed.), Academic Press, New York, 1982, pp. 69-109.

9
Mobile Phase Control

PAMELA A. PERRONE[*] University of Rhode Island, Kingston, Rhode Island

I. Ion-Pair Chromatography 162
   A. Introduction 162
   B. Applications 164
   C. Processes of ion pairing 171
   D. Conclusion 173
II. Ion Suppression 175
   References 176

In Chap. 7 the great separation potential of high performance liquid chromatography (HPLC) was discussed. Excellent separations are obtained due to the solute-solvent interactions as well as the solute-stationary phase interactions. By varying the mobile phase parameters, the capacity factor and/or selectivity may be changed. The resolving power is enhanced by control of the chemical equilibrium. This control of the chemical equilibrium is especially important in the separation of nucleic acid constituents. To increase the resolution of these components, the techniques of ion exclusion, ion suppression, and ion

---

[*]*Current affiliation*: The Perkin-Elmer Corporation, Norwalk, Connecticut.

pairing have been employed. Although ion exclusion was used in the early HPLC separations of the nucleotides, it is not generally used at present. Ion suppression, by adjusting the pH of the mobile phase, shifts the chemical equilibrium to the nonionic state. However, ion pairing permits the separation of components in both the ionized and nonionized forms simultaneously. Therefore, the latter technique is of special interest in the separation of the bases, nucleosides, and nucleotides.

I. ION-PAIR CHROMATOGRAPHY

A. Introduction

Ion-pair chromatography (IPC) is a technique which, by modification of the partitioning process, permits the separation of charged compounds normally unretained by the chromatographic conditions. Among the variety of names in the literature which have been used to describe this method are ion pairing [1], paired-ion [2], soap [3], solvent-generated (dynamic) ion-exchange [4], detergent-based cation-exchange [4], solvophobic-ion [5], hetaeric [6], surfactant [7], zwitterion-pair [8], and ion-association [9] chromatography. When first introduced by Schill [10], IPC involved coating the stationary support (silica gel or cellulose) with a reagent capable of forming ion pairs with the sample. In its present form a pairing agent is added to the mobile phase. This pairing agent, acting as a counterion, combines with the charged sample component to form a neutral ion pair which is retained longer by the chromatographic system. This reaction may be expressed as

$$\text{sample}^{\pm} + \text{counterion}^{\mp} \rightleftharpoons [\text{sample}^{\pm}\text{counterion}^{\mp}]^{0} \text{ pair}$$

Theoretically neutral sample components will not be affected by the counterion. Generally the pairing agent is a large organic ionic molecule. Classically, for strongly basic samples, alkyl sulfonates are used; while for strongly acidic samples, quaternary amines have been chosen. To optimize an ion-pairing separation, numerous mobile phase parameters may be varied. These parameters include pH, type of counterion (i.e., length of alkyl group), concentration of counterion, percent organic modifier, and temperature. A summary of these effects on both capacity factor (k') and selectivity is given in Table 1.

Thus IPC has several advantages over other HPLC separation modes. First, the separation of compounds with widely divergent properties (i.e., ionic and nonionic) in one chromatographic run is possible. Second, mechanically held stationary phases, which tend to be less stable than chemically bonded stationary phases, are not required, and hence counterion bleed from the stationary phase is eliminated. Third, IPC may be utilized with either normal or reversed-phase operation, thereby offering great versatility. Finally, by the use of a counterion

Table 1 Summary of Mobile Phase Effects in Reversed-Phase Ion-Pair Chromatography

Variable	Effect on k'	Effect on solvent selectivity
Counterion		
Increase concentration	Increase[a]	No effect[b]
Change counter-ion	Change	Small effect
Increase pH		
Sample anions	Increase[a]	Large effect
Sample cations	Decrease[a]	Large effect
Organic solvent		
Increase concentration	Decrease[a]	Small effect[b]
Change organic solvent	Change	Change
Temperature	Increase[a]	Large effect

[a] In normal-phase IPC, the effects are reversed.
[b] Larger effect for partly ionized sample molecules.
Source: Ref. 11.

with a high molar absorptivity, the detection limits of the compounds of interest may be enhanced and, more important, nonultraviolet-absorbing ionic components, via the ion pair formed, may be monitored [10, 12,13]. For a more complete discussion of the basics of IPC, the reader should consult the excellent review articles [7,14,15] which have been written on the subject or recently published texts on modern HPLC [11,16].

Because of their charge, nucleotides have traditionally been analyzed by ion-exchange chromatography [17-19]. However ion exchange has the disadvantages listed in Table 2. By virtue of their ionic character, nucleotides should be prime candidates for IPC. In addition, because nonionic compounds are not affected by the pairing agent, IPC has the potential for simultaneously analyzing the purine and pyrimidine bases, nucleosides, and nucleotides present in biological fluids. In this

Table 2 Disadvantages of Ion-Exchange Chromatography

1. Columns tend to be less stable than other LC columns.
2. Columns tend to be less efficient.
3. Choice of packing material is limited.
4. Long reequilibration times are necessary in gradient programming.

chapter the ion pairing of nucleic acid constituents will be presented. The current theories, applications, and advantages and disadvantages will be included in this discussion.

B. Applications

Since the nucleotides exist as charged compounds at pH values greater than 3, they are ideally suited for IPC. Due to their acidity, quaternary amines are generally employed as the pairing agent. The first separation of nucleotides by IPC was reported in 1977 by Hoffman and Liao [5]. Using tetra-n-butyl-ammonium hydrogen sulfate as the pairing agent, a separation of the mono-, di-, and triphosphates of cytosine, uracil, adenine, and guanine was achieved; however, complete resolution of all 12 ribonucleotides could not be obtained, despite different buffer compositions (Figs. 1 and 2).

It was not until several years later that other nucleotide separations by IPC were reported. Because of the important role of the adenine nucleotides in energy metabolism, Juengling and Kammermeier [20] developed a 10-min assay for AMP, ADP, and ATP. Ingebretsen et al. [21], also interested in monitoring the energy charge, reported

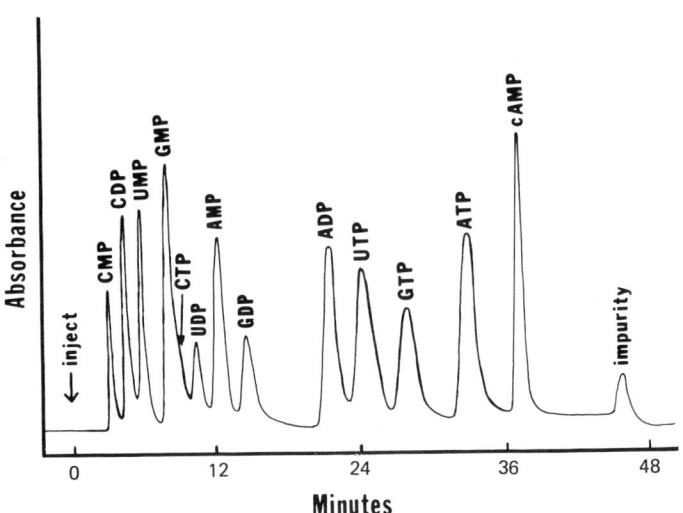

Figure 1. Gradient elution separation of ribonucleotides. Mobile phase: (A) 0.25 M tetra-n-butylammonium hydrogen sulfate (TBHS), 0.050 M KH_2PO_4, 0.080 M NH_4Cl buffer at pH 3.90. (B) 0.025 M TBHS, 0.10 M KH_2PO_4, 0.20 M NH_4Cl buffer at pH 3.40, 30% methanol. Operating conditions: 40-min gradient program. (Reprinted with permission from Ref. 5. Copyright 1977 American Chemical Society.)

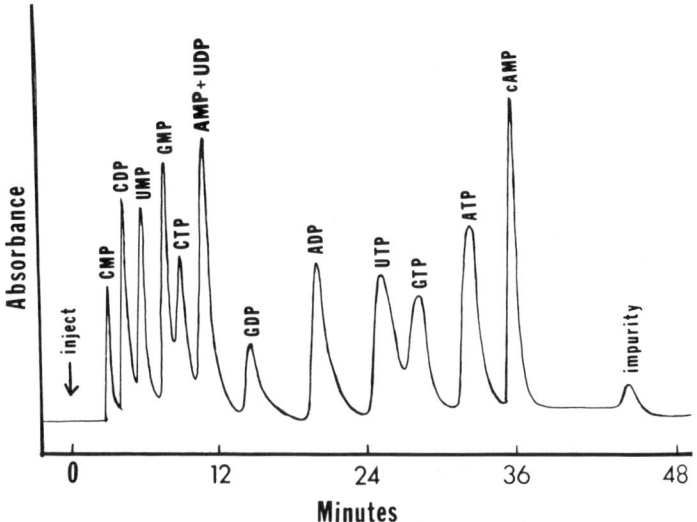

Figure 2. Gradient elution separation of ribonucleotides. Mobile phase: (A) 0.025 M TBHS, 0.050 M KH_2PO_4, 0.070 M NH_4Cl buffer at pH 3.90. (B) Same as in Fig. 1B. Operating conditions: 40-min gradient program. (Reprinted with permission from Ref. 5. Copyright 1977 American Chemical Society.)

conditions for the selective retention of the adenine nucleotides as a group relative to the mono-, di-, and triphosphates of guanine, uracil, and cytosine, using tetrabutylammonium hydrogen sulfate. Schwenn and Jender [22] separated the adenine nucleotides from their sulfated derivatives using 2-propanol and tetrabutylammonium hydroxide adjusted to pH 9.4; however, care should be exercised when employing these conditions, because high pH is known to dissolve the silica matrix of C18 columns. Gilbert [23], performing soap chromatography, separated nucleotides of adenine and guanine. Soap chromatography requires a long-chain hydrophobic pairing agent in the mobile phase; the agent used in this work was decyltrimethylammonium bromide. Van Haastert [24] developed a series of rules for the optimization of the separation of nucleobases, nucleosides, and nucleotides. In this systematic study polarity, pH, ion concentration, and polarity of the buffer ions of the mobile phase were varied. Walseth et al. [25], investigating several quaternary amines, reported a separation of the 5'-ribonucleoside monophosphates. Figure 3 illustrates the effects of alkyl chain length of the pairing agent and percent organic modifier in the mobile phase on retention. Knox and Jurand [8] chromatographed the nucleotides using the zwitterionic pairing agent, 11-aminoundecanoic

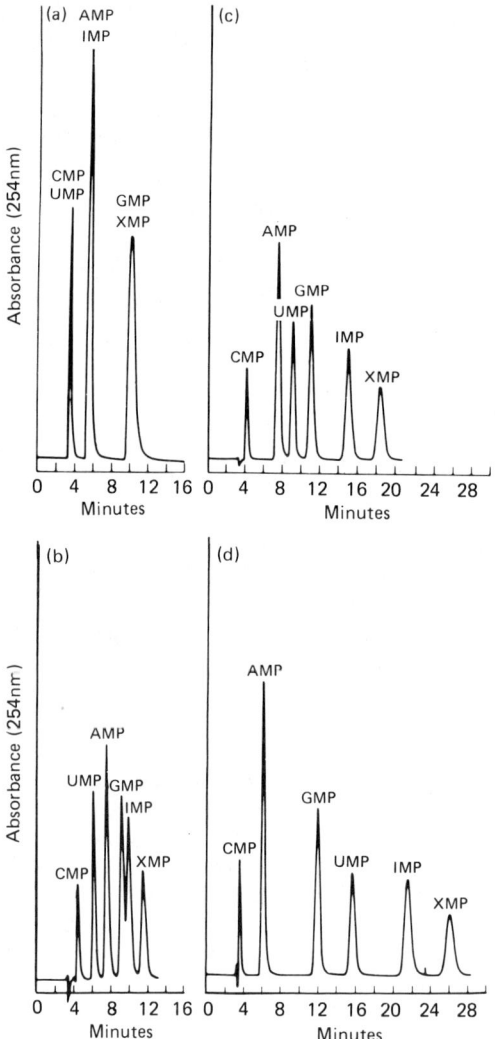

Figure 3. Effects of various ion-pair reagents on the separation of ribonucleotide monophosphates on a Spherisorb ODS column (4.6 × 250 mm). The standard injection was 7.5 μl of a solution that was 0.5 mM with respect to each of the following nucleotides: AMP, GMP, IMP, XMP, CMP and UMP. All separations were carried out at a flow rate of 1 ml/min at room temperature with detection at 254 nm (2 absorbance units full scale) and the following solvents, which were adjusted to pH 2.5 with formic acid. Standards were chromatographed individually and in various combinations to determine peak identities. (a) Water; (b) 20 mM TMA; (c) 5 mM TEA, 2% methanol; (d) 2.5 mM TBA, 8% methanol. (From Ref. 25.)

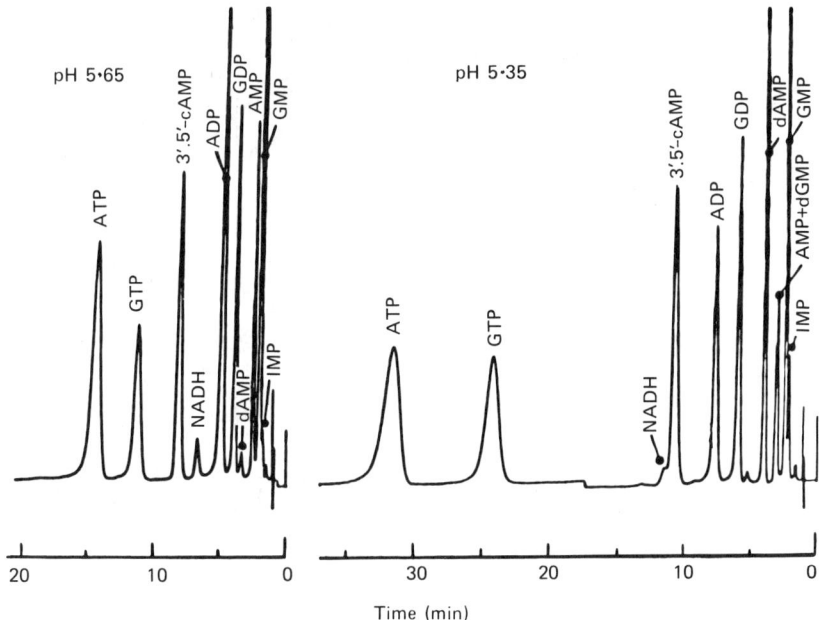

Figure 4. Separation of nucleotides by zwitterion-pair chromatography showing the effect of pH change. Packing material: ODS-Hypersil. Eluent: water-methanol (88:12, v/v) made 75 mM in phosphate and 1.25 mM in 11-aminoundecanoic acid (C11AA). (From Ref. 8.)

acid (C11AA). In their investigation the effect of pH variation of the mobile phase is evident (Fig. 4). The first separation of isomeric monoribonucleotides by HPLC was reported by Al-Moslih et al. [26]. Keeping pace with the technological developments of HPLC, a radially compressed reversed-phase column was used by Rao et al. [27] to separate the platelet nucleotides isocratically. Analysis time was more than halved by switching from an analytical C18 column to a Radial-Pak. Darwish and Prichard [28] also investigated the use of radially compressed columns. Figure 5 compares the separation obtained from a standard column and a radially compressed column. Most recently, El Rassi and Horváth [29] have used zwitterionic detergents, i.e., betaines, to achieve a separation; however, the retention moduli for these detergents were found to be lower than those obtained under comparable conditions with n-alkyl-trimethylammonium salts having the same alkyl chain length. Reversed-phase IPC has also been used for the separation of oligonucleotides [30-33]. For a more thorough discussion of oligonucleotides, see Chap. 11.

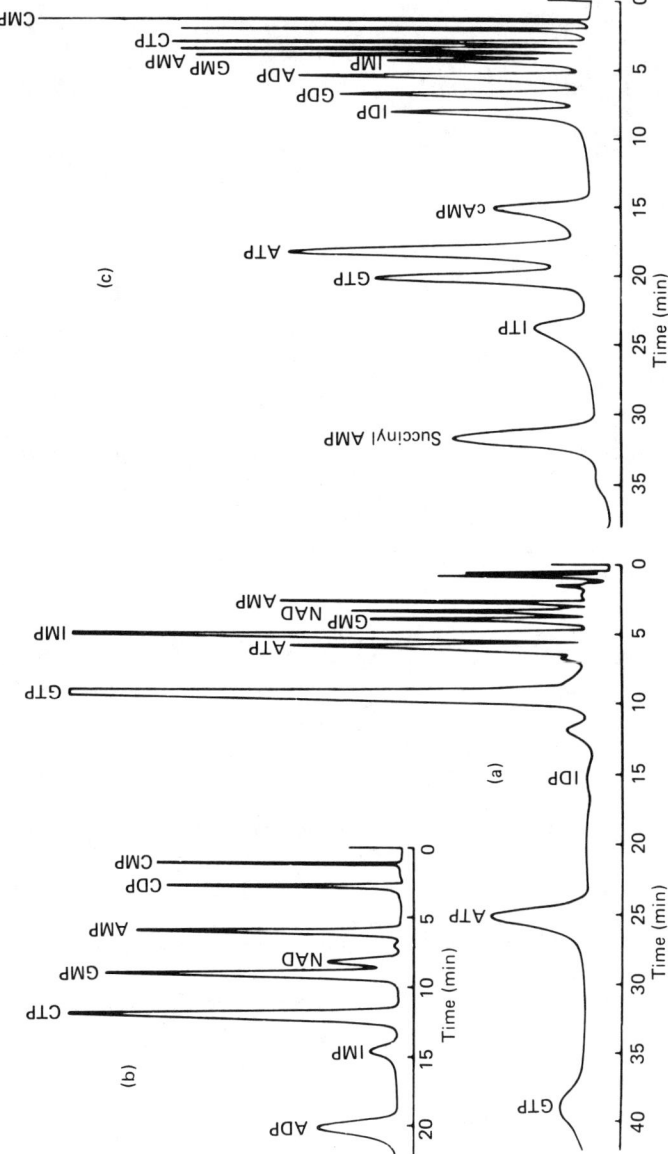

Figure 5. Isocratic elution in the absence (a and b) and presence (c) of acetonitrile using both columns and showing the difference in retention. Initial mobile phase, 65 mM KH_2PO_4, 0.9 mM TBAP, pH 3.2. (a) Tissue extract with added GMP, IMP, and GDP on μBondapak C_{18} column; flow rate 2.8 ml/min. (b) Separation of standard nucleotides on RadialPak A column; flow rate 3.5 ml/min. (c) Separation of nucleotide standards on RadialPak A column in the presence of 3.3% acetonitrile, flow rate 2.0 ml/min. (From Ref. 28.)

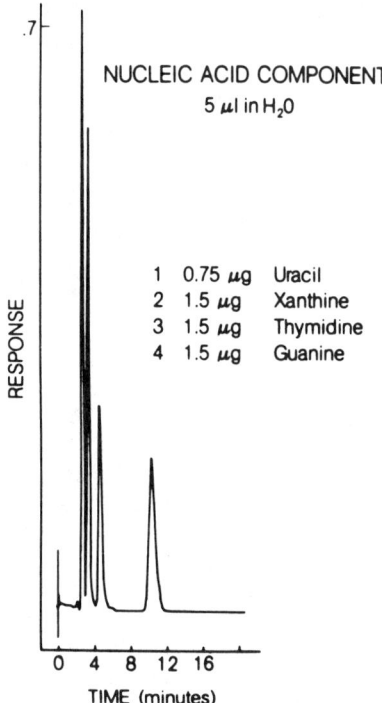

Figure 6. Separation of nucleic acid components on a Zorbax ODS (25 cm × 4.6 mm i.d.). Eluent: 0.005 M heptanesulfonic acid, pH 2.8 / methanol (85:15). Flow: 1.0 ml/min. Detection: 254 nm (0.16 aufs). Temperature: ambient. (From Ref. 35.)

In addition to the nucleotides, purine and pyrimidine bases have been separated by IPC. Because these compounds are basic, alkyl sulfonates or sulfates are generally employed as the pairing agents. Ehrlich and Ehrlich [34], unable to resolve cytosine, 5-methylcytosine, guanine, thymine, and uracil by reversed phase, developed an ion-pairing separation using heptane sulfonate. As uracil and thymidine do not ionize under acidic conditions, their retention was virtually unaffected by the addition of the pairing agent. Popovich [35] chromatographed uracil, xanthine, thymidine, and guanine in 12 min by IPC (Fig. 6). Csárnyi et al. [36], using octyl sulfate as the pairing agent, separated the 5-alkyluracils and purine bases in the hydrosylate of enzymatically synthesized nucleic acids. The 5-alkyl substituents on uracil included hydrogen, methyl, ethyl, n-propyl, n-butyl, n-pentyl, and n-hexyl. Brown et al. [37] reported an isocratic separation of the intermediate metabolites of uric acid in the purine salvage pathway.

In 6 min, levels as low as 5 ng of uric acid, hypoxanthine, and xanthine may be determined.

The separation and quantitation of certain bases, nucleosides, and nucleotides are of great importance because of their use as therapeutic drugs. Ion-pair chromatography has been used in several such applications. For example, Day et al. [38], developed an assay for mercaptopurine, an antineoplastic agent, with a minimum detection limit of 0.2 µg/ml using heptanesulfonic acid as the pairing agent. Allopurinol, an azapurine, is used in the treatment of gout. Chromatographic conditions for its determination were investigated by Voelter et al. [39]. Using tetrabutylammonium hydroxide, hypoxanthine, xanthine, uric acid, orotic acid, allopurinol, and its metabolite, oxipurinol, were resolved. Gelijkens and De Leenheer [40] have separated the nucleosides and nucleotides of 5-fluorouracil (FU), an agent used in the treatment of solid tumors, in the presence of quaternary ammonium ions. Au et al. [41] have also developed an assay for 5-fluorouracil, its analog, 5'-deoxy-5-fluorouridine (5-d FUR), and their nucleosides and nucleotides. Using a two-step elution with a mixture of tetraethylammonium and tetrabutylammonium pairing ions, a separation of the desired 10 compounds was achieved (Fig. 7). Shinohara et al. [42] determined the levels of 4-aminopyridine in serum using sodium heptanesulfonate; the minimum detection limit was 1 ng/ml. 4-Aminopyridine has been used clinically for reversing the effects of nondepolarizing neuromuscular blocking agents and in the treatment of myasthenia and botulism. Danielson and Huth [43], by measuring the levels of ATP, have developed an assay for creatine kinase, an important enzyme in the diagnosis of myocardial infarction and progressive muscular dystrophy.

In the analysis of nucleotides a unique form of "ion-pairing" has been attempted. Instead of the more traditional pairing agents (i.e., quaternary amines), metal ions are used. Chow and Grushka [44] have been able to separate nine nucleotides in less than 15 min. The column consisted of a $Co(en)_3^{3+}$ moiety bonded to a diamine group which in turn was bonded to silica gel. The packing material may be represented as

$$R-Si-O-Si-(CH_2)_3-N-Co-N$$

The nucleotides are retained by forming outer-sphere complexes with the cobalt moiety; however, very long retention was observed for the triphosphates. Analysis time could be shortened by the addition of

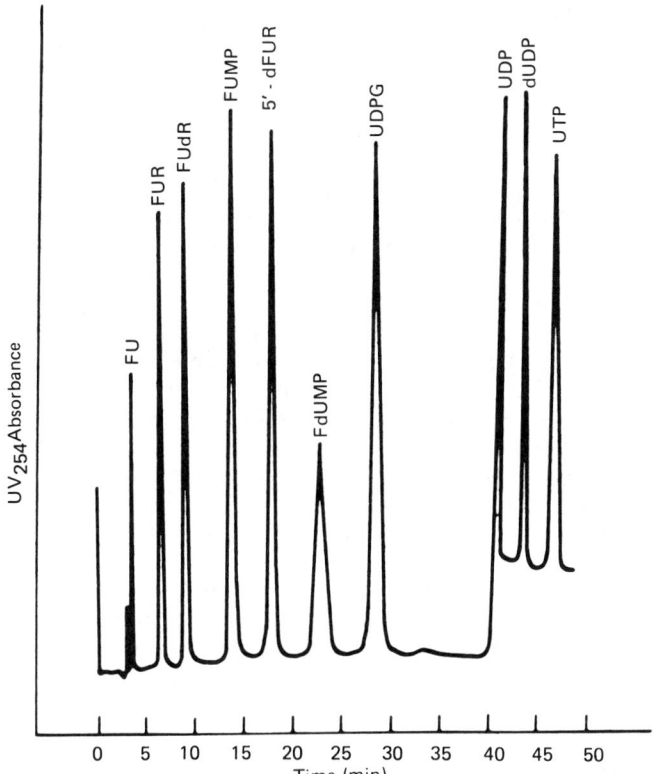

Figure 7. Separation of authentic chemicals of FU, 5'-dFUR, FU nucleosides and nucleotides, and uracil nucleotides. Eluting buffer from 0 to 30 min: 0.1 mM C_{16}, 2.5 mM C_8, and 2% methanol in 2 mM sodium acetate-1.5 mM phosphate buffer (pH 6.0) (A); from 30 to 50 min: A + 30 mM phosphate. (From Ref. 41.)

magnesium sulfate to the mobile phase. Magnesium forms inner-sphere complexes with the nucleotides.

C. Processes of ion pairing

From the variety of names given to describe ion pairing, one can surmise that numerous processes have been postulated to describe the mode of retention in IPC. In the "ion pairing" hypothesis [6,45-47], ion-pair formation is thought to occur in the aqueous mobile phase. This ion pair then partitions onto the bonded hydrophobic stationary phase. A "dynamic ion-exchange" process [3-5,48-54] has also been advanced. Under this theory the counterion is first adsorbed onto the stationary phase, basically creating an ion-exchange system prior to ion-pair formation. The charged sample components then form ion pairs onto the "charged" stationary phase. In another proposal, the concept of pairs of ions rather than ion pairs as the adsorbing species is advocated [9,55]. A charged primary ion layer is formed on the stationary phase surface; a secondary layer of opposite charge is formed on top of the primary layer. As both of these layers are in rapid dynamic equilibrium, the charged ions are constantly adsorbing and desorbing. When a charged functional group from a sample passes through the column, it is adsorbed into the primary layer due to electrostatic and van der Waals forces. Thus a charge deficiency is created in this primary layer. To restore the electrostatic equilibrium, another ion must be adsorbed into the primary layer. Hence a pair of ions rather than an ion pair is adsorbed. In a fourth process, attempts were made to combine the ion-pairing and dynamic ion-exchange hypotheses [56]; however, this did not adequately explain all the experimental observations. The most recent proposal is the "dynamic complex exchange" model [56]. Ion pairs are formed in the mobile phase and bind to a stationary phase which is coated with the ion-pairing reagent. A metathetical exchange of analyte between the ion bound to the stationary phase and the ion pairs formed in the mobile phase is assumed. Thus IPC has been explained by a number of processes. A general consensus on one process has yet to be achieved, and one may expect several more proposals before ion pairing is totally understood.

In the investigation of nucleic acid constituents by IPC, the question of the operative retention process has been explored. Hoffman and Liao [5] have concluded that for their nucleotide separation the dynamic ion-exchange process explains their observations. They first noted that the effects of the ion-pairing agent were seen even after its removal from the mobile phase. Therefore some adsorption of the pairing agent onto the stationary phase had to occur. Second, the order of elution of the components followed the predicted order for an ion-exchange separation rather than a reversed-phase system. For a reversed-phase system, the greater the number of phosphate groups, the more hydrophilic the solute, hence the faster the compound should elute. Thus ATP should elute first, then ADP and AMP, respectively. This order was not observed. Finally, the order of nucleotide retention, except for AMP, correlated with the order predicted for a positively charged surface—the smaller the negative charge, the less it would be retained.

Knox and Jurand [8], in their investigation of nucleotides, used a unique type of pairing agent and thus suggested a different process. For their separation, a zwitterionic pairing agent, 11-aminoundecanoic acid (C11AA) is used. The order of elution is the monophosphates, which are the least retained, followed by the diphosphates and finally the triphosphates. The enhanced retention by C11AA and its pH dependence is explained by the formation of quadropolar ion pairs between the agent and the nucleotides. The reaction between the zwitterionic solutes (the nucleotides between pH 2 and 4 have negative charges from the phosphates and positive charges from the cationic N atoms) is described as

$$S^{\pm} + P^{\mp} \rightleftharpoons (S^{\pm}P^{\mp})$$
aqueous   organic

This model is unique because both positive and negative charges on both components of the ion pair are involved. In the more traditional form of IPC, only one charge is present on the pairing agent, and it is attracted to an oppositely charged sample component.

In a recent report of the separation of 5-fluorouracil and its analogs by RP-IPC, Au et al. [41] support the ion-pairing retention process. Arguing that the concentration of the pairing agent used in their experiment was not sufficient to saturate the stationary-phase adsorption sites, they state that the process must be partitioning of an already formed ion pair to the stationary phase.

As these examples attest, there is disagreement in the literature over the IPC retention process of nucleic acid constituents, much as there is in the overall picture of IPC processes. More research is necessary in order to explain how this technique alters the retention times.

D. Conclusion

As discussed earlier, IPC has several advantages, the most significant being the ability to separate charged components. It is, however, not without its disadvantages. There are two serious limitations due to the size of the pairing agent. First, because of its abrasive nature, the pairing agent, when present in high concentration, has lead to some dissolution of the silica support, thus shortening column lifetime. Generally, the use of a precolumn can help to alleviate this problem. Second, as the length of the alkyl group increases, irreversible adsorption of the pairing agent onto the column may result. Hence reproducibility may be poor and column regeneration impossible. Despite these shortcomings, ion pairing is a very useful technique. Tables 3 and 4 summarize the applications of IPC to the analysis of nucleotides and bases, respectively. While IPC would appear to be an ideal tool for improving nucleotide separation, its potential is still virtually unexplored. In the future one may expect to see much more research on the ion pairing of nucleic acid constituents in order to determine the mode of retention and to further improve the ability to separate these compounds.

Table 3 Summary of Reversed-Phase IPC of Nucleotides

Mobile phase	Counterion	Program	Reference
Phosphate buffer/NH_4Cl	TBA	Gradient	5
Acetonitrile/phosphate buffer	TBA	Isocratic	20
MeOH/phosphate buffer	TBA	Isocratic	21
2-Propanol	TBA	Isocratic	22
MeOH/phosphate buffer	DTMA	Isocratic	23
MeOH/water	TMA,TEA,TBA	Isocratic	25
MeOH/phosphate buffer/water	C11AA	Isocratic	8
Phosphate buffer	TBA	Isocratic	26
Acetonitrile/water	TBA	Isocratic	27
Acetonitrile/phosphate buffer	TBA	Isocratic/gradient	28
Phosphate buffer	n-Decylbetaine	Isocratic	29

(NH_4Cl = ammonium chloride; MeOH = methanol; TBA = tetrabutylammonium, TMA = tetramethylammonium, TEA = tetraethylammonium ions; C11AA = 11-aminoundecanoic acid; DTMA = decyltrimethylammonium bromide).

Table 4 Summary of Reversed-Phase IPC of Bases

Mobile phase	Pairing agent	Program	Reference
Phosphate buffer	Heptane sulfonate	Isocratic	34
Methanol	Heptane sulfonic acid	Isocratic	35
Methanol/phosphate buffer	Octyl sulfate	Gradient	36
Acetate buffer	Heptane sulfonic acid	Isocratic	37

II. ION SUPPRESSION

Whereas ion pairing is the technique, other than ion-exchange chromatography, used for the analysis of strong acids or bases, ion suppression is the method chosen for weak acids or bases. Many ionic compounds may be expressed by the following equilibrium:

nonionic ⇌ ionic

By properly adjusting the pH of the mobile phase, the equilibrium can be driven to the left. This technique is known as ionic suppression. The reader is referred to Refs. 11 and 15 for a review of this method. The nucleosides and bases are relatively weak acids (high pK_{aa} values) and weak bases (low pK_{ab} values). Through proper selection of pH, in between the pK_{aa} and pK_{ab}, these compounds are neutral and hence can be readily analyzed by reversed-phase. The nucleotides, on the other hand, are strong acids. At a pH of 2.0, the charge on the phosphate moiety is proportional to the number of phosphate groups. Thus the monophosphates have one negative charge, the diphosphates two, and the triphosphates three. At a pH of 7.0 and above, the nucleotides gain an additional negative charge from the secondary phosphate dissociation. By adjusting the pH of the eluent to a low value, one can suppress this secondary dissociation. An example of the use of ion suppression in the analysis of nucleotides is the work of Whitehouse and Greenstock [57,58]. Using a C8 column and an eluent of 0.6 M ammonium phosphate at pH 3.5, the 2'-/ and 3'- monoribonucleotides of cytosine, uracil, guanine, and adenine were separated [57]. With the same column and eluent but adjusting the pH to 4.25, the deoxynucleotides of these compounds, in addition to deoxythymidine and 5-bromodeoxyuridine, 5'-monophosphoric acid, were chromatographed [58]. Another example was reported by Walseth et al. [25]. Using a reversed-phase column and a mobile phase of water adjusted to pH 2.5 with formic acid, a separation of CMP, UMP, AMP, and XMP was obtained; however, IMP and GMP were only partially resolved. To obtain a better separation of these compounds, ion pairing was used.

Ion suppression, where applicable, has distinct advantages over ion pairing. An expensive pairing agent is not required. Column lifetimes are greater as long as one does not exceed the pH limitations of the silica backbone. Longer lifetime is also achieved by the absence of an abrasive pairing agent and the irreversible adsorption of that agent. However, ion suppression is incompatible with strongly acidic or basic samples, and ion pairing must be the method of choice unless one prefers ion-exchange chromatography.

REFERENCES

1. S. Eksborg, P. O. Lagerström, R. Modin, and G. Schill, J. Chromatogr., 83, 99 (1973).
2. Paired-Ion Chromatography, an Alternative to Ion Exchange, Waters Assoc., Milford, Mass., December 1975.
3. J. H. Knox and G. R. Laird, J. Chromatogr., 122, 17 (1976).
4. J. C. Kraak, K. M. Jonker, and J. F. K. Huber, J. Chromatogr., 142, 671 (1977).
5. N. E. Hoffman and J. C. Liao, Anal. Chem., 49, 2231 (1977).
6. Cs. Horváth, W. Melander, I. Molnár, and P. Molnár, Anal. Chem., 49, 2295 (1977).
7. E. Tomlinson, T. M. Jefferies, and C. M. Riley, J. Chromatogr., 159, 315 (1978).
8. J. H. Knox and J. Jurand, J. Chromatogr., 203, 85 (1981).
9. B. A. Bidlingmeyer, S. N. Deming, W. P. Price, Jr., B. Sachok, and M. Petrusek, J. Chromatogr., 186, 419 (1979).
10. S. Eksborg and G. Schill, Anal. Chem., 45, 2092 (1973).
11. L. R. Snyder and J. J. Kirkland, Introduction to Modern Liquid Chromatography, 2nd ed., John Wiley & Sons, New York, 1979.
12. J. Crommen, B. Fransson, and G. Schill, J. Chromatogr., 142, 283 (1977).
13. M. Denkert, L. Hackzell, G. Schill, and E. Sjögren, J. Chromatogr., 218, 31 (1981).
14. R. Gloor and E. L. Johnson, J. Chromatogr. Sci., 15, 413 (1977).
15. B. A. Bidlingmeyer, J. Chromatogr. Sci., 18, 525 (1980).
16. M. T. W. Hearn, in Advances in Chromatography, vol. 18 (J. Calvin Giddings, Ed.), Marcel Dekker, New York, 1980.
17. W. E. Cohn, Science, 109, 377 (1949).
18. A. Floridi, C. A. Palmerini, and C. Fini, J. Chromatogr., 138, 203 (1977).
19. M. McKeag and P. R. Brown, J. Chromatogr., 152, 253 (1978).
20. E. Juengling and H. Kammermeier, Anal. Biochem., 102, 358 (1980).
21. O. C. Ingebretsen, A. M. Bakken, L. Segadal, and M. Farstad, J. Chromatogr., 242, 119 (1982).
22. J. D. Schwenn and H. G. Jender, J. Chromatogr., 193, 285 (1980).
23. M. T. Gilbert, in Current Developments in the Clinical Applications of HPLC, GC and MS (A. M. Lawson, C. K. Lin, and W. Richmond, Edits.), Academic Press, New York, 1980.
24. P. J. M. Van Haastert, J. Chromatogr., 210, 229 (1981).
25. T. F. Walseth, G. Graff, M. C. Moos, Jr., and N. D. Goldberg, Anal. Biochem., 107, 240 (1980).
26. M. I. Al-Moslih, G. R. Dubes, and A. N. Masoud, J. High Res. Chromatogr. and Chromatogr. Commun., 4, 173 (1981).

27. G. H. R. Rao, J. D. Peller, and J. D. White, J. Chromatogr., 226, 466 (1981).
28. A. A. Darwish and R. K. Prichard, J. Liq. Chromatogr., 4, 1511 (1981).
29. Z. El. Rassi and Cs. Horváth, Chromatographia, 15, 75 (1982).
30. J. B. Crowther, R. Jones, and R. A. Hartwick, J. Chromatogr., 217, 479 (1981).
31. A. Sokolows, N. Balgobin, S. Josephso, and J. B. Chattopa, Chem. SCR, 18, 189 (1981).
32. W. Jost, K. Unger, and G. Schill, Anal. Biochem., 119, 214 (1982).
33. J. B. Crowther, J. P. Caronia, and R. A. Hartwick, Anal. Biochem., 124, 65 (1982).
34. M. Ehrlich and K. Ehrlich, J. Chromatogr. Sci., 17, 531 (1979).
35. D. J. Popovich, Selective Detection in Liquid Chromatography, Tracor Chromatography, Appl. 79-10.
36. Á. H. Csárnyi, M. Vajda, and J. Sági, J. Chromatogr., 204, 213 (1981).
37. N. D. Brown, J. A. Kintzios, and S. E. Koetitz, J. Chromatogr., 177, 170 (1979).
38. J. L. Day, L. Tterlikkis, R. Neimann, A. Mobley, and C. Spikes, J. Pharm. Sci., 67, 1027 (1978).
39. W. Voelter, K. Zech, P. Arnold, and G. Ludwig, J. Chromatogr., 199, 345 (1980).
40. C. F. Gelijkens and A. P. De Leenheer, J. Chromatogr., 194, 305 (1980).
41. J. L. S. Au, M. G. Wientjes, C. M. Luccioni, and Y. M. Rustum, J. Chromatogr., 228, 245 (1982).
42. Y. Shinohara, R. D. Miller, and N. Castagnoli, Jr., J. Chromatogr., 230, 363 (1982).
43. N. D. Danielson and J. A. Huth, J. Chromatogr., 221, 39 (1981).
44. F. K. Chow and E. Grushka, J. Chromatogr., 185, 361 (1979).
45. D. P. Wittmer, N. O. Nuessle, and W. G. Haney, Jr., Anal. Chem., 47, 1422 (1975).
46. D. Westerlund and A. Theodersen, J. Chromatogr., 144, 27 (1977).
47. C. P. Terweij-Groen, S. Heemstra, and J. C. Kraak, J. Chromatogr., 161, 69 (1978).
48. J. H. Knox and J. Jurand, J. Chromatogr., 125, 89 (1976).
49. P. T. Kissinger, Anal. Chem., 49, 883 (1977).
50. J. L. M. van de Venne, J. H. H. Hendrikx, and R. S. Deedler, J. Chromatogr., 167, 1 (1978).
51. J. H. Knox and J. Jurand, J. Chromatogr., 149, 297 (1978).
52. R. P. W. Scott and P. Kucera, J. Chromatogr., 175, 51 (1979).

53. A. P. Konijnendijk and J. L. M. van de Venne, in *Advances in Chromatography* (A. Zlatkis, Ed.), Chromatography Symposium, Dept. of Chem., Univ. of Houston, Houston, Tex., 1979.
54. J. H. Knox and R. A. Hartwick, J. Chromatogr., *204*, 3 (1981).
55. S. N. Deming and R. C. Kong, J. Chromatogr., *217*, 421 (1981).
56. W. R. Melander and Cs. Horváth, J. Chromatogr., *201*, 211 (1980).
57. R. P. Whitehouse and C. L. Greenstock, J. Chromatogr., *243*, 376 (1982).
58. R. P. Whitehouse and C. L. Greenstock, J. Liq. Chromatogr., *5*, 2085 (1982).

Part III
APPLICATIONS

10
Nucleic Acids

RICHARD C. SIMPSON University of Rhode Island,
Kingston, Rhode Island

   I. Introduction 181
  II. Classical Chromatography 182
      A. Applications 182
      B. Disadvantages 183
 III. HPLC 184
      A. Ion exchange 184
      B. Ion exchange/reversed phase 185
      C. Reversed-phase ion pairing 187
      D. Size exclusion 187
      E. Hydroxylapatite 188
 IV. Summary 190
  V. Conclusion 191
     References 191

I. INTRODUCTION

Due to the importance of nucleic acids in biochemical systems, scientists have long been interested in methods for nucleic acid separation and/or purification from various other materials found in biological systems. Classical separation methods of the past have included such techniques as density gradient centrifugation [1] and electrophoresis [2]. However, today's rapidly growing field of genetic engineering is producing an ever-increasing demand for faster and more efficient separation techniques.

One technique which has the potential to meet these demands is liquid chromatography. Although classical forms of liquid chromatography have been used extensively in the past, they possess certain disadvantages not found in high performance liquid chromatography (HPLC). As a result, HPLC promises to continue growing in popularity and will gradually replace the more restrictive classical techniques.

The next few paragraphs provide a brief review of some typical applications of various forms of classical chromatographic techniques, along with their inherent disadvantages. The remainder of the chapter concentrates on the newer HPLC techniques which have become available in the last few years.

II. CLASSICAL CHROMATOGRAPHY

A. Applications

Chromatography using an agarose medium has been used to separate various transfer RNAs (tRNAs) by utilizing mobile phase salt concentration gradients [3-5]. Affinity chromatography has also been used to achieve separation of tRNAs [6-8]. Boronate chromatography has been employed for the separation of O-methylribose nucleosides and aminoacyl tRNAs [9]. Retention was due to the interaction of borate with the cis-2',3'-hydroxyl groups of ribonucleosides or with the 3'-adenosine in unacylated tRNAs. Elution was accomplished by a salt and pH gradient.

Anion-exchange chromatography has been used extensively. Isolations of specific tRNAs [10-11], and DNA [12] were accomplished using diethylaminoethyl (DEAE)-cellulose columns. In addition, packings such as DEAE-dextran [13], 2-hydroxypropylamino(QAE)-dextran [14], and triethylaminoethyl(TEAE)-cellulose [15] have been used to separate various oligonucleotides. Finally, a benzoylated diethylaminoethyl (BD)-cellulose column has been used to separate various tRNAs [16].

A form of calcium phosphate known as hydroxylapatite has been widely used as a packing for nucleic acid separations. Retention is due to the attractions between the packing's calcium sites and the phosphate groups in the nucleic acids. This material has often provided better separations than can be obtained by use of cellulose-based ion-exchange packings. Some of the many applications of hydroxylapatite include separation of DNA from RNA [17], separation of single-stranded DNA from double-stranded DNA [18], and the isolation of extrachromosomal DNA [19,20].

Another popular classical technique is known as RPC-5 chromatography. This form of chromatography uses inert polychlorotrifluoroethylene beads as a support material. The beads are coated with trialkylmethylammonium chloride. The alkyl chain length lies predominantly in the C_8-C_{10} range. The resultant packing material possesses both ionic sites (quaternary ammonium ions) and hydrophobic sites (inert resin). Thus retention of compounds is due to ion-exchange and/or reversed-phase mechanisms [21].

Some typical uses of RPC-5 chromatography include separation of various tRNAs [22-26], DNA restriction fragments [27], and oligonucleotides [28].

It is interesting to note that siliconized glass beads have also been coated with trialkylmethylammonium chloride for use in fractionation of tRNAs [29].

B. Disadvantages

Most of these methods have certain disadvantages which are not present in HPLC. For example, cellulose-, agarose-, or dextran-based packings are soft and cannot be subjected to high pressures. This results in severe limitations in mobile phase flow rates and the subsequent time required for a given separation to be achieved.

All of these methods require that the bulk packing be mixed with a solvent to produce a slurry. This step is often necessary to swell or hydrate the packings prior to use. Additionally, since the columns are generally slurry-packed by the user, extra time is required to prepare the column and it is difficult to reproduce results on a column-to-column basis.

Some techniques, such as affinity chromatography or RPC-5 chromatography, require the user actually to manufacture the packing material due to lack of commercial availability.

In RPC-5 chromatography, the stationary phase is mechanically coated onto the support. As a result, it has been shown that in mobile phases which are low in ionic strength (less than 0.2 M) or which contain organic solvents, the stationary phase can be readily stripped from the support [30]. RPC-5 material cannot tolerate high pressures and often results in poor sample recovery [31]. It has also been suggested that reproducibility is poor in some separations [30].

Many of these techniques use off-line detection methods, employing fraction collectors and some sort of detector; thus the errors in quantitation are increased and more of the user's time is required during the separation process.

As a result of these limitations, some of which are quite severe, HPLC appears to be a viable alternative technique for many separations involving nucleic acids. Researchers are beginning to use HPLC more frequently and are obtaining encouraging results.

III. HPLC

The advent of HPLC has revolutionized the field of chromatography. Modern column technology has provided packings which are reproducible, versatile, selective, stable at high pressures, and highly efficient. Modern solvent delivery systems are capable of operating at high pressures and delivering extremely accurate flow rates of up to several milliliters per minute, thus yielding fast and highly efficient separations of compounds, some of which were impossible to obtain by classical chromatographic techniques. Therefore, it is advantageous to apply HPLC to the separation of nucleic acids and other closely related compounds. The next section will explore some of these HPLC methods.

A. Ion exchange

Since nucleic acids contain anionic sites, ion exchange is a logical separation technique. Pearson and Regnier have described the preparation and use of an anion-exchange column for the separation of oligonucleotides [30]. The material is silica modified with polyethyleneimine, which provides the ion-exchange sites. Their work describes the effects of silica type and stationary phase thickness on the separation process. The paper also demonstrates that the silica surface can be modified either by a batch reaction or in situ. Elution was accomplished by a salt concentration gradient.

Colpan et al. have described the preparation of a silica-based ion-exchange packing and its use for purification of RNA [32]. Spherical silica (10 μm diameter, 500 Å pore size) was reacted with (1,2-epoxy-3-propylpropoxy)-trimethylsilane. It was then condensed with N,N-dimethylaminoethanol to yield a weak anion-exchange packing. The structure of the bonded-phase packing is shown in Fig. 1.

Using this packing and a salt concentration gradient, the resolution of four different RNAs was achieved in less than 50 min. The separation obtained by this method is shown in Fig. 2.

$$\equiv Si-O-Si-(CH_2)_3-O-CH_2-CH(OH)-CH_2-O-CH_2-CH_2-\overset{\oplus}{N}H(CH_3)_2$$

Figure 1. Structure of DMA-500 weak anion-exchange material. (From Ref. 32.)

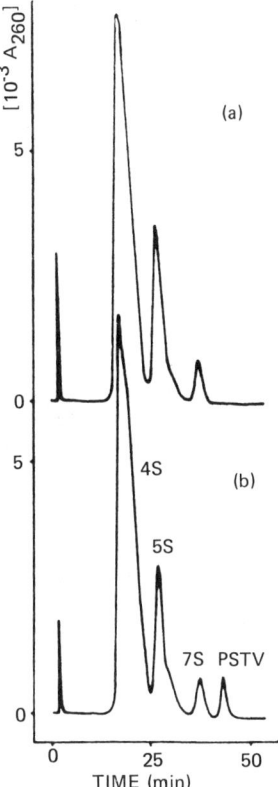

Figure 2. Analytical HPLC of (a) noninfected and (b) PSTV-infected plant RNA extract. 25 μg RNA in 20 μl were injected into a 1/2-in. × 4-cm DMA-500 column, eluted with a linear gradient of 0.25 M KCl to 0.66 M KCl in 50 min, in 5 M urea, 20 mM potassium phosphate, pH 6.5, 0.2 mM EDTA, at 3 mL/min, 15 bar, and ambient temperature. (From Ref. 32.)

Thompson and Blakesley have recently reported a "nucleic acid chromatography system" based on ion exchange [33]. Though limited information is available, they do report that use of a salt gradient allows separation of a wide variety of nucleic acid species.

B. Ion exchange/reversed phase

Bischoff et al. have described the preparation of a unique stationary phase for the separation of various tRNAs [31]. The stationary phase was an aggregate of C_{18} and trioctylmethylammonium chloride. The

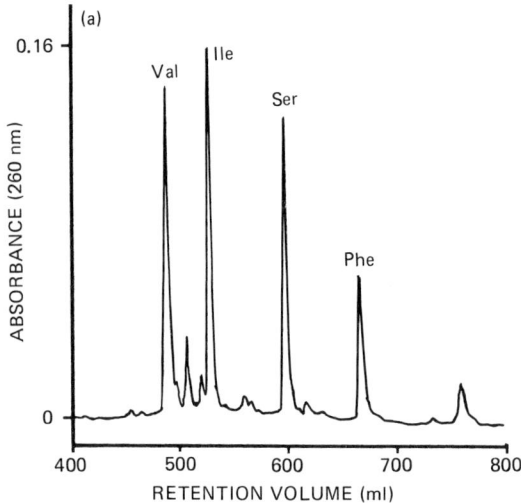

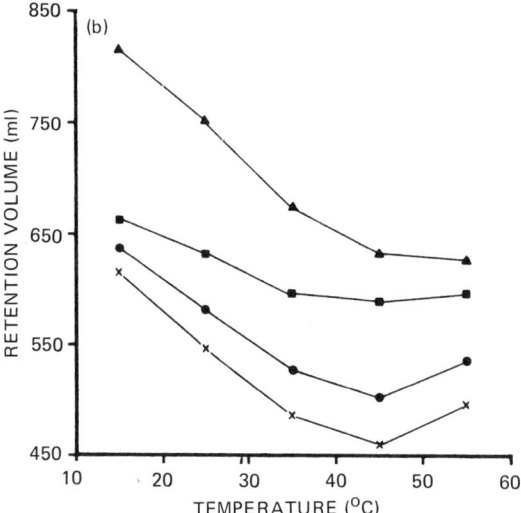

Figure 3. (a) Separation of four specific tRNAs at 35°C. (b) Temperature dependence of the retention volumes of the tRNAs specific for valine (x), isoleucine (●), serine (■), and phenylalanine (▲). Column: 6.2 × 250 mm trioctylmethylammonium chloride-treated ODS-Hypersil. Flow rate: 1 ml/min. Gradient: buffer A, 0.5 M ammonium acetate pH 4.5; buffer B, 5.0 M ammonium acetate pH 6.0; 100% A to 60% A– 40% B in 1080 min. (From Ref. 31.)

support material was 5-μm spherical silica. This packing retains nucleic acids by both ion-exchange and reversed-phase mechanisms. Elution of tRNAs was accomplished by a salt concentration/pH gradient. A typical separation of tRNAs achieved by this technique is shown in Fig. 3a.

Additionally, the authors demonstrated the effect of column temperature on the retention volume of tRNAs. A graph of these data is illustrated in Fig. 3b. The results are as expected; i.e., increased temperature causes a decrease in retention volume. This behavior is observed in reversed-phase systems as well as ion-exchange systems.

It is interesting to note the similarity between this method and RPC-5 chromatography. Both methods use a combination of reversed-phase and ion-exchange retention mechanisms. Both methods have hydrophobic sites (C_{18} or polychlorotrifluoroethylene) and ionic sites (trialkylmethylammonium). Due to these similarities, perhaps many previously established RPC-5 separations can be readily accomplished by this newer method described by Bischoff et al.

C. Reversed-phase ion pairing

Recently Nguyen et al. have reported the separation of tRNA from mRNA by use of reversed-phase ion-pairing HPLC [34]. For the reader who is unfamiliar with this technique, Chap. 9 provides a brief review of the theory, along with references to several other reviews on the subject. Nguyen's method employs two C_{18} columns in series. The mobile phase contains tetrabutylammonium phosphate as the ion-pairing reagent. Elution of the RNAs is accomplished by use of a methanol gradient. The authors report that this method requires only 40 min to obtain the same efficiency of a several-hour sucrose centrifugation. Additionally, the mobile phase and ion-pairing reagent are volatile and easily removed from collected fractions. Currently underway are studies observing the effects of various solvents, solute concentrations, and gradient profiles. The authors hope to improve the efficiency of the separation through use of these variables.

D. Size exclusion

High performance size exclusion (HPSEC), or gel filtration, chromatography has also been used in nucleic acid separations. The packings used may be composed of porous silica or synthetic polymers. They are smaller, more rigid, and more porous than classical SEC materials. Also, their pore size distribution is narrower and better controlled. Because of these factors it is possible to achieve rapid and efficient SEC separations, using high pressures and flow rates, which cannot be duplicated by classical techniques.

Uchiyama et al. reported the separation of three low-molecular-weight RNAs using a silica gel packing [35]. The mobile phase was a neutral phosphate buffer containing sodium dodecyl sulfate (SDS). Complete separation of the analytes was obtained in approximately 30 min.

HPSEC was used successfully by Himmel et al. to separate DNA from DNA fragments and DNA linkers following ligation reactions [36]. They also noted an interesting interaction between DNA and the packing material in the presence of Mg(II) ions, and have not yet determined the exact nature of the interaction. To eliminate it, ethylenediaminetetraacetate was added to the mobile phase to chelate any Mg(II) ions present.

Thrall and Spelsberg have used HPSEC in a variety of DNA and RNA fractionations [37]. They employed a silica gel with a bonded ether stationary phase. Their work noted the effect of different silica pore sizes on the resolution of various nucleic acid species. The results of their study indicated that this packing material results in somewhat poor fractionation.

For preparative work, Graeve et al. used HPSEC for the fractionation of high-molecular-weight RNAs [38]. They have also been able to separate complex mRNA mixtures while still retaining the biological activity of the analytes. The investigators examined the use of both denaturing and nondenaturing mobile phases. They found that by using a denaturing mobile phase, the resolution obtained was superior to that obtained by use of sucrose gradient centrifugation or a previously reported reversed-phase ion-pairing technique [34]. The separation was accomplished in 4 hrs, resulting in the chromatogram shown in Fig. 4.

The study also indicated that high-molecular-weight RNAs were not degraded during the chromatographic process. This is supportive of results from another study involving HPSEC of DNA. This previous study indicated that no significant shearing of DNA occurred during the chromatographic separation [36]. This is an important factor to consider when evaluating the retention of the analytes' biological integrity.

E. Hydroxylapatite

Recently, hydroxylapatite has become commercially available for use as a HPLC packing material [39]. Special preparation and packing methods permit its use at elevated pressures and flow rates. In addition, the capacity and resolution are much higher than in classical hydroxylapatite chromatography. Thus, hydroxylapatite may now be used in preparative or analytical-scale HPLC separations of nucleic acids. The advantage in using this new method is that classical hydroxylapatite separations may now be transferred with little or no modification to the

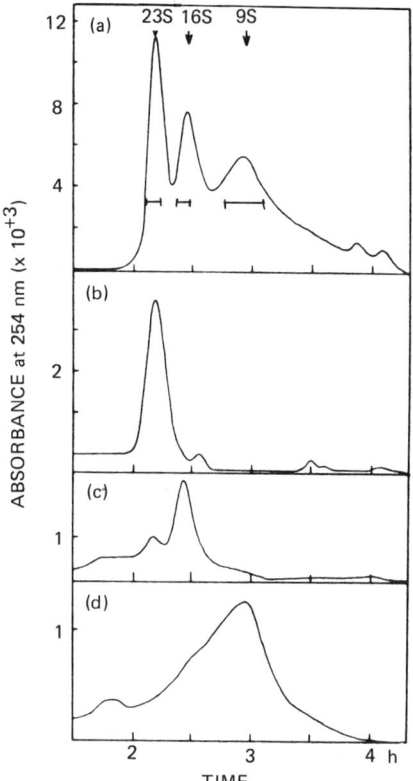

Figure 4. HPLC gel filtration of high-molecular-weight RNAs. (a) Separation of a mixture of 16S and 23S prokaryotic rRNA (8 µg) and 9S globin mRNA (6 µg) under nondenaturing conditions on a TSK G 4000 SW column. (b-d) A part of the indicated fractions of profile A was rechromatographed. (From Ref. 38.)

faster HPLC technique. An example of a hydroxylapatite HPLC separation is the resolution of single-stranded DNA from double-stranded DNA in 12 min. The chromatogram obtained is illustrated in Fig. 5.

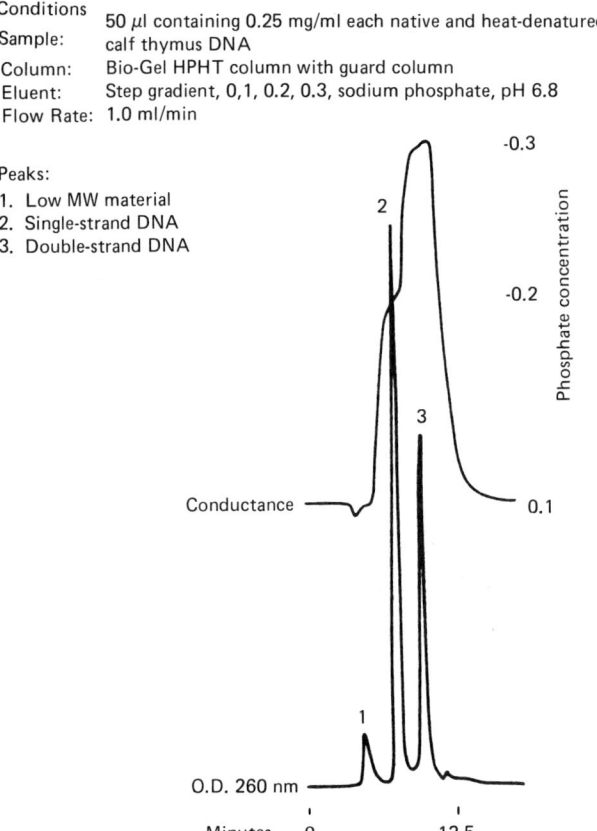

Figure 5. Separation of single-stranded and double-stranded DNA with Bio-Gel HPHT. (From Ref. 39 with permission of Bio-Rad Laboratories. Bio-Gel is a registered trademark of Bio-Rad Laboratories.)

IV. SUMMARY

Table 1 provides a quick summary of the HPLC methods for nucleic acid separations.

Table 1 Summary of HPLC Nucleic Acid Separations

Type of HPLC	Mobile phase program	Reference
Ion exchange	Salt gradient	30
Ion exchange	Salt gradient	32
Ion exchange	Salt gradient	33
Ion exchange/reversed-phase	Salt/pH gradient	31
Reversed-phase/ion pairing	Methanol gradient	34
Size exclusion	Isocratic	35
Size exclusion	Isocratic	36
Size exclusion	Isocratic	37
Size exclusion	Isocratic	38
Hydroxylapatite	Salt gradient	39

V. CONCLUSION

HPLC is only beginning to be utilized in nucleic acid separations. This is evidenced by the relatively few HPLC applications developed to date. However, its advantages over classical chromatographic techniques are many and varied. They include higher operating pressures, higher flow rates, more stable packings, more selective packings, more efficient separations, faster separations, higher capacity, and more flexibility. All of these factors, combined with future advances in state-of-the-art technology, ensure that HPLC will continue to develop into a powerful separation technique in nucleic acid research. It is hoped that researchers will be willing to abandon established, but outdated, chromatographic methods for the newer methods of the future.

REFERENCES

1. J. Draper, M. R. Davey, J. P. Freeman, and E. C. Cocking, Experientia, 38, 101 (1982).
2. U. E. Loening, Biochem. J., 102, 251 (1967).
3. J. M. Egly, J. L. Plassat, and E. Boschett, J. Chromatogr., 243, 301 (1982).

4. M. Spencer and M. M. Binns, J. Chromatogr., 238, 297 (1982).
5. M. Spencer, J. Chromatogr., 238, 307 (1982).
6. C. M. Garcia and R. P. Singhal, Biochem. Biophys. Res. Commun., 86, 697 (1979).
7. D. J. Gross and L. S. Parkhurst, J. Biol. Chem., 253, 7804 (1978).
8. M. M. Smith and R. C. C. Huang, Proc. Nat. Acad. Sci. U. S., 73, 775 (1976).
9. R. P. Singhal, R. K. Bajaj, C. M. Buess, D. B. Smoll, and V. N. Vakharia, Anal. Biochem., 109, 1 (1980).
10. Y. Kawade, T. Okamoto, and Y. Yamamoto, Biochem. Biophys. Res. Commun., 10, 200 (1963).
11. J. D. Cherayil and R. M. Brock, Biochemistry, 4, 1174 (1965).
12. A. F. Wu and T. T. Wu, Prep. Biochem., 121, 1 (1982).
13. M. Staehelin, Progr. Nucleic Acid Res., 2, 169 (1963).
14. H. Schott, J. Chromatogr., 237, 429 (1982).
15. J. Satava, O. Mikes, and P. Strop, J. Chromatogr., 180, 31 (1979).
16. I. C. Gillam and G. M. Tener, in *RNA Protein Synthesis* (K. Moldave, Ed.), Academic Press, New York, 1981, p. 43.
17. D. W. Sutton and J. D. Kemp, Biochemistry, 15, 3153 (1976).
18. Y. Miyazawa and C. A. Thomas, Jr., J. Mol. Biol., 11, 223 (1965).
19. M. Shoyab and A. Sen, J. Biol. Chem., 253, 6654 (1978).
20. M. Shoyab and A. Sen, Meth. Enzymol, 68, 199 (1979).
21. R. L. Pearson, J. F. Weiss, and A. D. Kelmers, Biochim. Biophys. Acta, 228, 770 (1971).
22. B. Wittig and S. Wittig, FEBS Lett., 105, 254 (1979).
23. D. Hatfield, A. Diamond, and B. Dudock, Proc. Nat. Acad. Sci. U. S., 79, 6215 (1982).
24. M. A. Francis and B. Dudock, J. Biol. Chem., 257, 11195 (1982).
25. M. A. Kashdan, I. L. Pirtle, J. L. Calagan, H. J. Vreman, and B. S. Dudock, J. Biol. Chem., 255, 8831 (1980).
26. B. Roe, K. Marcu, and B. Dudock, Biochim. Biophys. Acta, 319, 25 (1973).
27. A. Landy, C. Foeller, R. Reszelbach, and B. Dudock, Nucleic Acids Res., 3, 2575 (1976).
28. R. D. Wells, S. C. Hardies, G. T. Horn, B. Klein, J. E. Larson, S. K. Nevendorf, N. Panayotatos, R. K. Patient, and E. Seling, Meth. Enzymol., 65, 327 (1980).
29. T. Narihara, Y. Fujita, and T. Mizutani, J. Chromatogr., 236, 513 (1982).
30. J. D. Pearson and F. E. Regnier, J. Chromatogr., 255, 137 (1983).

31. R. Bischoff, E. Graeser, and L. W. McLaughlin, J. Chromatogr., 257, 305 (1983).
32. M. Colpan, J. Schumacher, W. Bruggemann, H. L. Sanger, and D. Riesner, Anal. Biochem., 131, 257 (1983).
33. J. A. Thompson and R. W. Blakesley, Fed. Proc., 42, 1953 (1983).
34. P. N. Nguyen, J. L. Bradley, and P. M. McGuire, J. Chromatogr., 236, 508 (1982).
35. S. Uchiyama, T. Imamura, S. Nagai, and K. Konishi, J. Biochem., 90, 643 (1981).
36. M. E. Himmel, P. J. Perna, and M. W. McDonell, J. Chromatogr., 240, 155 (1982).
37. C. Thrall and T. C. Spelsberg, in *Biological/Biomedical Applications of Liquid Chromatography I* (G. Hawk, Ed.), Marcel Dekker, New York, 1979, p. 283.
38. L. Graeve, W. Goemann, P. Foldi, and J. Kruppa, Biochem. Biophys. Res. Commun., 107, 1559 (1982).
39. Bio-Rad Laboratories, *The Liquid Chromatographer*, March 1983, p. 6.

11
Oligonucleotides

JONATHAN B. CROWTHER* Rutgers, The State University of New Jersey, Piscataway, New Jersey

I. Introduction 195
II. Separation Methods 198
   A. Open-column and low-pressure methods 198
   B. High performance methods 200
III. Identification and Detection of Oligonucleotides 207
   A. Optical and radioactive detection 207
   B. Absolute methods for determining oligomer sequence 208
   C. Methods for determining relative base composition 208
IV. Future Developments in HPLC of Oligonucleotides 209
   References 210

I. INTRODUCTION

Recent advances in recombinant DNA research have inspired keen interest in the development of improved separation and identification methods for analysis of the shorter chains (from dimers to about 60-mers) of deoxy- and ribooligonucleotides.

---

*Current affiliation: Cornell University, Ithaca, New York.

Oligonucleotides of this length are either prepared synthetically or can be produced from enzymatic digestion of DNA or RNA. In the enzymatic production of oligonucleotides, restriction enzymes, known to cleave DNA and RNA at specific base sites, are employed to digest these materials into oligomer fragments of various chain lengths and base compositions. However, since enzymatic production of oligonucleotides is limited by the various restriction enzymes and DNA material available, it is suitable for the production of relatively few sequences and chainlengths. Synthetic production of these polymers, on the other hand, is more versatile. New methods of oligomer synthesis, both solid phase and solutions, have reduced the time required for synthesis of oligonucleotides while increasing yields. These advances are due mainly to the use of modern chromatographic techniques in the analysis of synthetic intermediates and final product purification.

Traditionally, nucleotides, nucleosides, and their short polymers have been separated using ion-exchange column chromatography [1,2], DEAE-cellulose [3], thin-layer [4,5] or gel electrophoresis [7]. With the development of high efficiency bonded support materials and improved instrumentation in the late 1960s [6], high performance liquid chromatography (HPLC) has evolved into the method of choice for rapid and efficient separations of these compounds. In recent years the separations of nucleosides on reversed-phase materials [8], the separation of nucleotides on anion exchangers [9], nucleobases on cation exchangers [10], and the simultaneous separation of nucleotides and nucleosides using ion-pair/reversed-phase HPLC [11] have been reported. HPLC separations of nucleosides, nucleotides, and bases were recently reviewed by Zakaria and Brown [12]. Although oligonucleotides consist of nucleotides, the chromatographic properties differ significantly as the chain length increases. The separation of the oligonucleotides thus becomes a unique problem, distinct from the separation of the shorter monomers. This chapter reviews the state of the art of separations of oligomers from dimers up through small DNA fragments (i.e., less than 60 bases). A systematic survey of transfer RNA (tRNA) and larger DNA separations will not be presented here, since the separation problems encountered are distinctly different from those of the shorter fragments.

The area of short oligonucleotide separations is one of excitement and growing significance. The synthesis and purification of pure and reliably sequenced oligonucleotides is essential for the advancement of research in such areas as molecular cloning, the synthesis of genes, and fetal abnormality probes. Furthermore, the oligonucleotides can be used as probes to improve our understanding of genes themselves.

Classically, the three most popular approaches to the synthesis of oligonucleotides have been the phosphodiester [13], phosphotriester [14], and the phosphite/phosphotriester chemistries [15]. The main differences among these three approaches lie in the method and degree to which the nucleotides are derivatized in order to prevent undesirable side reactions during the synthesis of the longer nucleotide chain. Using the "protected" nucleotide monomers or dimers, oligomer chains of 30 or more base units can be readily synthesized either manually or with the automated assistance of "gene machines" now available commercially.

Figure 1 shows a C-G dimer and the same molecule in one of its protected forms [11]. Substituents, which are base-labile, protect the free hydroxyl in the sugar moiety and the amino groups on the aglycone. The phosphate linkage protecting groups, which form the phosphodiester, can also be removed upon treatment with base. The acid-liable dimethoxytrityl groups are used to eliminate polymerizations of the nucleotides and to increase the lipophilic properties of the molecule, which aids product purification. Thus, throughout the synthesis of these biomolecules, different and complex separation mechanisms must be evaluated in order to provide adequate purification of both the organic-soluble protected oligomer and the water-soluble deprotected counterpart.

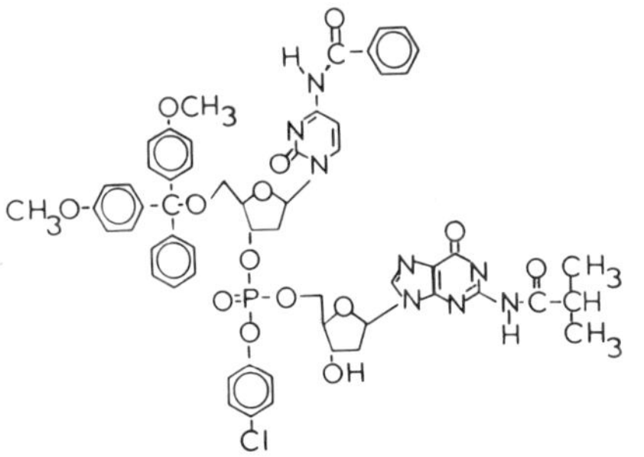

C - G PROTECTED

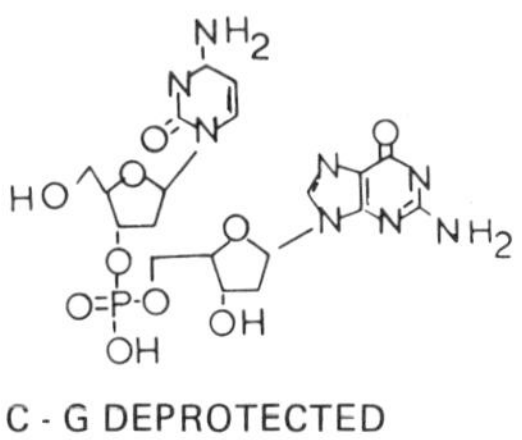

C - G DEPROTECTED

Figure 1. Structure of the deoxy dimer cytidyly (3'-5') guanosine (CG) and its protected form showing several of the various protecting groups (Ref. 11) used routinely during synthesis of oligonucleotides to increase reaction yields.

II. SEPARATION METHODS

A. Open-column and low-pressure methods

Much of the original open-column chromatographic work on oligonucleotides was performed on ion-exchange sorbents such as DEAE-cellulose and DEAE-Sephadex [3,15,16]. The celluloses have been modified to improve their selectivity, as demonstrated in the work of Pace and Pace [17]. They used boronated celluloses (DBAE-cellulose) to form cis-diol complexes of monoribonucleotides and small ribooligomers. The oligomers can thus be eluted by mild acid or a competing diol. Larger oligomers (mono to 11-mers) were separated by Satava et al. [18] using DEAE-Spherons 300, a reticulated rigid gel which has a very low ion-exchange capacity with a weakly hydrophobic surface. The Spheron 300 particles provided better resolution for these oligomers than the DEAE-Sephadex. Furthermore, the rigid particles enabled analysis to be performed at higher pressures.

De Rooij et al. [19] separated and purified protected phosphotriester oligonucleotides of chain lengths up to 12 bases on Sephadex LH-60. The surface of LH-60 was found to have both hydrophilic and hydrophobic properties, which necessitated the addition of 5% tetrahydrofuran to the mobile phase. Purification of similarly protected oligomers on pure silica columns resulted in transesterfication of the aryl protecting groups, and recoveries were found to be substantially lower. Grzeskowiak et al. [20] found Sephadex LH20 to be useful in the purification of protected phosphodiesters.

An optimum blend of the hydrophobic, hydrophilic, and ion-exchange properties for separations of oligomers and RNA materials was found in a series of stationary phases developed in 1971 by Pearson et al. [21]. These materials, coined RPC-5, used particles of polychlorotrifluoroethylene, Plaskon 2300 (Allied Corporation, Morristown, N.J.), as the support. These rigid materials were coated with Adogen 464 (Archer Daniels Midland Co., Decatur, Ill.), a C_8-C_{10} trialkyl-methylammonium compound. The base, which was adsorbed to the support, imparted some weak ion-exchange properties to the support, achieving a fortuitous blend of hydrophobic/hydrophilic/ionic properties which were well suited for larger oligonucleotides. RPC-5 was found to be quite suitable for separations of DNA restriction fragments [22], RNA fragments [23], and polymeric nucleotides [24].

Despite the popularity of Plaskon 2300, the supplier of this support material has discontinued the product. A suitable substitute has recently been reported by Shum and Crothers [23], who used Kel F powder manufactured by 3M, Inc. (St. Paul, Minn.), as a support material for RPC-5 chromatography.

Using this Kel-F material coated with Adogen, an excellent comparison between the merits of RPC-5 chromatography and cellulose was presented by Shum and Crothers. As shown in Fig. 2, RPC-5 is

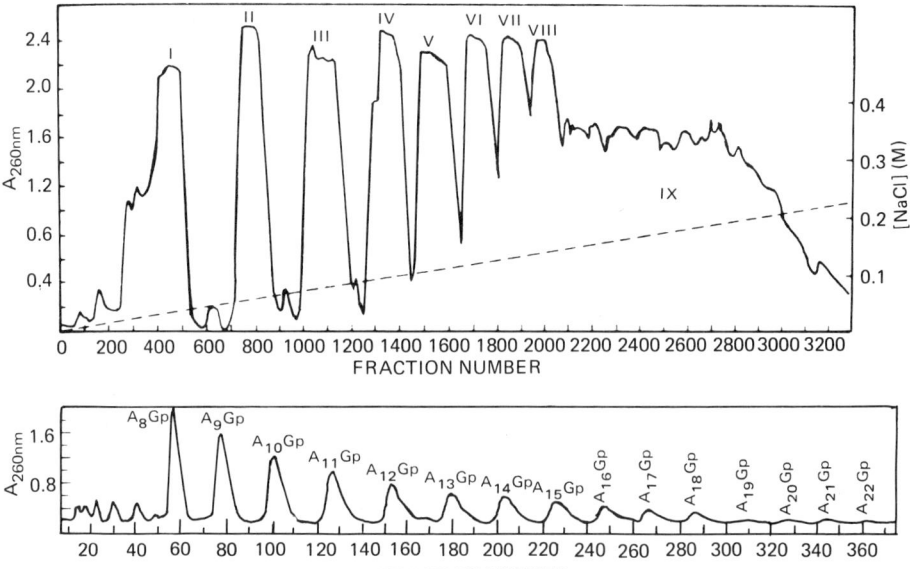

Figure 2. (Top) Large-scale TEAE-cellulose anion-exchange chromatography of the products of exhaustive T_1 ribonuclease digestion of poly (A,G). Column size = 3.2 × 120 cm; gradient volume = 40 liters, from 0.0 to 0.4 M NaCl, in 0.005 M Tris, 7.0 M urea, pH 7; flow rate = 300 ml/hr, fraction size = 12 ml, T = 4°C. Peak identification, I = Gp, II = ApGp, III = (Ap) Gp, . . . , IX = (Ap)nGp, N ⩾ 8. (Bottom) RPC-5 reversed-phase chromatography of the pooled fraction labeled IX above. Column = 1.6 × 100 cm, gradient volume 12 liters, from 0.4 to 0.8 NaCl, in 5 mM sodium cacodylate, pH 7, flow rate = 70 ml/hr, fraction size = 12 ml, room-temperature elution. (From Ref. 23.)

capable of separating extended chain lengths, while TEAE-cellulose is limited to separations of about 15 bases [23]. Others [22,23,25] have reported similar success with separations at low pH of both the ribo- and deoxyribooligonucleotides having chain lengths of up to 60 bases. Selsing et al. [24] achieved a basic pH separation of dG oligomers up to 30-mers.

Separations using RPC-5 are not without limitations. Since RPC-5 is a coated polymer, nucleic acids can be eluted only by salt gradients buffered between pH 4 and 8. These solvent conditions help prevent stationary phase bleed-off from the column. However, even under the best conditions, adsorbed Adogen will sometimes be found in the effluent, thus creating problems in preparative purification methods. The column itself must be periodically unpacked and recoated, unless a presaturated mobile phase is employed. Usher [26] found that uncoated Kel-F may be used with mobile phases having relatively volatile solvents and amine salts at high concentrations. Usher's approach has the advantage of minimizing solute contamination upon collection and lyophilization.

B. High performance methods

Celluloses and polymeric-based supports such as those described above are useful for the separation of delicate oligomers because of their mild surface properties and unique selectivities. However, most of these polymers are not rigid enough to withstand substantial flow rates and pressures; therefore, they are generally unsuited as high performance packings in modern HPLC. Silica-based supports are almost universally used in modern HPLC packings [27] because few other support materials have been found to possess the desirable properties of silica. Much of the current popularity of silica supports for high performance liquid chromatography is due to the rigid nature of the gel, the large surface area, and the good mass transfer properties.

Initial analytical separations of oligonucleotides on silica supports were performed on nonbonded silica. Since protected oligomers are organic-soluble, they can often be purified on silica columns, typically using chloroform/methanol mobile phases [28,29]. Unfortunately, this active surface can result in the methanol-catalyzed [19] or TEA [28] conversion of triesters to diesters and subsequent low yields, typically ca. 65% for smaller oligomers and much less for oligomers of 10 or more. Thus, in order to increase recoveries, the surface properties of silica must be modified.

Before the advent of porous silica with good mass transfer characteristics, pellicular materials, notably pellicular ion exchangers, were designed to provide efficient analyses. Basically, these pellicular materials consist of solid glass spheres with a surface of polymerized silica which can be bonded using established silane procedures. The loading capacities of these materials are low, but this deficiency is offset by the ease of column packing.

Using pellicular packings, AS pellionex SAX and AL Pellionex WAX (Whatman, Inc., Clifton, N.J.), Komiya et al. [30] achieved excellent results in separation of the products of RNase digestions of ribooligomers. The SAX packing material is a strong cation exchanger (quarternary amine) and the WAX material is a weak anion exchanger (R-NH2). The WAX column provided good separations of oligomers to a 10-mer, chiefly on the basis of chain length but also on the basis of composition. Vandenberghe and associates [31,32] also achieved excellent separations at neutral pH on Pellionex WAX and reported selectivity of the material toward chain length on enzymatic digestion products of RNA.

The Pellionex materials have recently been discontinued; however, primary amine columns bonded onto porous silica are now marketed by several commercial suppliers of bonded silica materials. Brown and associates [33] found that propylamine columns, although somewhat unstable in a phosphate buffer system, are adequate to resolve several forms of 2-5 oligo(A). Wakizaka et al. [34] reported excellent separations of di- and trinucleotides on Lichrosorb NH2 column (E. M. Industries Co., Cincinnati, Ohio). Furthermore, McLaughlin et al. [35] reported RNase separations of tridecamers using aminopropyl-bonded silica (Zorbax-NH2, Dupont, Wilmington, Del.), as shown in Fig. 3. The column exhibited some deterioration over a 4-month period and required high salt gradients to elute the higher-molecular-weight oligomers. Dizdaroglu et al. [36] have reported separations of oligodeoxypentamers using a Micro pack AX-10 (Varian Assoc.) amine column. Finally, Jost et al. [37] synthesized a novel amino/hydroxyl-bonded

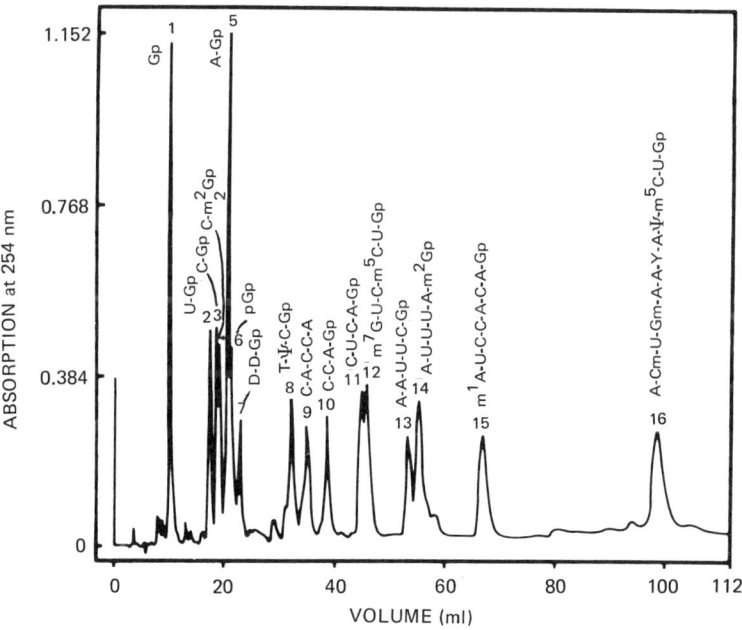

Figure 3. Separation of an RNase T_1 digestion column, 4.6 × 250 mm, Zorbax-NH2 (7 μm); buffers: A, 0.05 M KH_2PO_4; B, 10% methanol (v/v) in 0.9 M KH_2PO_4; flow rate 2 ml/min; detector, 254 nm, 1.28 aufs; temperature 35°C. (From Ref. 35.)

stationary phase having properties similar to those observed in commercial primary amine columns. Their work included separations of (U) oligomers up to nonamers.

Strong anion-exchange resins are often preferred to weak anion exchangers since their ion exchange properties are more easily understood. A stationary phase of Partisil SAX (Whatman, Inc.) has been reported by several authors to be useful in both initial purification of synthetic oligomers [38,41] as well as for final efficient fractionation of base-protected oligomers [39]. This separation of base-protected oligomers was performed using a linear salt gradient in 20% ethanol. Gait et al. [40] used a SAX column for the initial preparative purification of a 27-mer synthesized on a solid support (Fig. 4). The 27-mer was separated using a linear gradient of phosphate buffer pH 6.3 in 60:40 formamide/water. The formamide was reported to increase greatly column lifetime as well as to improve resolution. Crea et al. [43], in the synthesis of 29 oligodeoxyoligomers, used a Permaphase

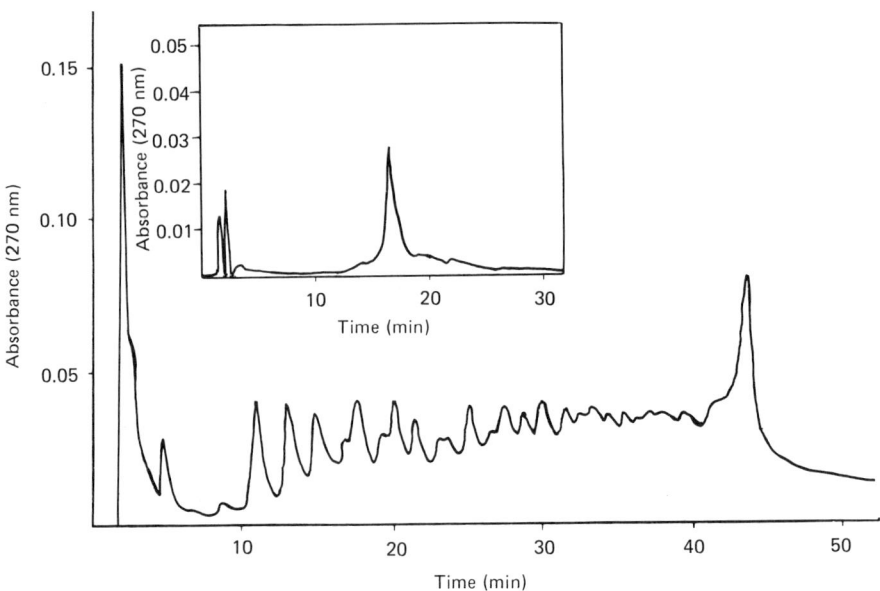

Figure 4. Analytical ion exchange HPLC of 27-mer, d(G-G-G-T-C-T-A-G-A-A-T-T-C-A-A-G-C-T-T-G-G-A-T-C-C-T-C), gradient 0.001 M KH_2PO_4 to 0.3 M KH_2PO_4 in 6/4 formamide/water, pH 6.3 in 50 min. Temperature 41°C. Inset: Reversed-phase HPLC of 27-mer. Buffer A = 0.1 M ammonium acetate and buffer B = 0.1 M ammonium acetate/acetonitrile (2:8 v/v), gradient 10-25% buffer B in 45 min. (From Ref. 40.)

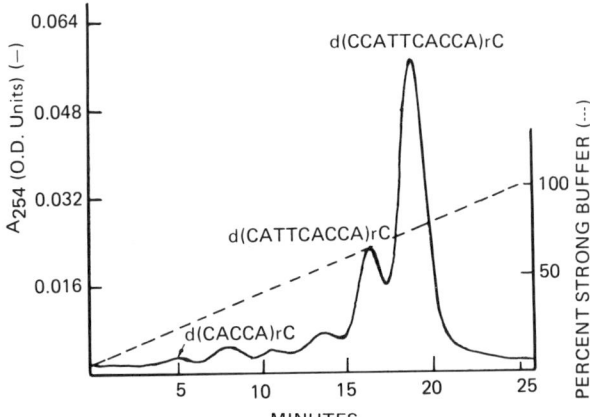

Figure 5. Deoxyribonucleotides 6-11 bases long. Column PEI 6-Li-Chrosorb Si 100 (10 μm). Flow rate 2 ml/min, temperature 25°C. Weak buffer: 0.05 KH_2PO_4, pH 7.0, 30% methanol. Strong buffer: 1.0 M NH_4Cl in weak buffer. (From Ref. 42.)

AAX (DuPont) anion exchanger for the purification of the deprotected synthetic fragments. During the stepwise synthesis of longer oligonucleotides, anion by-products are shorter oligomers. These shorter oligomers have fewer internucleotide phosphate linkages and thus elute prior to the longer oligonucleotide of interest.

Silica-bonded ion exchangers are noted for their poor efficiencies and instability. Furthermore, the pH working range is only between 2 and 6.5. Several resin manufacturers are currently marketing highly cross linked and stable ion-exchange resins suitable for HPLC. Another approach is that of Alpert and Regnier [42], of coating silica with a polyethyleneimine. Upon cross-linking, a stable anion exchanger is formed with a reasonably high exchange capacity. Within this system, oligomers are separated according to chain length (Fig. 5.)

Due to the lypophilic as well as the ionic character of oligonucleotides, these complex molecules can be separated on reversed-phase materials as well as the classic mode of ion exchange. Reversed-phase bonded silica is the most popular support material used in modern HPLC. Typically, these materials consist of a octyl or octyldecyl ligand bonded via siloxane linkages to a silica support. Reversed-phase materials are found to be more stable and are more efficient then bonded ion exchangers, due either to the bonding chemistries or to the mechanism of retention itself. Thus, reversed-phase materials are an attractive complement to ion-exchange chromatography of the oligonucleotides. Gait et al. [40] reported the use of reversed-phase materials for the final purification of fractions collected from SAX columns (Fig. 4, inset). Using 0.1 M ammonium acetate in a acetonitrile gradient, excellent separations of 14, 20, and 27 oligomers were observed on a µBondapack C_{18} column (Waters Assoc., Milford, Mass.). It was observed [44,45] that the oligonucleotide separations occur as a function both of chain length and base composition. Others have found reversed-phase materials suitable for separation of 2-15 oligomers [11,29,44,45,47].

Reversed-phase separations of fully and partially protected oligomers are possible. Chow et al. [47] have separated the dimethoxytrityl partially deportected derivatives of oligomers using a 0.1 M triethylammonium acetate buffer in 20-30% acetonitrile. Fritz et al. [45] discussed the effects of protecting groups on chromatographic behavior of oligomers. In a companion article, Jones et al. [46] discussed the design of oligonucleotide protecting groups that enhance separations on reversed-phase materials. Since bulky protecting groups, such as dimethoxytrityl, increase the hydrophobicity of the oligomer, separations of protected longer-chained oligomers are limited because the organic content of the mobile phase must be increased to a point where selectivity diminishes.

Crowther et al. [11] showed separations on a polar bonded stationary phase (Whatman PAC) for these organic-soluble protected oligomers. This phase was found to be highly suited for oligomer purification while using an easily evaporated mobile phase of methylene chloride and methanol. Figure 6 shows a preparative separation of a tetramer and its synthetic by-product using this chromatographic system. With the PAC stationary phase and an aqueous mobile phase, Hagemeir et al. [48] were able to separate the dinucleoside polyphosphates.

Most of the separation modes discussed so far have been designed to resolve oligomers according to chain length or base composition. Often, it is advantageous to separate these polymers according to size. In size exclusion chromatography (SEC), separation is not performed by interaction with the stationary phase; in fact, stationary-phase interaction must be prevented. Separation is accomplished by the mean path length a particular size molecule travels through a porous bed. Thus, the larger molecules elute first, while those molecules which can enter the pores elute later. Molko et al. [49] reported the size separation of oligomers on an I-125 Protein Analysis Column (Waters Associates). Their results show, in both preparative and analytical modes, that the oligomers of interest (the largest) is eluted first. Thus the order of elution is reversed to that observed for ion-exchange columns. Furthermore, volatile salts are used for size exclusion of these polymers and thus are more suited for preparative chromatography. In ion-exchange chromatography, high concentrations of nonvolatile salts are often needed to elute the compound of interest. Figure 7 illustrates the retention volume of oligodeoxynucleotides as a function of chain length. Using the mobile phase as

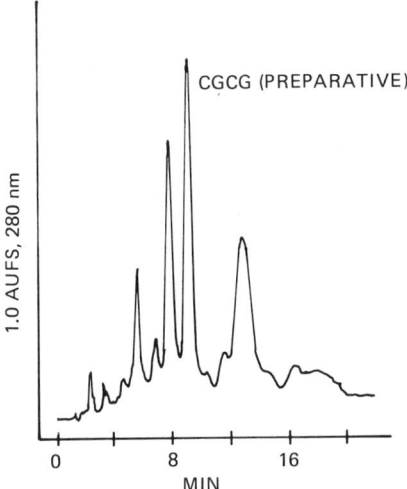

Figure 6. Injection of 2 mg of crude C-G-C-G protected tetramer, using a Whatman Magnum-9 PAC column. Flow rate = 6.0 ml/min of a mobile phase of (7.5:92.5) methanol/methylene chloride. (From Ref. 11.)

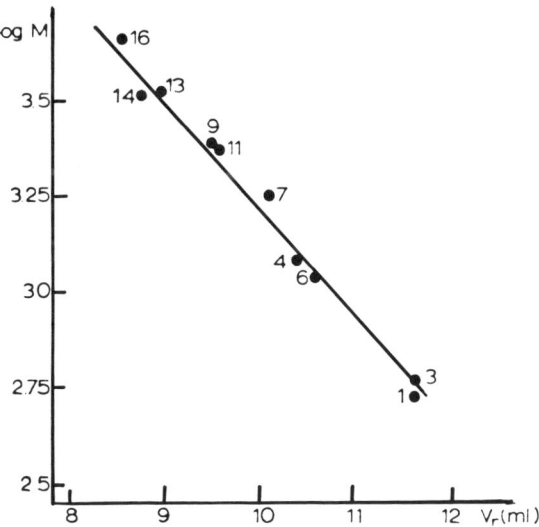

Figure 7. Logarithms of molecular weights of oligodeoxyribonucleotides as a function of their retention volumes on the Waters I-125 Protein Analysis Column. Solvent: 0.1 M triethyammonium acetate (TEAA), pH 7.0. Flow rate 1.04 ml/min. (From Ref. 49.)

described, adequate resolution of oligomers to a 15-mer were reported. The addition of triethylammonium acetate presumably is used to interact with active sites of the column and "cover" these sites to prevent interactions with the oligomers. Thus the concentration of TEAA and pore size characteristics of the silica are the only parameters that must be optimized. Robison et al. [50] used a series of three columns varying in pore size to separate restriction enzyme fragments of endonuclease HAE III as shown in Fig. 8. They also separated polymers of base-hydrolyzed poly A on an I-60 column using similar mobile phase conditions (Fig. 9.)

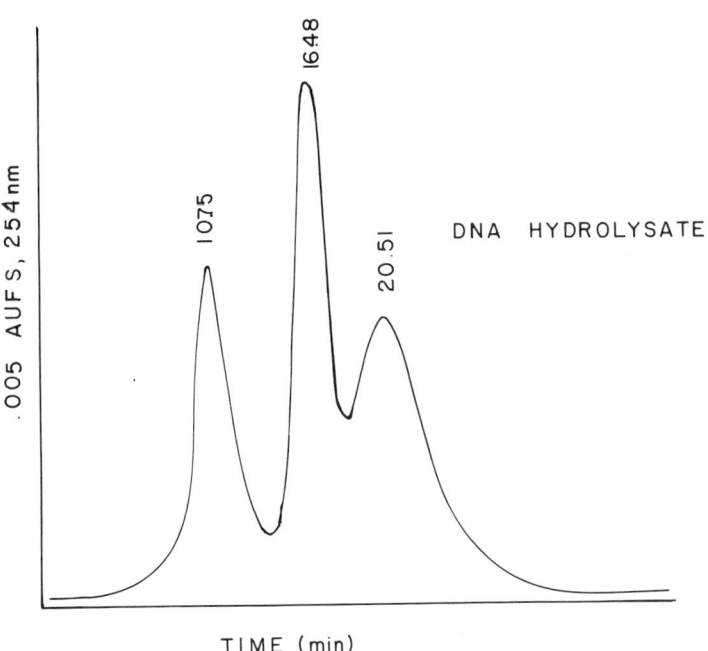

Figure 8. Size exclusion analysis of restriction fragments of an Endonuclease HAE III digestion of DNA using three Protein Analysis Columns in series (I-60, I-125, I-250) and a mobile phase as in Fig. 7. Flow rate = 1.5 ml/min.

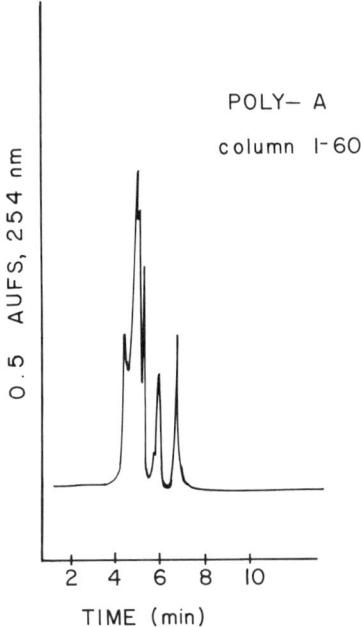

Figure 9. Size exclusion analysis of the base hydrolysis products of Poly-A. Mobile phase condition as in Fig. 8. Column: Protein Analysis I-60.

III. IDENTIFICATION AND DETECTION OF OLIGONUCLEOTIDES

A. Optical and radioactive detection

Due to the extremely high extinction coefficients at 254 nm (Hg I line) [2,51,52], oligonucleotides are usually monitored using a fixed-wavelength detector. By incorporating an appropriate phosphore, the oligomers can be monitored at 280 nm as well. Multiwavelength detectors, although more expensive, allow the monitoring at various wavelengths as well as spectral scans of eluting peaks. It was reported, however, that due to the complexity of these molecules, limited information could be expected to be derived from the ultraviolet (UV) scans [11]. Webster and Whaun [53] used both a UV detector and a flow-through radioactive detector to identify and quantify radiolabeled oligonucleotides.

Several off-line methods have been found more suitable for oligomer characterization. Tyson and Wickstrom [54] used optical detection followed by fraction collection to determine the radioactivity of the radiolabeled trinucleoside diphosphates.

B. Absolute methods for determining oligomer sequence

Exact identification of a collected oligomer can be accomplished using the popular method developed by Maxam and Gilbert [7] and the method proposed by Sanger and Coulson [55]. Both methods are based on electrophoretic techniques requiring labeling with ^{32}P. Additional equipment as well as radiolabeled reagents are necessary for the analysis. Except for analysis of the final product, the use of time-consuming "exact" sequence methods is unnecessary, since knowledge of chain length and/or base composition is adequate to describe the sequence when used in conjunction with prior knowledge of the anticipated oligomer sequence.

A new instrumental technique of oligomer sequence analysis is the use of mass spectrometry. Budzikiewicz [63] used several different ionization methods, and field desorption is his method of choice. Although the method is limited to oligomers of chain lengths of about 10, the method may be of value to analyze oligomers with modified or rare bases. Grotjahn et al. [64] used fast atom bombardment (FAB) mass spectrometry to reveal the sequence of a 10-mer. The method is fast (about 1 hr) and required only about 1 O.D. (ca. 10 nmol.) per analysis.

C. Methods for determining relative base composition

Several methods have been devised to enzymatically or chemically digest the oligomer into its respective nucleoside/nucleotide constituents. Thus, in knowing the "response factor" for each component, a base ratio can be defined. Clark and Trebilcock-Guzman [56] describe an improved method for determining base composition of tRNA. Mischke and Wickstrom [57], meanwhile, have reported the nanogram sensitivity of both DNA and tRNA digestion products using reversed-phase columns, while Kuo et al. [58] quantitatively separated major and modified bases of DNA.

Pon and Ogilvie [51] separated some ribonucleoside and nucleotides on a single weak anion-exchange column. This separation enabled them to determine composition of oligodimers. Dizdaroglu et al. [59] were able to separate and sequence the partial enzymatic digestion fragments of pyrimidine deoxynucleotides on a MicroPak AX-10 weak anion exchanger (Varian Assoc.). Reliable sequencing of pentamers of dC and dT were recorded; however, the partial digestion using phosphodiesterase I (PDE I) and phosphodiesterase II (PDE II) were both time-consuming and required several chromatographic injections for complete analysis.

Crowther et al. [60] developed a simplified method to analyze enzymatic digest (PDE I) of both deoxyribooligomers and ribooligomers in order to determine both chain length and base composition. By developing an ion-pair separation of nucleoside and nucleotide components of the oligomer in a single chromatographic run, the chain length and base composition can be determined according to Eq. (1) if the leading oligomer is dephosphorolated [61]:

$$\ell = \sum_{i=1}^{n} \frac{\text{moles}_{NTDi}}{\text{moles}_{NSD}} + \frac{\text{moles}_{NSD}}{\text{moles}_{NSD}} \tag{1}$$

where

n = number of base types in oligomer
ℓ = calculated chain length

A chain-length determination of up to a 10-mer was reported. Through statistical analysis it was determined that at 1% precision, a chain length of 30 could be determined. The following expression was derived to express the reliability of the method:

$$Z\left[4RSD^2\left(1 + \frac{4(\ell-1)}{n}\right)^2\right]^{1/2} \leq \pm \Delta \ell \tag{2}$$

where

ℓ = chain length
RSD = precision
n = number of base types
Z = chosen confidence level

Finally, Ho and Gilham [62] also performed a simultaneous separation of nucleosides and nucleotides on polystryrene; however, elution times were about 8 hr per separation.

IV. FUTURE DEVELOPMENTS IN HPLC OF OLIGONUCLEOTIDES

The current trends in separations of oligonucleotides appear to be focused on the development of support materials specifically to resolve these oligomers. Some of the most successful separations of higher-molecular-weight chains have been performed on stationary phases possessing both ionic and hydrophobic qualities. Jost et al.

[65] reported the use of ion-pair reagents to modify dynamically the reversed phase to simulate these hydrophobic/ionic interactions. Unfortunately, in preparative separations, ion-pair reagents contaminate the oligomer fraction, necessitating an additional purification step. Thus it would appear that the design of a chemically bonded stationary phase which would allow selective interactions with oligomers without the presence of high salt concentrations or ion-pair reagents would be beneficial to both analytical and preparative separations of oligomers. This principle has been shown to be applicable to ion-pair separations of nucleoside/nucleotides, and it provides a degree of flexibility not offered by dynamic surface modification methods [66]. Thus, by tailoring the polarity of the silica surface by varying the reverse-phase and ion-exchange ligand densities, separations according to chain length and base composition can be designed. Finally, the pore structure and surface area of the silica must be matched to give an optimum phase ratio of the stationary phase with suitably large pores to facilitate efficient mass transfer.

In conclusion, the next few years will be active in the development of efficient separations of the oligomers. With the development of new solid-phase reaction systems, methods of product purification must keep abreast. Furthermore, methods must be improved to reduce the time required to synthesize these genetic fragments and increase the reaction yields. The development of novel stationary phases with unique selectivities will undoubtedly continue to be an area of major effort. As the entire field of recombinant DNA and nuclcic acid research continues to advance, increasing reliance will be put upon the development of more selective and efficient separation methods for the oligonucleotides.

REFERENCES

1. M. J. Gait and R. C. Sheppard, Nucleic Acids Res., 4, 1135 (1977).
2. C. F. Crampton, F. R. Rankel, A. M. Benson, and A. Wade, Anal. Biochem., 1, 249 (1960).
3. R. V. Tomlinson and G. M. Tener, Biochemistry, 2, 697 (1963).
4. H. Domdey and H. J. Gross, Anal. Biochem., 98, 346 (1979).
5. E. Y. Chen and B. A. Roe, Anal. Biochem., 89, 45 (1978).
6. I. Halasz and I. Sebastion, Angew. Chem. Int. Ed., 8, 453 (1969).
7. A. M. Maxam and W. Gilbert, Proc. Nat. Acad. Sci. U. S., 74, 560 (1977).
8. R. A. Hartwick and P. R. Brown, J. Chromatogr., 126, 679 (1976).
9. R. A. Hartwick and P. R. Brown, J. Chromatogr., 112, 651 (1975).

10. J. C. Kraak, C. X. Ahn, and J. Fraanje, J. Chromatogr., 209, 369 (1981).
11. J. B. Crowther, R. Jones, and R. A. Hartwick, J. Chromatogr., 217, 479 (1981).
12. M. Zakaria and P. R. Brown, J. Chromatogr., 226, 267 (1981).
13. H. G. Khorana, Pure Appl. Chem., 17, 349 (1968).
14. C. B. Reese, Tetrahedron, 34, 3143 (1978).
15. R. Wu, C. P. Bahl, and S. A. Narang, Prog. in Nucleic Acid Res. Mol. Biol., 21, 101 (1978).
16. C. B. Reese, R. C. Titmas, and L. Yau, Tetrahedron Lett., 30, 2727 (1978).
17. B. Pace and N. R. Pace, Anal. Biochem., 107, 128 (1980).
18. J. Satava, O. Mikes, and P. Strop, J. Chromatogr., 180, 31 (1979).
19. J. F. M. De Rooij, R. Arentzen, J. A. J. Den Hartog, G. Van Der Marel, and J. H. Van Boom, J. Chromatogr., 171, 453 (1979).
20. K. Grzeskowiak, R. W. Adamiak, and M. Wiewiorowski, Nucleic Acids Res., 8, 1097 (1980).
21. R. L. Pearson, J. F. Weiss, and A. D. Kelmers, Biochim. Biophys. Acta, 228, 770 (1971).
22. S. C. Hardies and R. D. Wells, Proc. Nat. Acad. Sci. U. S., 73, 3117 (1976).
23. B. W. Shum and D. M. Crothers, Nucleic Acids Res., 5, 2297 (1978).
24. E. Selsing, J. E. Larson, and R. D. Wells, Anal. Biochem., 99, 213 (1979).
25. D. A. Usher and J. A. Rosen, Anal. Biochem., 92, 276 (1979).
26. D. A. Usher, Nucl. Acids Res., 6, 2289 (1979).
27. R. E. Majors and M. J. Hopper, J. Chromatogr. Sci., 12, 767 (1974).
28. K. L. Agarwal and F. Riftina, Nucleic Acids Res., 5, 2809 (1978).
29. P. S. Miller, D. M. Cheng, N. Dreon, K. Jayaraman, L. Kan, E. E. Leutzinger, S. M. Pulford, and P. O. P. Ts'O, Biochemistry, 19, 4688 (1980).
30. H. Komiya, K. Nishikawa, K. Ogawa, and S. Takemura, J. Biochem., 86, 1081 (1979).
31. A. Vandenberghe, C. Van Broeckhoven, and R. DeWachter, Arch. Int. Physiol. Biochem., 87, 848 (1979).
32. A. Vandenberghe, L. Nelles, and R. De Wachter, Anal. Biochem., 107, 369 (1980).
33. R. E. Brown, P. J. Cayley, and I. M. Kerr, Meth. Enzymol., 79, 208 (1981).
34. A. Wakizaka, K. Kurosaka, and E. Okuhara, IRCS Med. Sci. Libr. Compend., 6, 48 (1978).

35. L. W. McLaughlin, F. Cramer, and M. Sprinzl, Anal. Biochem., *112*, 60 (1981).
36. M. Dizdaroglu, M. G. Simic, and H. Schott, J. Chromatogr., *185*, 403 (1979).
37. W. Jost, K. K. Unger, R. Lipecky, and H. G. Gassen, J. Chromatogr., *185*, 403 (1979).
38. M. J. Gait, M. Singh, R. C. Sheppard, M. D. Edge, A. R. Greene, G. R. Heathcliffe, T. C. Atkinson, C. R. Newton, and A. F. Markham, Nucleic Acids Res., *8*, 1081 (1980).
39. M. J. Gait and R. C. Sheppard, Nucleic Acids Res., *4*, 4391 (1977).
40. M. J. Gait, H. W. D. Matthes, M. Singh, B. S. Sproat, and R. C. Titmas, Nucleic Acids Res., *10*, 6243 (1982).
41. M. Dizdaroglu and W. Hermes, J. Chromatogr., *171*, 321 (1979).
42. A. J. Alpert and F. E. Regnier, J. Chromatogr., *185*, 375 (1979).
43. R. Crea, A. Kraszewski, T. Hirose, and K. Itakura, Proc. Natl. Acad. Sci. U. S., *75*, 5765 (1978).
44. G. D. McFarland and P. N. Borer, Nucleic Acids Res., *7*, 1067 (1979).
45. H.-J. Fritz, R. Belagaje, E. L. Brown, R. H. Fritz, R. A. Jones, R. G. Lees, and H. G. Khorana, Biochemistry, *17*, 1257 (1978).
46. R. A. Jones, H.-J. Fritz, and H. G. Khorana, Biochemistry, *17*, 1268 (1978).
47. F. Chow, T. Kempe, and G. Palm, Nucleic Acids Res., *9*, 2807 (1981).
48. E. Hagemeier, S. Bornemann, K.-S. Boos, and E. Schlimme, J. Chromatogr., *237*, 174 (1982).
49. D. Molko, R. Derbyshire, A. Guy, A. Roget, R. Teole, and A. Boucherle, J. Chromatogr., *206*, 493 (1981).
50. K. J. Robison, J. Magram, and R. A. Hartwick, Paper 189, Pittsburgh Conference on Analytical Chemistry and Applied Spectroscopy, March 1982.
51. R. T. Pon and K. K. Ogilvie, J. Chromatogr., *205*, 202 (1981).
52. J. H. Van Boom and J. F. M. De Rooy, J. Chromatogr., *131*, 169 (1977).
53. H. K. Webster and J. M. Whaun, J. Chromatogr., *209*, 283 (1981).
54. R. W. Tyson and E. Wickstrom, J. Chromatogr., *192*, 485 (1980).
55. F. Sanger and A. R. Coulson, J. Mol. Biol., *94*, 441 (1975).
56. I. Clark and M. A. Trebilcock-Guzman, J. Biochem. Biophys. Meth., *1*, 287 (1979).
57. C. F. Mischke and E. Wickstrom, Anal. Biochem., *105*, 181 (1980).

58. K. C. Kuo, R. A. McCune, C. W. Gerhke, R. Midgett, and M. Ehrlich, Nucleic Acids Res. *8*, 4763 (1980).
59. M. Dizdaroglu, M. G. Simic, and H. Schott, J. Chromatogr., *188*, 273 (1980).
60. J. B. Crowther, J. P. Caronia, and R. A. Hartwick, Anal. Biochem., *124*, 65 (1982).
61. J. P. Caronia, J. B. Crowther, and R. A. Hartwick, J. Liq. Chrom. *6*, 1673 (1983).
62. N. W. Y. Ho and P. T. Gilham, Biochim. Biophys. Acta, *308*, 53 (1978).
63. H. Budzikiewicz, Pure Appl. Chem., *50*, 159 (1978).
64. L. Grotjahn, R. Frank, and H. Blocker, Nucleic Acids Res., *10*, 4671 (1982).
65. W. Jost, K. Unger, and G. Schill, Anal. Biochem., *119*, 214 (1982).
66. J. B. Crowther and R. A. Hartwick, Chromatographia, *16*, 349 (1982).

12
Free Nucleotides in Biological Samples

MALCOLM McKEAG University of Rhode Island,
Kingston, Rhode Island

I. Introduction 215
II. Biological Matrices 236
III. Sample Preparation 238
IV. Chromatography 239
   A. Instrumentation 239
   B. Ion exchange 240
   C. Reversed phase and ion pairing 240
V. Abbreviations 242
   References 243

I. INTRODUCTION

Interest in separating and quantifying hydrolysates of nucleic acids and free nucleotides in biological samples has contributed significantly to the development of high performance liquid chromatography (HPLC). In this chapter applications of HPLC to research on free nucleotides will be described. The data necessary to reproduce the ion-exchange (IE) chromatographic separations are summarized in Table 1, and reversed-phase (RPLC) separations in Table 2. The important aspects of these data are discussed in the text.

Table 1 Ion Exchange

Compounds of interest	Sample preparation	Column	Anal. time mode[b] (slope)	Chromatographic conditions	Detector[a]	Refs.
1. Cultures						
a. Bacteria						
N-tide[c] pools	0.3 M CCl$_3$COOH or HClO$_4$/Alamine 336-Freon TF soln.	Aminex A-27 or A-28	75; G (concave)	(A) 0.025 M C$_6$H$_5$N$_3$O$_7$ pH 8.2 (B) 0.5 M C$_6$H$_5$N$_3$O$_7$ pH 8.2 0–100% (B) 0.6 ml/min	254 nm	45
N-tide pools	1.0 M CH$_2$O$_2$/ lyophilization	Partisil PXS 10/25 SAX	48; G (slight concave)	(A) 0.040 M CH$_3$COONH$_4$ pH 6.0 (B) 3.0 M CH$_3$COONH$_4$ pH 6.0 45-min gradient 1.0 ml/min	254 nm	70
b. Yeast						
N-base, N-side, N-tide pools	5% HClO$_4$(c)/ 10 M KOH	Aminex A-14	225; G (linear)	(A) 0.1 M CH$_3$(NH$_2$)C(CH$_3$)CH$_2$OH 0.1 M NaCl pH 9.90 (B) 0.1 M CH$_3$(NH$_2$)C(CH$_3$)CH$_2$OH	254 nm	29

c. Fungal

N-tide pools	10% CCl$_3$COOH/ C$_2$H$_5$OC$_2$H$_5$ (H$_2$O sat.) pH adj. to 6.8 with KOH	AS-Pellionex-SAX	45; G (linear)	0.4 M NaCl pH 10.00 0-100% (B) 1.67 ml/min	254 nm 280 nm	51

d. Mammalian cell cultures

N-tide pools	HClO$_4$ or CCl$_3$COOH	Pellicular anion-exchange resin	65; G (linear)	Prepurif.: Bio-gel P-2, H$_2$O (A) 2.5 M KH$_2$PO$_4$ pH 3.6 (B) 1.0 M KH$_2$PO$_4$ 0.25 M KCl 0.4 ml/min	254 nm	11
				(A) 0.015 M KH$_2$PO$_4$ (B) 0.25 M KH$_2$PO$_4$ 2.2 M KCl 0-100% (B) 0.2 ml/min		
Ribo N-tide pools	12% CCl$_3$COOH (c)/ C$_2$H$_5$OC$_2$H$_5$ (H$_2$O sat.) (Ref. 81)	Partisil PXS 10/25 SAX	35; G (linear)	(A) 0.005 M NH$_4$H$_2$PO$_4$ pH 2.8 (B) 0.75 M NH$_4$H$_2$PO$_4$ pH 3.7 0-100% (B); 40 min 2.0 ml/min	254 nm 280 nm	67
Purine bases, N-sides, N-tides	0.4 M HClO$_4$ (c)/ unspecified neutralization	Aminex A-25	120; G (linear)	(A) 0.08 M Na$_2$B$_4$O$_7$ 0.05 M NH$_4$Cl pH 9.10	254 nm ^{14}C, ^{3}H (on line)	6

Table 1 (Continued)

Compounds of interest	Sample preparation	Column	Anal. time mode[b] (slope)	Chromatographic conditions	Detector[a]	Refs.
Deoxy N-tide pools	0.5 M HClO$_4$ (c)/ 4 M KOH, 0.4 M KH$_2$PO$_4$	Partisil PXS 10.25 SAX	30; I	(B) 0.01 M Na$_2$B$_4$O$_7$ 0.5 M NH$_4$Cl pH 9.00 0-100% (B) 0.5 ml/min	254 nm	31
N-bases, N-sides, N-tides	0.8 M HClO$_4$ (c)/ KOH, pH adj. 6-8	Aminex A-25	160; G (linear)	0.4 M (NH$_4$)$_2$HPO$_4$ pH 3.25 10% CH$_3$CN 2.0 ml/min (A) 0.07 M Na$_2$B$_4$O$_7$ 2.5% CH$_3$CH$_2$OH 0.045 M NH$_4$Cl pH 9.15 (B) 0.01 M Na$_2$B$_4$O$_7$ 0.5 M NH$_4$Cl pH 8.80 15 min (A); 0-100% (B) 0.5 ml/min 65°C	254 nm ^{14}C (on line)	68
Uridine deoxy N-tides FdUMP	0.5 M HClO$_4$ (c)/ KOH periodate/ methylamine elimination of ribo N-tides	Partisil PXS 10/25 SAX	33; G (linear)	(A) 0.02 M Na$_3$PO$_4$ pH 3.3 (B) 0.75 M Na$_3$PO$_4$ pH 3.3	254 nm 280 nm ^{14}C, ^{3}H (off line)	25

Sample	Extraction	Column	Length (cm); G/I (gradient/isocratic)	Mobile phase	Detection	Ref.
N-tide pools	$HClO_4$/unspecified neutralization	Partisil PXS 10/25 SAX	50; G (linear)	(A) 0.01 M KH_2PO_4 pH 3.4 (B) 0.8 M KH_2PO_4 pH 4.3 Isocratic (A) 0–80% (B) 1.6 ml/min 0–100% (B) 1.8 ml/min	254 nm ^{14}C, ^3H (on line)	100
Ribo N-tides: Diphosphates	3% $HClO_4$/0.72 M KOH, 0.60 M $KHCO_3$ 100 µg/ml EDTA	µBondapak NH_2	10; I	0.075 M $NH_4H_2PO_4$ pH 3.15 4.0 ml/min	254 nm	76
Triphosphates	(Ref. 45 as mod. by Ref. 14) DEPC added (see text)		30; I	0.125 M $NH_4H_2PO_4$ pH 3.15 4.0 ml/min		
N-tide pools	5% $HClO_4$ (Ref. 45)	Partisil PXS 10/25 SAX	75; G (linear)	(A) 0.010 M KH_2PO_4 pH 4.0 (B) 0.5 M KH_2PO_4 pH 4.0 12 min (A); 0–100% (B) 50 min 2.0 ml/min	254 nm ^{32}P (off line)	86
2. Tissues						
N-tide pools	$HClO_4$/lyophilization	Pellicular anion-exchange resin	75; G (linear)	(A) 0.04 M $HCOONH_4$ pH 4.35 (B) 1.5 M $HCOONH_4$ pH 4.35	260 nm	40

Table 1 (Continued)

Compounds of interest	Sample preparation	Column	Anal. time mode[b] (slope)	Chromatographic conditions	Detector[a]	Refs.
N-tide pools	Frozen tissue pulverization, titrated with 50% CH_3CH_2OH (-20°C), 0.6 N $HClO_4$/KOH to pH 5.0-7.0, lyophilization	Pellicular anion-exchange type LFS	45; G (linear)	0.2 ml/min 71°C (A) 0.01 M KH_2PO_4 0.001 M H_3PO_4 (B) 0.05 M KH_2PO_4 0.44 M KCl 10 min (A) 0-100% (B) 0.4 ml/min	254 nm	82
N-tide pools	(Ref. 11)	Permaphase AAX	30; G (linear)	(A) 0.005 M KH_2PO_4 pH 3.0 (B) 0.5 M KH_2PO_4 pH 4.3 2 min (A) 0-100% (B) 1.0 ml/min	254 nm	38
N-tides: Monophosphates	(Ref. 45)	MicroPak AX-10	10; I	0.01 M KH_2PO_4 pH 3.3 2.0 ml/min	254 nm 259 nm 260 nm	28
Adenine N-tides			25; I	0.01 M KH_2PO_4 pH 2.85 0.75 ml/min		

Sample	Extraction	Column	Length; gradient	Elution	Detection	Ref.
N-tide pools			60; G (linear)	(A) 0.01 M KH_2PO_4 pH 2.85 (B) 0.75 M KH_2PO_4 pH 4.40 5–100% (B) 0.5 ml/min		
N-tide pools	10% CCl_3COOH, 20% CH_2OH (c) / $C_2H_5OC_2H_5$ (H_2O sat.), pH adj. with Tris-Cl, Microtube analysis.	AS-Pellionex-SAX	55; G (linear)	(A) 0.004 M KH_2PO_4 pH 6.5 (B) 0.18 M KH_2PO_4 0.13 M KCl pH 4.2 0–100% (B) 0.36 ml/min	254 nm ^{14}C (off line)	56
Pyrimidine N-tides Monophosphates	1.0 M $HClO_4$ (heat) pH adj. 3–4, KOH	Partisil PXS 10/25 SAX	20; I	0.05 M HCOOH 0.001 M Na_3PO_4 pH 4.1 1.0 ml/min	254 nm 280 nm	63
Triphosphates	0.5 N $HClO_4$ / 10 N KOH		40; I	0.4 M Na_3PO_4 pH 3.3 1.5 ml/min		
N-tide pools	0.6 M $HClO_4$ (Ref. 45 as mod. by Ref. 14)	μBondapak NH_2	54; G (linear)	Prepurifi.: Chelex 100 (Cu^{2+}-loaded) column (A) 0.005 M KH_2PO_4 pH 3.0 (B) 0.5 M KH_2PO_4 pH 4.0 1.3 ml/min		87

Table 1 (Continued)

Compounds of interest	Sample preparation	Column	Anal. time mode[b] (slope)	Chromatographic conditions	Detector[a]	Refs.
N-tide pools	$HClO_4/KOH$; charcoal adsorption; CCl_3COOH/amine-Freon	Partisil PXS 10/25 SAX	50; G (slight concave)	(A) 0.007 M KH_2PO_4 pH 4.2 (B) 0.007 M KH_2PO_4 pH 4.5 2.2 M KCL 10 min. (A) 1.6 ml/min, 0-100% B, 55 min, 3.0 ml/min	260 nm	77
Cyclic AMP derivatives + degradation products	(Ref. 95 – ion exchange)	Partisil PXS 10/25 SAX	10; I	0.005 M $(C_2H_5)_3NHOOCH$ pH 3.0 2.0 ml/min	254 nm 3H (off line)	96
		Partisil PXS 10/25 SCX	10; I	0.003 M $(C_2H_5)_3NHCOOH$ 65% CH_3CN 2.0 ml/min		
N-tide pools	Frozen tissue, 0.1 M $HClO_4$ (c)/ 3 N KOH	Partisil PXS 10/25 SAX	80; G (linear)	(A) 0.007 M KH_2PO_4 pH 4.0 (B) 0.25 M KH_2PO_4 0.50 M KCl pH 4.5 0-100% (B) 1.0 ml/min	254 nm	74

3. Blood					
Adenine: N-tides	1.0 N HClO₄ (c) plus sonication/ 1.0 N KOH	AS-Pellicular-SAX	14; I	0.15 M KH₂PO₄ pH 3.3 1.0 ml/min	254 nm	75
N-tide pools			55; G (linear)	(A) 0.005 M KH₂PO₄ (B) 1.0 M KH₂PO₄ 0–100% (B) 1.0 ml/min		
N-tide pools	CCl₃COOH/ C₂H₅OC₂H₅	Partisil PXS 10/25 SAX	85; G (linear)	(A) 0.007 M KH₂PO₄ pH 4.0 (B) 0.25 M KH₂PO₄ 0.25 M KCl 15 min (A) 0–100% (B) 1.5 ml/min	254 nm	36
N-side and N-tide pools	1.0 N HClO₄ (c)/ 1.0 N KOH	Partisil PXS 10/25 SAX	75; G (linear)	(A) 0.0015 M H₃PO₄ (B) 1.0 M NH₄H₂PO₄ pH 4.8 iso (A) 0–100% (B) 1.0 ml/min	254 nm	26
N-tide pools	12% CCl₃COOH (Ref. 81)	Partisil PXS 10/25 SAX	30; G (linear)	(A) 0.01 M NH₄H₂PO₄ pH 2.8 (B) 0.5 M NH₄H₂PO₄ pH 4.8 0–100% (B) 2.0 ml/min	260 nm	73

Table 1 (Continued)

Compounds of interest	Sample preparation	Column	Anal. time mode[b] (slope)	Chromatographic conditions	Detector[a]	Refs.
Adenine N-tides	0.4 M HClO$_4$ (c)/ 1.0 N KOH	Micropack NH$_2$	50; G (linear)	(A) 0.005 M (NH$_4$)H$_2$PO$_4$ pH 3.2 (B) 0.5 M (NH$_4$)H$_2$PO$_4$ pH 4.15 0-50% (B) 0.42 ml/min	254 nm	20
N-tide pools	(Ref. 45)	Partisil PXS 10/25	55; G (linear)	(A) 0.03 M Na$_3$PO$_4$ pH 4.0 (B) 0.75 M Na$_3$PO$_4$ pH 4.0 0-100% (B) 1.0 ml/min	260 nm 280 nm	57
N-tide pools	CCl$_3$COOH (Ref. 45 as mod. by Ref. 14)	Partisil PXS 10/25 SAX	51; G (linear)	(A) 0.01 M KH$_2$PO$_4$ pH 3.90 (B) 0.25 M KH$_2$PO$_4$ 0.5 M KCl pH 4.50 iso. (A) 2-100% (B) 1.5 ml/min	280 nm	90
N-tide pools	12% CCl$_3$COOH/ amine-Freon (Ref. 45 as mod. by Ref. 97)	Partisil PXS 10/25 SAX	55; G (linear)	(A) 0.007 M KH$_2$PO$_4$ 0.007 M KCl pH 4.0	254 nm	60

Sample	Preparation	Column	% Recovery; Gradient	Mobile Phase	Detection	Ref.
Adenine N-tides	12% CCl$_3$COOH/no neutralization (Ref. 61)	Dianion CDR-10	70; G (linear)	(B) 0.25 M KH$_2$PO$_4$ 0.50 M KCl pH 5.0 5 min (A) 0-100% (B) 2.0 ml/min	254 nm	85
Adenine N-tides	6% CCl$_3$COOH (c) / no neutralization	Partisil PXS 10/25 SAX	50; G (slight concave)	(A) H$_2$O (B) 6.0 M CH$_3$COONH$_4$ 0-100% (B) 0.7 ml/min 22-70°C: 0-30 min then maintain	254 nm 280 nm	88
N-tide pools	0.4 M HClO$_4$/4.0 N KOH, 1.0 M KH$_2$PO$_4$	Partisil PXS 10/25 SAX	96; G (linear, multistep)	(A) 0.005 M KH$_2$PO$_4$ pH 3.2 (B) 0.5 M KH$_2$PO$_4$ pH 3.9 0-100% (B) 2.0 ml/min	254 nm 280 nm	1
				(A) 2% CH$_3$CN/98% H$_2$O (v/v) (B) 0.05 M KH$_2$PO$_4$ 2% CH$_3$CN pH 3.35 (C) 0.25 M KH$_2$PO$_4$ 0.50 M KCl 2% CH$_3$CN pH 5.25		

Table 1 (Continued)

Compounds of interest	Sample preparation	Column	Anal. time mode[b] (slope)	Chromatographic conditions	Detector[a]	Refs.
				80-7.5% (A) 0-90 min 20-7.5% (B) 0-85% (C)		
4. Urine						
N-tide pools	Urine applied directly	Aminex BRX 150; G (linear)		(A) 0.01 M KH_2PO_4 pH 3.25 (B) 1.0 M KH_2PO_4 pH 4.2 0-100% (B) 0.2 ml/min 70°C	254 nm	13

[a] UV-visible detector unless otherwise noted. Wavelength of UV detector given.
[b] Eluent Mode
   I = isocratic
   G = gradient
[c] Abbreviations
   N-tides = nucleotides
   N-sides = nucleosides

Table 2 Reversed Phase

Compounds of interest	Sample preparation	Column	Anal. time mode[b] (slope)	Chromatographic conditions	Detector	Refs.
1. Cultures						
a. Mammalian Cell Cultures						
Uridine N-tides, FdUMP	(see Ref. 25–ion exchange)	Partisil 10-ODS-2	38; I	0.15 M CH_3COONa pH 5.5 0.7 ml/min	254 nm 280 nm ^{14}C, ^3H (off line)	25
N-sides, N-tides	0.33 M CH_2O_2 (c)/ no neutralization	μBondapak C_{18}	25; I	0.2 M $(NH_4)H_2PO_4$ pH 5.1 1.0 ml/min	254 nm	92
2. Tissues						
Adenine N-tides	Perfusate: cells ppt. by centrifugation, conc. by evaporation	μBondapak C_{18}	25; I (2-step)	(A) 0.05 M $(NH_4)H_2PO_4$ pH 6.0 (B) 0.05 M $(NH_4)H_2PO_4$ 5% CH_3CN pH 6.0 6 min (A) 18 min (B) 2.0 ml/min	254 nm ^{14}C (off line)	3
Cyclic N-tides	0.4 M $HClO_4$/no neutralization	μBondapak C_{18}	25; G (linear)	(A) 0.02 M KH_2PO_4 pH 3.7	254 nm	54

Table 2 (Continued)

Compounds of interest	Sample preparation	Column	Anal. time mode[b] (slope)	Chromatographic conditions	Detector	Refs.
N-tide pools	(see Ref. 87–ion exchange)	μBondapak C_{18}	30; G (linear)	(B) 60% CH_3OH/40% H_2O 1.5 ml/min (A) 0.2 M KH_2PO_4 pH 4.0 (B) 0.2 M KH_2PO_4 20% CH_3OH pH 4.0 2.0 ml/min	254 nm	87
CoA precursors	Protein ppt. with CH_3OH and centrifugation	μBondapak C_{18}	50; G (linear)	(A) 0.02 M KH_2PO_4 pH 4.0 (B) 0.02 M KH_2PO_4 45% CH_3OH pH 8.25 15 min (A) 0–20% (B) : 35 min 1.4 ml/min	254 nm ^{14}C	35
Adenylate pyridine di N-tides	3.0 M $HClO_4$ (c)/no neutralization	μBondpak C_{18}	35; G (multistep)	(A) 0.10 M KH_2PO_4 pH 6.0 (B) 0.10 M KH_2PO_4 5% CH_3OH/H_2O (v/v) pH 6.0		42

Compound	Sample preparation	Column	Gradient/Eluent	Detection	Ref.
Reduced pyridine di-n-tides	0.5 M KOH in 50% (v/v) CH CH OH plus 35% (w/v) ClCs (c) (Ref. 94)		8.5 min 100% (A), 100-80% (A): 0.5 min., hold 6 min 80-0% (A): 5 min, hold 20 min 1.0 ml/min 50-75% (B) 1.0 ml/min	260 nm 340 nm	
3. Blood					
Purine N-tides	(Ref. 45 as mod. by Ref. 14)	C_{18}RP, 5 µm 12; I	0.35 M Na_3PO_4 pH 5.5 2.0 ml/min	254 nm	55
Purine N-sides and N-tides	0.56 N $HClO_4$/ 10 N KOH	µBondapak C_{18} 35; G (concave)	(A) 0.06 M K_2HPO_4 0.04 M KH_2PO_4 pH 6.0 (B) 0.08 M K_2HPO_4 0.053 M KH_2PO_4 25% CH_3OH pH 6.0 0-100% (B) 1.0 ml/min	254 nm	83
Adenine N-tides		20; I (2-step)	10 min (A); 10 min (B) 1.0 ml/min		

Table 2 (Continued)

Compounds of interest	Sample preparation	Column	Anal. time mode[b] (slope)	Chromatographic conditions	Detector[a]	Refs.
4. Standards						
Pyridine di-N-tide stds		LiChrosorb RP-18 (5 μm)	80; I	0.1 M $CaH_4O_8P_2$ 0.01% NaN_3 pH 7.0 1.0 ml/min	254 nm	34
Reversed-phase ion pairing						
1. Cultures						
a. Bacteria						
N-tide pools	(see Ref. 70–ion exchange)	Ultrasphere ODS	36; G (slight convex)	(A) 0.005 M TBAP 4% CH_4CN 0.03 M KH_2PO_4 pH 6.0 (B) 100% CH_3CN 0-40% (B): 60 min 1.5 ml/min	254 nm Fluoro.	70
b. Mammalian Cell Cultures						
Uridine N-tides, FdUMP	(See Ref 25–ion exchange)	LiChrosorb RP-18	40; I	0.005 M Na_3PO_4 0.005 M TBAB 10% CH_3OH 1.0 ml/min	254 nm 280 nm ^3H (off line)	25

2. Tissues

Adenine N-tides	Freeze-stop tissue, ground (c) 0.3 N HClO$_4$/KOH	LiChrosorb RP-8	10; I	22% (v/v) CH$_3$CN/H$_2$O 0.3% TABP 0.65% KH$_2$PO$_4$ pH 5.8 2.0 ml/min	254 nm	43
Sulfated adenine N-tides	(see Ref. 45–ion exchange)	LiChrosorb RP-18	14; I	9.4% CH$_3$CH$_2$CH$_2$OH 0.003 M TBAH pH 9.4 (1 N H$_3$PO$_4$) 1.5 ml/min	254 nm 280 nm ^{35}S (off line)	84
N-tide pools	(Ref. 45 as mod. by Ref. 97)	µBondapak	35; I	0.065 M KH$_2$PO$_4$ 0.009 M TBAP 3.3% CH$_3$CN pH 3.2 2.8 ml/min	254 nm	22
		RCM-C$_{18}$		(A) 0.065 M KH$_2$PO$_4$ 0.009 M TBAP pH 3.2 (B) (A) plus 6% CH$_3$CN 2.0 ml/min		
Cyclic AMP derivatives and degradation products	(see Ref. 95–ion exchange)	Machery Nagal RP-18	5; I	0.008 M Na$_2$HPO$_4$ 0.01 M TBAF pH 6.6 48% CH$_3$OH 1.5 ml/min	254 nm ^{3}H (off line)	95

Table 2 (Continued)

Compounds of interest	Sample preparation	Column	Anal. time mode[b] (slope)	Chromatographic conditions	Detector[a]	Refs.
3. Blood						
5'-Ribo N-side monophos.s	10% HClO$_4$/C$_2$H$_5$OC$_2$H$_5$ (H$_2$O sat.)	Spherisorb ODS, Ultrasphere ODS, μBondapak C$_{18}$	28; I	Prior sep. of n-tide classes on DEAE-cellulose 2.5 mM TBAH pH 2.5 (CH$_2$O$_2$) 12% CH$_2$OH 1.0 ml/min	254 nm	98
4. Standards						
Ribo N-tides, C-AMP		Spherisorb ODS	40; G (concave)	(A) 0.025 M TBHS 0.05 M KH$_2$PO$_4$ 0.07 M NH$_4$Cl pH 3.90 (B) 0.025 M TBHS 0.10 M KH$_2$PO$_4$ 0.20 M NH$_4$Cl pH 3.40 30% CH$_3$OH 0-100% (B) 1.0 ml/min	254 nm	39
Mono N-tides		Lichrosorb RP-18	30; I	0.1 M KH$_2$PO$_4$ 0.012 M TMAH pH 3.9 0.8 ml/min	260 nm	2

Analyte	Column	Mobile phase	Detection	Ref.
N-tides	ODS-Hypersil 35; I	0.075 M KH_2PO_4 1.25 mM C11AA pH 5.35 12% CH_3OH (v/v) (cond.s not given) same as above except pH 5.65	UV	50
N-tides	µBondapak C_{18}, Ultrasphere-Ion Pair C-18-1P 16; I 22; G (linear)	(A) 0.007 M KH_2PO_4 0.005 M TBAP pH 7.0 5% CH_3OH (v/v) (B) 0.007 M KH_2PO_4 0.50 M KCl pH 7.0 0.005 M TBAP 5% CH_3OH (v/v) 5-45% (B) 1.6 ml/min	254 nm	89
N-sides, N-tides of 5-FUra and 5-dFUrd	µBondapak C_{18} RCM-C_{18} Zorbax TMS C_{18} Zorbax TMS C_8 50; I (2-step)	(A) 0.002 M CH_3COONa 0.0015 M Na/K phosphate 0.1 mM TBHS pH 6.0 2.5 mM TEAB 2% CH_3OH (v/v) (B) (A) plus 30 mM phosphate 0-30 min (A) 30-50 min (B)	254 nm	4

Table 2 (Continued)

Compounds of interest	Sample preparation	Column	Anal. time mode[b] (slope)	Chromatographic conditions	Detector[a]	Refs.
N-tide triphosphates		LiChrosorb RP-18	80; I	0.05 M CaH$_4$O$_8$P$_2$ 0.01 M TBHS 1.5% CH$_3$CN 0.01% NaN$_3$ pH 6.5 1.0 ml/min	254 nm	34
N-sides, N-tide mono- and diphos.s				Same as above without CH$_3$CN		

[a]UV-visible detector unless otherwise noted. Wavelength of UV detector given.
[b]Eluent mode
  I = isocratic
  G = gradient
[c]Abbreviations
  N-tides = nucleotides
  N-sides = nucleosides

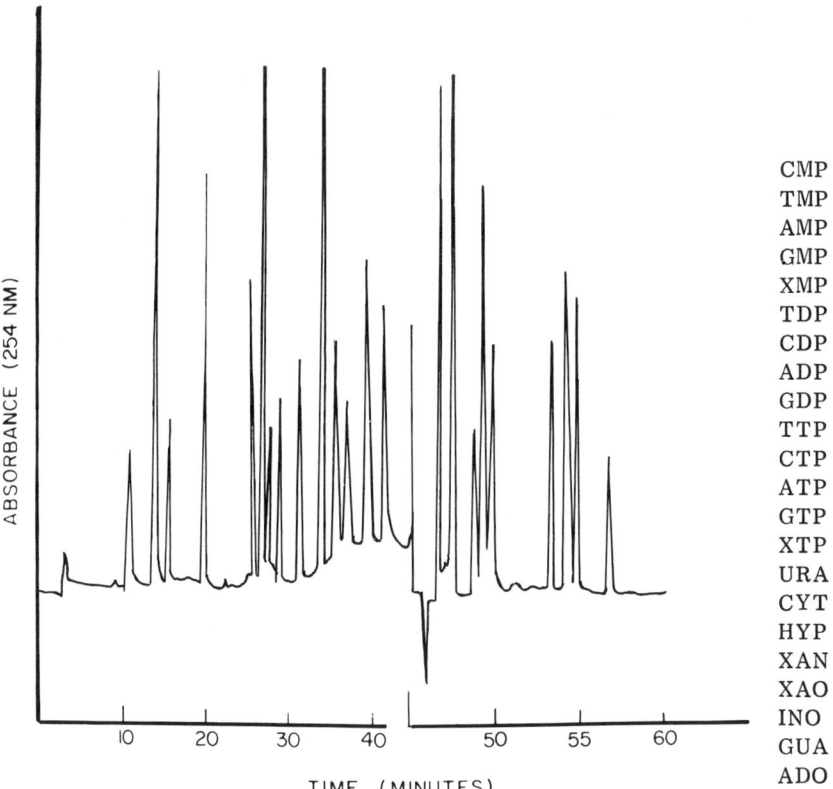

Figure 1. Separation of a standard ribonucleotide, ribonucleoside, and base mixture using column switching. The nucleotides were separated on a Partisil 10-SAX column (Whatman, Inc., Clifton, N. J.). A nonlinear gradient was used with a phosphate buffer from 0.005 KH_2PO_4, pH 5.5, to 0.5 KH_2PO_4, pH 5.5. Flow rate was 2 ml/min. After the elution of the XTP peak, the column was switched out of line and a radially compressed C_{18} column (Waters Assoc., Milford, Mass.) was switched in line. The nucleosides were then separated using a linear gradient elution from 0.005 M KH_2PO_4 to a 30:70 solution of methanol: water. (Unpublished chromatogram of Dr. Anne P. Halfpenny, from the laboratory of Dr. Phyllis R. Brown).

In mammalian cells the major constituents which comprise the nucleotide pools are the mono-, di-, and triphosphate ribonucleotides of adenine, guanine, hypoxanthine, uracil, thymine, and cytosine. These compounds, along with their bases and nucleosides, can be readily separated by HPLC using column switching with both reversed-phase and anion-exchange columns (Fig. 1). In addition, it is possible to separate the deoxyribonucleotides of these bases. However, separations of the deoxyribonucleotides in the presence of these ribonucleotide counterparts are more difficult because of the great concentration differential; the concentrations of the deoxy compounds are less than one-hundredth of the ribonucleotides. Concentrations of cyclic forms of AMP and GMP [15,26,48], sugar derivatives of UDP and ADP [17, 60], pyridine and flavin nucleotides [8,10,42] and their reduced forms [42], coenzyme A derivatives [21,35], and the monophosphates of xanthine and orotidine [6,10,100] have also been determined by HPLC.

In looking at nucleotide extracts of other species, the nucleotide profiles of the following extracts have been found to be very similar: cilicates [52,76], trematodes [22], yeast extracts [29,72], bacterial cultures [45,70], plant tissue [37,84,91], and fungal extracts [51]. The nucleotides were separated with anion-exchange HPLC.

A major portion of HPLC nucleotide research has been directed toward separating the maximum number of nucleotides recoverable, i.e., the nucleotide pool (column 1 of Tables 1 and 2). These pools are commonly called nucleotide profiles.

It is possible, however, to optimize HPLC conditions for the separation of a specific nucleotide or classes of nucleotides, e.g., cyclic nucleotides [27,39,48,54,99], pyridine nucleotides [8,34,42,58,62], coenzyme A precursors [21,35], 5'-substituted 2'-deoxyuridylates [30], nucleotide sugar complexes [17,47,51], and sulfated adenine nucleotides [84]. Particular attention has been paid to adenine nucleotides because of their importance in nucleic acid metabolism [18,24,44], cellular energy and homeostasis [43,56,74,75,83,88], and hormonal mechanisms [26].

II. BIOLOGICAL MATRICES

As is shown in Tables 1 and 2, many biological matricies have been analyzed for nucleotides by HPLC. Among the specific mammalian cell lines that have been studied are Chinese hampster ovary cells [92], murine sarcoma 180 and leukemia (15178) cells [11,82], mouse leukemia (1210) cells [25,67], human erythrocyte cultures [100], S-49 mouse lymphoma cells [31], skin fibroblasts [6,47,68], HeLa cells [47], and rat heart cells [86]. In research on blood cells, a great deal of work has been done on erythrocytes; however, investigations of nucleotide concentrations in platelets [26,73,75,98] and lymphocytes [1, 64,75] have also been carried out. An example of a chromatogram of the nucleotides in leukocytes is shown in Fig. 2. In addition to the formed elements of blood, the HPLC nucleotide profiles of liver [28, 29,32,40,95], muscle [56], brain [15,40,54,82], and granulation [74] tissues as well as spleen perfusate [3] have been reported in the literature.

Free Nucleotides in Biological Samples / 237

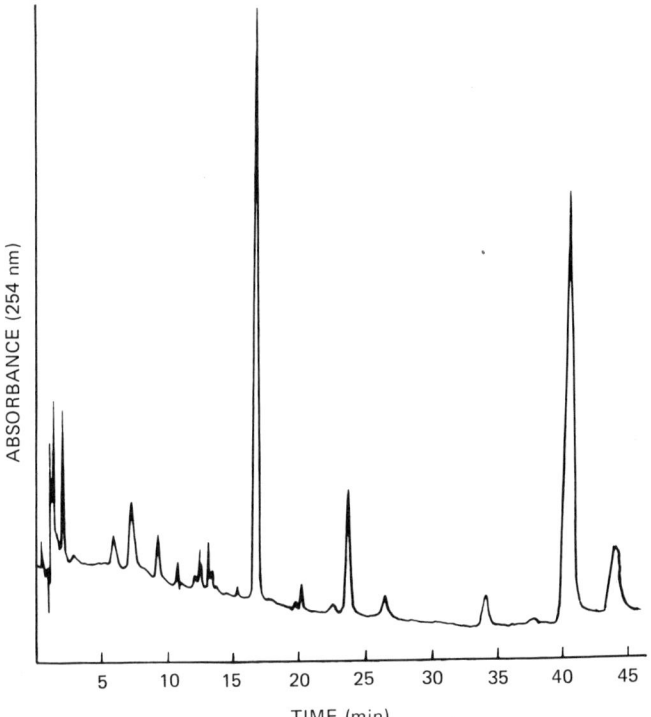

Figure 2. Anion-exchange separation of ribonucleotides in human leukocytes. Conditions for the analysis: column, strong anion-exchange column, Partisil-10-SAX (Whatman, Inc., Clifton, N. J.); linear gradient elution over period of 45 min; low concentration eluent, 0.007 M KH_2PO_4, pH 4.0; high concentration eluent, 0.25 M KH_2PO_4, 0.50 M in KCl, pH 4.5. Flow rate, 1.5 ml/min; Detector, UV detector set at 254 nm. (Unpublished chromatogram from the laboratory of Dr. Phyllis R. Brown.)

Research on free nucleotides can be divided into four main categories: (a) biological studies comparing nucleotide profiles of various species or the tissues within a species; (b) biomedical studies on the effect of diseases on nucleotide metabolism; (c) pharmacological studies of the effects of exogenous agents (e.g., chemotherapeutic, toxic, or carcinogenic) on nucleotide concentrations in physiological fluids or tissues; and (d) biochemical studies on the role and/or fate of nucleotides in metabolic pathways or cellular mechanisms. Examples of comparative studies are those conducted on species of fish [12,55] and mammals [12,57,60]. In biomedical research, the interrelationships of free nucleotide concentrations in cells with various disease states such as cancer, cardiovascular problems, immunodeficiencies, birth defects, etc., have been investigated [7,20] (see Chap. 19). In pharmacology studies, HPLC has been used to study the effects of fluoropurine derivatives [4,33,59,69], mercapto- and thiopurine derivatives [80,93], and arabinoside nucleotides [53,79] (see Chap. 17.) The effects on nucleotide pools of lead intoxication [90] and of enzyme inhibitors such as fluoroenylhydroxamic [17] and mycophenolic acids [19] have also been investigated. HPLC has been coupled with radionuclides to monitor with high sensitivity and specificity, the metabolic fate of compounds labeled with ^{14}C, ^{3}H, ^{35}S, and ^{32}P. Both off-line and on-line scintillation counters have been used in these studies of metabolic mechanisms [23,32,38].

III. SAMPLE PREPARATION

Sample preparation techniques have been developed to isolate some or all of the low-molecular-weight compounds of the nucleotide pools in cell matrices. These techniques are constantly being refined, as it has been found that the procedures themselves can cause degradation of the nucleotides. In addition, enzymes released from cells can cause metabolism of the nucleotides. Thus, sample preparation prior to the HPLC separation of nucleotides has the following requirements: (a) The integrity of the nucleotide pool must be preserved, thus the action of the enzymes present in the biological sample must be stopped immediately after the sample is collected; (b) the nucleotides must be completely released or extracted from the cells; (c) the high-molecular-weight compounds (usually proteins) must be removed from the sample, both to remove interferences with the analysis and to protect the column. As evident from Tables 1 and 2, acids have been traditionally used for lysing the cells and precipitating out both enzymatic and nonenzymatic proteins.

Individual researchers have modified the basic deproteinization procedures (see Chap. 3) to maximize isolation and preservation of the compounds of interest. During isolation of excised tissue, immediate freezing by the freeze-clamp method [87] or submerging in liquid N_2 [32,43,63,74,82] has been used to halt enzyme action prior to addition of the acid solution. Cells obtained from cultures [25,70,76] or a matrix such as blood [83,85] were also maintained in a cold environment until addition of a cold acid solution to retard the biochemical activity of the cells. Another way to inhibit enzymatic activity is the addition of sodium fluoride to freshly drawn blood. This method was used specifically to inhibit phosphatase activity [66]. Several aspects of deproteinization by the cold strong acid technique have been investigated: mode of addition [37,56,61], optimum acid concentration

[45,97], as well as use of an organic in conjunction with the acid [56, 82]. To avoid degradation of the nucleotides and possibly the HPLC column, optimal conditions were determined for subsequent removal of the acid by extraction [45,56,77,81] or neutralization of the acid with a base (see Tables 1 and 2). It has also been found that EDTA [76] or Tris-Cl [47,56], which can be added to adjust the pH of the solution, can decrease the amount of nucleotide hydrolysis. Although losses of 30-40% ribonucleotide triphosphates have been noted in samples in which the acid was neutralized by a base, these losses were minimized by mixing the acid extracts with dicarbonic acid diethylester (DEPC) prior to neutralization [76]. In perfusates where lysing of cells is not necessary, centrifugation [3] or addition of an organic solvent [35] has been used to precipitate proteinaceous material. Filtration prior to analysis by HPLC and additional chromatographic methods [15,47,51,87] have also been used for prepurification of certain classes of nucleic acid constituents. In addition, periodate and methylamine have been used to eliminate interfering ribonucleotides [25,31,78] in studies of deoxyribonucleotides.

IV. CHROMATOGRAPHY

A. Instrumentation

The basic HPLC equipment is discussed in Chaps. 4, 7, and 8. Although earlier researchers used a single pump with a mixing chamber to generate gradients [71], the majority of the chromatographic analyses listed in Tables 1 and 2 use a gradient system requiring two pumps. Today, most chromatographs are equipped with electronic programmers which control the pumps and types of gradients generated.

While ultraviolet (UV) absorbance, specifically in the range 250-280 nm, has been used extensively to monitor nucleotides, other wavelengths as well as other methods of detection have been employed. Reduced forms of pyridine nucleotides have been differentiated from oxidized forms by simultaneously comparing the absorbance at 340 nm with the absorbance at 280 nm [42,58,62] (Chap. 19). Straightforward application of fluorescence detection to nucleotides is limited; however, the flavin adenine dinucleotide complexes have enough fluorescence intensity to be monitored directly [70]. Post-column derivatization using sodium hydroxide and potassium hexacyanoferrate(III) has been used to detect thiamine nucleotides by fluorescence [49]. An automatic ashing technique for organic phosphate determination [9] and conductivity measurements [8] have also been applied to nucleotide analyses. Radioactive internal standards [3,56,95] and the incorporation of a radioactive label into nucleotide pools [25,35,84,86] have been monitored off-line. This procedure is time-consuming and costly. Although use of on-line radioactivity monitors has been reported [6,68,100], this type of detector will undoubtedly gain popularity now that several models are commercially available.

B. Ion exchange

Ion exchange is the oldest chromatographic method for separating nucleotides. Basically, an ionic group, usually a quarternary amine, is chemically bonded to a solid support. These supports can be a polystyrene (e.g., Aminex-A series) or a silica having a pellicular (e.g., Pellionex-SAX) or totally porous (e.g., Partisil PXS 10/25 SAX) nature. A metal ion complex, $Co(em)_3^{3+}$, bonded to silica has also been tested for effectiveness in separating nucleotides [16]. The ionic bonds are formed between the charged solutes (compounds of interest) and the ionic sites of the packing. In ion exchange the mobile phase is some type of aqueous buffer solution. Mobile phase parameters of polarity, pH, ion types, and concentration have been studied to determine the mechanisms of retention. Several fundamental rules regarding retention of nucleic acid fragments have been proposed for ion-exchange chromatography with silica-based packings [96]: (a) The ion concentration has an effect only on the ion-exchange properties. (b) The pH of the mobile phase determines the degree of protonation, and therefore the charge and polarity of the solutes. (c) Ion exchangers also have reversed-phase and normal-phase properties which are determined by the polarity of the mobile phase. (d) The polarity of the buffer ions have an indirect effect on ion-exchange chromatography through differential distribution of the buffer ions between the two phases.

Examination of Table 1 indicates that a variety of counterions have been used successfully to separate nucleotides. A majority of the ion-exchange analyses have utilized a pH range between 2 and 7.5, the range recommended for protecting the silica-based packings. While this range utilizes the pK_a's of many organic acids, the effectiveness of solute ionization on separation quality is greater with weak (e.g., μBondapak NH_2) rather than strong (e.g., Partisil PXS 10/25 SAX) anion exchangers [41]. Regardless of the salt used, the actual ionic strength (corrected for pH adjustments) has the greatest effect on the separation. Volatile salts have been used effectively where recovery of a purified nucleotide was necessary [5,70].

C. Reversed phase and ion pairing

Reversed phase has been applied mainly to the analysis of selected nucleotides, particularly when rapid analyses are required. Although nucleotides, nucleosides, and bases have been separated using reversed phase, there are problems involved in the use of reversed-phase liquid chromatography (RPLC) in resolving the weakly retained

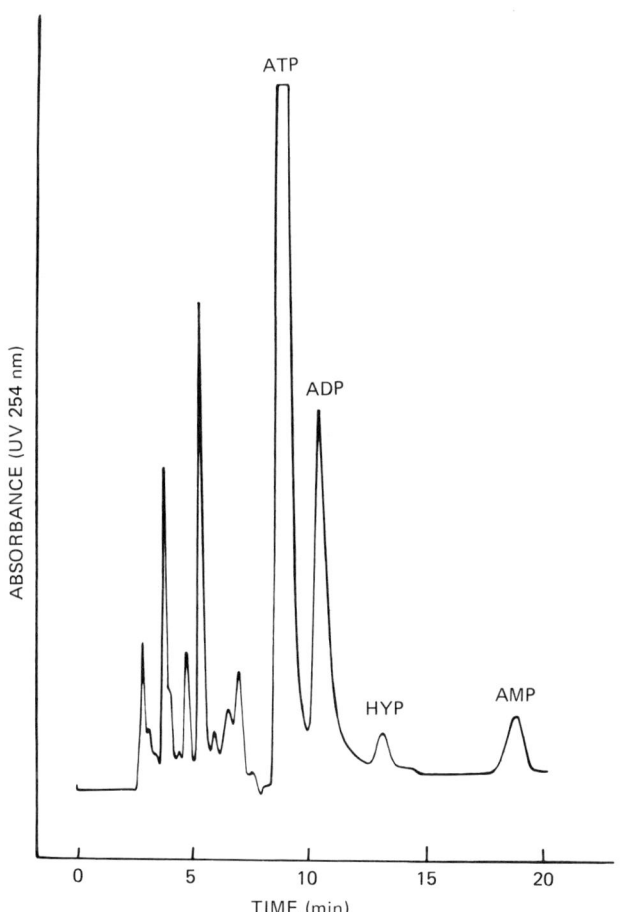

Figure 3. Reversed-phase separation of ribonucleotides and bases in human erythrocytic lysates. Conditions for analysis: column, Partisil ODS-3 IV (Whatman, Inc., Clifton, J. J.); guard column, Co Pel ODS (also Whatman); isocratic elution; eluent, 0.04 M KH_2PO_4, pH 6.0; flow rate, 1.0 ml/min.; temperature, ambient; detection, UV detector set at 254 nm. (Unpublished chromatogram from the laboratory of Dr. Phyllis R. Brown.)

nucleotides from other related compounds such as uric acid, hypoxanthine, and guanine. Thus column switching, as shown in Fig. 1, appears to hold the greatest potential for separating all these compounds in one analysis [101].

The majority of the HPLC separations of free nucleotides have been performed on strong anion-exchange columns; however, the ion-pairing technique with the reversed-phase (RP-IP) is now being used more widely. This technique is discussed in Chap. 9. Ion pairing is very useful in metabolic or pharmacological studies when it is desirable to determine concentrations of nucleosides and bases together with concentrations of nucleotides. Several reports of RPLC-IP are included in Table 2; however, these studies are limited mainly to the ion pairing of standards, as reports of investigations on the ion pairing of nucleotides in biological matrices are limited. With RP-IP a gradient program is not usually necessary to obtain an efficient separation of the nucleotide pools [22]. The choice of elution mode, i.e., isocratic or gradient, is dependent on the number and type of nucleotides to be resolved and the time required for the analysis. However, a gradient program must be used if the study requires resolution of nucleosides and their bases along with the nucleotides. A major advantage of using RPLC is the reduced time needed for an analysis. Separation of nucleotides, nucleosides, and their bases using ion-exchange gradient programs [6,29,26,65,68] required from 120 to 225 min. However, with the reversed-phase mode [83,92] with or without the ion-pairing technique [40], the analysis time for separating selected nucleosides and nucleotides can be as little as 30 min, as shown in Fig. 3. Systematic investigations of the effects of various parameters, i.e., pH, ion concentration, and polarity of the mobile phase, on the separation of nucleo-bases, nucleosides, and nucleotides have been carried out for reversed phase and ion pairing [4,33,50,95].

V. ABBREVIATIONS

C11A = 11-amino undecanoic acid
DEPC = dicarbonic acid diethylester
TMAH = tetramethyl ammonium hydroxide
TBHS = tetra-N-butyl ammonium hydrogen sulfate
TBAB = tetra-N-butyl ammonium bromide
TBAP = tetra-N-butyl ammonium phosphate
TBAF = tetra-N-butyl ammonium formate
TEAB = tetraethyl ammonium bromide
TEAH = tetraethyl ammonium hydroxide
TBAH = tetrabutyl ammonium hydroxide

REFERENCES

1. R. A. Abrau, J. M. vanBaal, J. A. J. M. Bakkeren, C. H. M. M. DeBruyn, and E. D. A. M. Schretlen, J. Chromatogr., 227, 45 (1982).
2. M. I. Al-Moslih, G. R. Dubes, and A. N. Masoud, J. High Res. Chromatogr. and Chromatogr. Commun., 4 (April), 173 (1981).
3. F. S. Anderson and R. C. Murphy, J. Chromatogr., 121(2), 251 (1976).
4. J. L. S. Au, M. G. Wientjes, C. M. Luccioni, and Y. M. Rustum, J. Chromatogr., 228, 245 (1982).
5. J. T. Axelson, J. W. Bodley, and T. F. Walseth, Anal. Biochem., 116, 357 (1981).
6. B. Bakay, E. Nissinen, and L. Sweetman, Anal. Biochem., 86(1), 65 (1978).
7. B. Bakay, E. Nissinen, L. Sweetman, U. Francke, and W. L. Nyhan, Pediatr. Res., 13, 1365 (1979).
8. C. Bernofsky and W. J. Gallagher, Anal. Biochem., 67(2), 611 (1975).
9. S. P. Bessman, P. J. Geiger, T. C. Lu, and E. R. B. McCabe, Anal. Biochem., 59, 533 (1974).
10. D. P. Brenton, K. H. Astrin, M. K. Cruickshank, and J. E. Seegmiller, Biochem. Med., 17(3), 231 (1977).
11. P. R. Brown, J. Chromatogr., 52, 257 (1970).
12. P. R. Brown, R. P. Agarwal, J. Gell, and R. E. Parks, Jr., Comp. Biochem. Physiol., 43B, 891 (1972).
13. C. A. Burtis, N. M. Munk, and F. R. McDonald, Clin. Chem., 16, 667 (1970).
14. S. C. Chen, P. R. Brown, and D. M. Rosie, J. Chromatogr. Sci., 15(6), 218 (1977).
15. A. Chiu and D. Eccleston, Anal. Biochem., 78(1), 148 (1977).
16. F. K. Chow and E. Grushka, J. Chromatogr., 185, 361 (1979).
17. Y. M. Chow, H. R. Gutman, and A. H. Potter, Biochem. Biophys. Acta, 585(1), 154 (1979).
18. J. K. Christman, Anal. Biochem., 119, 38 (1982).
19. M. B. Cohen, J. Maybaum, and W. Sadeé, J. Chromatogr., 198, 435 (1980).
20. M. S. Coleman, J. Donofrio, J. J. Hutton, and L. Hahn, J. Biol. Chem., 253, 1619 (1978).
21. B. E. Corkey, M. Brandt, R. J. Williams, and F. R. Williamson, Anal. Biochem., 118, 30 (1981).
22. A. A. Darwish and R. K. Prichard, J. Liq. Chromatogr., 4(9), 1511 (1981).
23. B. M. Dean and D. Perrett, Biochem. Biophys. Acta, 437(1), 1 (1976).
24. J. C. Donofrio, J. Meier, and J. J. Hutton, Cell Immunol., 42(1), 79 (1979).

25. R. Dreyer and E. Cadman, J. Chromatogr., *219*, 273 (1981).
26. L. D'Souza and H. I. Glueck, Thromb. Heam., *38*(4), 990 (1977).
27. S. P. Dutta, A. Mittelman, and G. B. Chheda, Anal. Biochem., *72*(2), 649 (1976).
28. E. H. Edelson, J. G. Lawless, C. T. Wehr, and S. R. Abbott, J. Chromatogr., *174*, 409 (1979).
29. A. Floridi, C. A. Palmerini, and C. Fini, J. Chromatogr., *138*, 203 (1977).
30. C. Garrett, A. L. Pogoletti, and D. V. Santi, Anal. Biochem., *79*(1-2), 602 (1977).
31. C. Garrett and D. V. Santi, Anal. Biochem., *99*(2), 268 (1979).
32. T. Gasser, J. D. Moyer, and R. E. Handschumacher, Science, *213*, 777 (1981).
33. C. F. Gelijkens and A. P. DeLeenkeer, J. Chromatogr., *194*, 305 (1980).
34. E. Hagele, J. Finke, P. Lehmann, and M. Grabl, Fresenius Z. Anal. Chem., *311*, 419 (1982).
35. O. Halvorsen and S. Skrede, Anal. Biochem., *107*, 103 (1980).
36. R. A. Hartwick and P. R. Brown, J. Chromatogr., *112*, 651 (1975).
37. H. W. Heldt, A. R. Portis, R. M. Lilley, and A. Mosbach, Anal. Biochem., *101*(2), 278 (1980).
38. R. A. Henry, J. A. Schmit, and R. C. Williams, J. Chromatogr. Sci., *11*, 358 (1973).
39. N. E. Hoffman and J. C. Liao, Anal. Chem., *49*(4), 2231 (1977).
40. C. Horvath, B. Preiss, and S. Lipsky, Anal. Chem., *39*, 1422 (1967).
41. E. L. Johnson and R. Stevenson, *Basic Liquid Chromatography*, Varian Associates, Inc., 1978.
42. D. P. Jones, J. Chromatogr., *225*, 446 (1981).
43. E. Juengling and H. Kammermeier, Anal. Biochem., *102*(2), 358 (1980).
44. M. J. Kessler, J. Liq. Chromatogr., *5*(1), 111 (1982).
45. J. X. Khym, Clin. Chem., *2*(9), 1245 (1975).
46. J. X. Khym, J. Chromatogr., *124*(2), 415 (1976).
47. J. X. Khym, J. W. Bynum, and E. Volkin, Anal. Biochem., *77*(2), 446 (1977).
48. J. X. Khym, J. Chromatogr., *151*, 421 (1978).
49. M. Kimura, T. Fujita, S. Nishida, and Y. Itokawa, J. Chromatogr., *188*(2), 417 (1980).
50. J. H. Knox and J. Jurand, J. Chromatogr., *203*, 85 (1981).
51. C. Y. Ko, H. M. McNair, and J. R. Vercello, Carbohydrate Res., *58*(2), 453 (1977).
52. G. Krauss and H. Reinbothe, Anal. Biochem., *78*(1), 1 (1977).
53. W. Kreis, C. Gordon, C. Gizoni, and T. Woodcock, Cancer Treatment Rep., *61*(4), 643 (1977).

54. A. M. Krstulovic, R. A. Hartwick, and P. R. Brown, Clin. Chem., 25(2), 235 (1979).
55. C. Leray, Comp. Biochem. Physiol., 64B, 77 (1979).
56. C. Lush, Z. H. A. Rahim, D. Perrett, and J. R. Griffiths, Anal. Biochem., 93, 227 (1979).
57. N. S. Magnuson and L. E. Perryman, Comp. Biochem. Physiol., 67B, 205 (1980).
58. S. A. Margolis and R. Schaffer, J. Liq. Chromatogr. 2(6), 837 (1979).
59. J. Maybaum, F. K. Klein, and W. Sadee, J. Chromatogr., 188, 149 (1980).
60. M. McKeag, P. R. Brown, and J. D. Sallis, Comp. Biochem. Physiol., 70B, 541 (1981).
61. R. P. Miech and M. C. Tung, Biochem. Med., 4, 435 (1970).
62. J. R. Miksic and P. R. Brown, J. Chromatogr., 142, 641 (1977).
63. J. D. Moyer and R. E. Handschumacher, Cancer Res., 39, 3089 (1979).
64. B. Munch-Peterson, G. Tyrsted, and B. Dupont, Exp. Cell. Res., 79, 249 (1973).
65. F. Murakani, S. Rokushika, and H. Hatano, J. Chromatogr., 53, 584 (1970).
66. G. S. Nakai and C. G. Craddock, Cancer Res., 25, 575 (1965).
67. J. A. Nelson, L. M. Rose, and L. L. Bennett, Jr., Cancer Res., 36, 1375 (1976).
68. E. Nissinen, Anal. Biochem., 106(2), 497 (1980).
69. R. E. Parks, Jr., and P. R. Brown, Biochemistry, 12(7), 3294 (1973).
70. S. M. Payne and B. N. Ames, Anal. Biochem., 123, 151 (1982).
71. D. Perrett, J. Chromatogr., 124(2), 187 (1976).
72. K. H. Pfluger, Anal. Biochem., 81(1), 136 (1977).
73. S. H. Pross, T. W Klein, and C. W. Fishel, Proc. Soc. Exp. Biol. Med., 154, 508 (1977).
74. D. Pruneau, E. Wulfert, M. Pascal, and C. Baron, Anal. Biochem., 119, 274 (1982).
75. G. H. R. Rao, J. G. White, A. A. Jachimowicz, and C. J. Withop, Jr., J. Lab. Clin. Med., 12, 839 (1974).
76. M. P. Reinhart and M. J. Koroly, Anal. Biochem., 119, 392 (1982).
77. T. L. Riss, N. L. Zorich, M. D. Williams, and A. Richardson, J. Liq. Chromatogr., 3(1), 133 (1980).
78. E. J. Ritter and L. M. Bruce, Biochem. Med., 21(1), 16 (1979).
79. L. M. Rose and R. W. Brockman, J. Chromatogr., 133(2), 335 (1977).
80. E. M. Scholar, P. R. Brown, and R. E. Parks, Jr., Cancer Res., 32, 259 (1972).

81. E. M. Scholar, P. R. Brown, R. E. Parks, Jr., and P. Calabresi, Blood, 41(6), 927 (1973).
82. H. W. Schmukler, J. Chromatogr. Sci., 10, 38 (1972).
83. P. D. Schweinsberg and T. L. Loo, J. Chromatogr., 181, 103 (1980).
84. J. D. Schwenn and H. G. Jender, J. Chromatogr., 193, 285 (1980).
85. K. Seta, M. Washitake, T. Anmo, N. Takai, and T. Okuyama, J. Liq. Chromatogr., 4(1), 129 (1981).
86. E. S. Sharp and R. L. McCarl, Anal. Biochem., 124, 421 (1982).
87. N. M. Shaw, E. G. Brown, and R. P. Newton, Biochem. Soc. Trans., 7, 1250 (1979).
88. R. C. Smith, Comp. Biochem. Physiol., 69B, 505 (1981).
89. R. J. Smith, I. R. C. S. Med. Sci.: Biochem.; Biomed. Techn.; Cell and Membrane Biol., 9, 963 (1981).
90. M. S. Swanson, C. R. Angle, S. J. Stohs, and K. S. Rovang, Res. Commun. Chem. Pathol. Pharmacol., 27, 353 (1980).
91. T. Tashiro and E. Fujita, Nippon Shokuhin Kogyo Gakkaishi, 28(11), 588 (1981).
92. M. W. Taylor, H. V. Hershey, R. A. Levine, K. Coy, and S. Olivelle, J. Chromatogr., 219, 133 (1981).
93. D. M. Tidd and S. Dedhar, J. Chromatogr., 145(2), 237 (1978).
94. M. E. Tischler, D. Friedrichs, K. Coll, and J. R. Williamson, Arch. Biochem. Biophys., 184, 222 (1977).
95. P. J. M. VanHaastert, J. Chromatogr., 210, 229 (1981).
96. P. J. M. VanHaastert, J. Chromatogr., 210, 241 (1981).
97. D. A. Van Haverbeke and P. R. Brown, J. Liq. Chromatogr., 1(4), 507 (1978).
98. T. F. Walseth, G. Groff, M. C. Moss, Jr., and N. D. Goldberg, Anal. Biochem., 107, 240 (1980).
99. D. M. Watterson, D. B. Iverson, and L. J. Vaneldik, J. Biochem. Bio., 2(3), 139 (1980).
100. H. K. Webster and J. M. Whaum, J. Chromatogr., 209, 283 (1981).
101. I. Molnar, Abstract #, 8th International Liquid Column Symposium, New York, May 1984.

13
Free Nucleosides and Their Bases in Physiological Fluids

KATSUYUKI NAKANO PL Medical Data Center, Tondabayashi, Osaka, Japan

I. Introduction 247
II. Applications of HPLC to Physiological Fluids Analysis 248
   A. Serum and plasma 248
   B. Saliva 254
   C. Urine 258
   References 264

I. INTRODUCTION

Physiological fluids, such as serum, saliva, and urine, are very complex matrices both in the quality and quantity of their components. For example, urine is thought to contain thousands of components with comparatively low molecular weight, of which about 40 purine and pyrimidine metabolites are known [1]. On the other hand, serum or plasma contains a more limited number of proteins, peptides, and low-molecular-weight compounds.

The analysis of purines and pyrimidines and their metabolites is very important for the understanding of metabolic disorders and related pathological states. Disorders of nucleic acid metabolism found so far include gout, Lesch-Nyhan syndrome, xanthine uria, orotic acid-uria, uracil uria, adenosine deaminase deficiency, and purine nucleoside phosphorylase deficiency. Recently, concentrations of minor components such as modified (alkylated or methylated) nucleosides and bases have been investigated with keen interest for their clinical significance as biochemical markers in cancer detection.

During the last decade, analytical methods using high performance liquid chromatography (HPLC) have been developed as a powerful and practical tool for the determination of nucleosides, bases, and their metabolites in body fluids. These methods might be categorized into the following approaches for clinical use: (a) a profile analysis and (b) a selective analysis.

A profile analysis aims at determining simultaneously as many as possible compounds in body fluids with one or more HPLC analyses, thus characterizing normal or pathological states by metabolic profiles. Recently, some studies that directly utilize the chromatographic patterns in urine have been reported for diagnostic purposes for cancer [2] and kidney functions [3].

In a selective analysis, specific compounds in body fluids are marked for the HPLC analysis and determination. Several methods to isolate or prefractionate the specific compounds from body fluids prior to the analytical HPLC have been developed.

In this chapter, both approaches for the HPLC analysis of free nucleosides and bases in physiological fluids will be discussed, with special emphasis on the present state of the art in biochemical and clinical applications of HPLC for purine and pyrimidine metabolites.

II. APPLICATIONS OF HPLC TO PHYSIOLOGICAL FLUIDS ANALYSIS

A. Serum and plasma

1. Profile analysis

In the late 1960s, ion-exchange HPLC analyses of the ultraviolet (UV)-absorbing materials and nucleosides [4] were carried out, and some of the purine and pyrimidine metabolites in serum were studied. However, when serum samples were analyzed using HPLC anion exchangers, the only nucleic acid metabolites found were uric acid, hypoxanthine, and uracil [5,6]. With the developments of more efficient packing materials and instruments, the resolution of nucleosides and bases has increased greatly and the time required for the analysis decreased.

Comprehensive studies of nucleosides and bases in serum or plasma have been performed by Brown and her collaborators. They first established the separation conditions of some standard nucleosides and bases using microparticle chemically bonded reversed-phase packing (μBondapak C_{18}) and a methanol gradient in phosphate buffer [7,8].

Since there is no ionized group in nucleosides and bases, the reversed-phase partition mode of HPLC was particularly suited for this separation.

Krstulovic et al. [9] identified nucleosides and bases in serum and plasma by the following techniques: peak height ratios (UV 254 nm/ 280 nm), fluorescence, coinjection of standards, and enzymatic peak shift. From this study, eight purine and pyrimidine metabolites, uric acid, hypoxanthine, uridine, xanthine, guanine, xanthosine, inosine, and guanosine, as well as other UV-absorbing constituents, tryptophan, theobromine, and caffeine, were identified in a serum sample from a normal subject. The identity of these serum constituents was also confirmed by the stopped-flow UV-scanning technique [10,11].

Hartwick et al. [12] further reported a comprehensive investigation of HPLC separation of 86 compounds including nucleosides, their bases and other low-molecular-weight UV-absorbing compounds. Figure 1 shows a typical chromatogram of a normal serum sample analyzed by the above separation conditions [13]. The changes in peak areas of serum sample incubated at 25°C (removed from the clot) over a 10-hr period were examined, and the drastic changes of adenosine, inosine, hypoxanthine, and xanthine levels were observed, indicating the intracellular and extracellular enzyme reactions in purine metabolism. Thus, standard protocol conditions for processing and storage of serum samples prior to HPLC analyses were proposed on the basis of the above findings [13] and the investigations for sample preparation techniques, especially concerning the deproteinization methods [14]. Quantitative analyses were also performed to determine the normal levels of 12 compounds identified in serum from 31 normal subjects [13], and to compare serum profiles of humans and dogs [15].

Reversed-phase, radially compressed, flexible-walled columns were also used by Assenza and Brown [16] for the rapid separation of UV-absorbing compounds in serum with high flow rates (> 2.0 ml/min). The radially compressed columns were found to reduce the analysis time of human serum by 50% compared to ordinary rigid-walled columns.

Recently, Zakaria and Brown [17] investigated the effects on plasma chromatograms of several anticoagulants such as heparin, acid-citrate-dextrose (ACD), or EDTA. The presence of UV-absorbing compounds which interfered with the determination of plasma constituents were found in ACD and EDTA solutions; however, there were no interferences in the chromatograms of plasma samples with heparin. It was also found that in all 15 sets of samples examined, the levels of xanthine, inosine, and particularly hypoxanthine were consistently higher in the serum than in the plasma. It was also found that unless heparinized plasma and serum layer were instantly separated from the

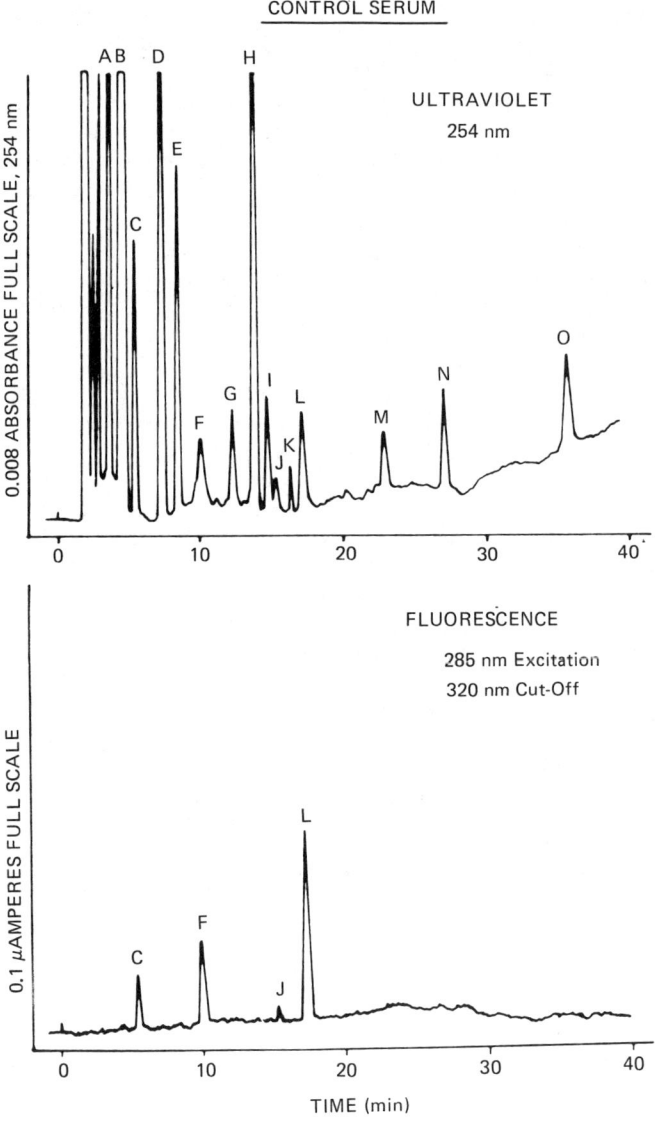

Figure 1. Chromatogram of the serum filtrate from a normal donor using both UV (254 nm, 0.008 aufs) and fluorescence (285 nm excitation, 320 nm cutoff) detection. Injection volume: 80 μl. Column: reversed-phase, 10-μm particle diameter, 4.6 × 300 mm. Eluents: low strength, 0.02 mol/liter KH_2PO_4, pH 5.6; high strength 60% methanol-water. Gradient: linear, 0–100% of high strength eluent in 87 min, slope 0.69% methanol/min. Flow rate: 1.5 ml/min. Temperature: ambient. (From Ref. 13.)

erythrocyte cells, the concentrations of hypoxanthine, xanthine, and inosine in their samples changed significantly due to the by-products of nucleotide catabolism in blood cells.

The reversed-phase HPLC techniques developed by Brown et al. have great advantages for clinical applications: (a) The simple deproteinization by ultrafiltration (Amicon Centriflo CF 25 membrane cone) and/or precolumn is used in sample preparation; (b) the injection volume of sample is only 50-80 µl; (c) the required analysis time is 35 min or less (see p. 65) by the use of a linear gradient of methanol; and (d) more than 12 constituents identified in serum or plasma are analyzed simultaneously. The applications of these HPLC systems to the metabolic profiles of several pathological states demonstrated that the nucleoside and base profiles in serum or plasma might be used for the detection of several types of cancerous conditions [18,19].

Wung and Howell [20] recently reported a HPLC method for the simultaneous determination of cancer chemotherapeutic agents such as 5-fluorouracil, allopurinol, and its metabolite oxipurinol together with endogenous compounds, such as uridine, hypoxanthine, xanthine, and uric acid, in plasma. Their separation was optimized by the reversed-phase mode of HPLC using phosphate buffer isocratically, and the levels of the endogenous compound in EDTA plasma and serum were quantified. These methods are especially effective for monitoring the drug levels in blood from cancer patients undergoing treatment with 5-fluorouracil and allopurinol.

Other profile analyses of plasma purines and pyrimidines using reversed-phase HPLC have been developed by Mcburney and Gibson [21]. Overall chromatographic separation was rather poor for the simultaneous determination of all these compounds in plasma; however, this method was suitable for the determination of hypoxanthine, xanthine, and oxipurinol levels, and was applied to the determination of these compounds in plasma from subjects with gout and renal failure, and in urine of patients with enzyme deficiencies.

Recently, Simmonds and Harkness [22] published a reversed-phase HPLC method for bases and nucleosides in plasma and in cells including erythrocytes, lymphocytes, and polymorphonuclear neutrophil leukocytes. However, in the normal plasma samples, only levels of hypoxanthine, xanthine, and uridine were quantified by this method.

The normal human serum or plasma levels of nucleosides, bases, and their metabolites, and the chromatographic conditions for profile analysis discussed in this section, are listed in Table 1.

2. *Selective analysis*

Several methods of selective HPLC analysis for certain nucleosides, bases, and their metabolites have been developed. Reversed-phase HPLC procedures for the selective determination of adenosine in human serum [23] and deoxyadenosine and adenosine in human plasma [24] in the presence of other nucleic acid components were reported. Rapid

Table 1 Normal Human Serum or Plasma Levels of Nucleosides, Bases, and Their Metabolites, and Chromatographic Conditions for Profile Analysis

Compounds	Samples[a]	Means ± S.D. or range (nmol/ml)	Stationary phase	Mobile phase	Ref.
Uric acid	S (F, 14)	171 ± 30	μBondapak C_{18} (Waters)	Linear gradient from 0.02 M KH_2PO_4 (pH 5.6) to 100% methanol-water (3:2) in 87 min	13
	S (M, 17)	295 ± 39			
Hypoxanthine	S (F, 14; M, 17)	7.16 ± 2.81			
Uridine	S (F, 14; M, 17)	3.17 ± 1.11			
Xanthine	S (F, 14; M, 17)	2.62 ± 1.04			
Inosine	S (F, 14; M, 17)	5.62 ± 2.87			
Guanosine	S (F, 14; M, 17)	0.881 ± 0.515			
Uric acid	EP	276 ± 55	μBondapak C_{18} (Waters)	0.05 M KH_2PO_4 (pH 4.60), isocratic	20
Hypoxanthine	EP	0.46 ± 0.21			
	S	3.20 ± 1.40			

Xanthine	EP	0.40 ± 0.27			
Uridine	S	1.40 ± 0.60			
	EP	4.50 ± 1.70			
Hypoxanthine	HP (10)	1.025 ± 0.9	μBondapak C$_{18}$ (Waters)	Linear gradient from 0.073 mM KH$_2$PO$_4$ (pH 5.8) to 100% methanol	21
Xanthine	HP (10)	4.9 ± 1.5			
Hypoxanthine	EP (M, 4)	1.50 ± 0.2	ODS-Hypersil (Shandon)	0.004 M KH$_2$PO$_4$ (pH 5.8) with 1% (v/v) methanol, isocratic	22
Xanthine	EP (M, 4)	0.46 ± 0.09			
Uridine	EP (M, 4)	3.2 ± 0.6			
Hypoxanthine	S (F, 6); S (M, 9)	13.7–19.8; 15.6–24.0	μBondapak C$_{18}$ (Waters)	Linear gradient from 0.02 M KH$_2$PO$_4$ (pH 5.7) to 40% methanol–water (3:2) in 35 min	17
	HP (F, 5); HP (M, 9)	3.84–12.1; 4.79–14.6			
	EP (F, 5); EP (M, 8)	2.18–6.41; 3.20–10.4			
	AP (F, 5); AP (M, 9)	0.94–5.40; 1.34–6.41			

[a] Abbreviations used: P, plasma; EP, EDTA plasma; HP, heparin plasma; AP, ACD plasma; S, serum; PS, pooled serum. Sex (M, male; F, female) and number of samples are shown in parentheses.

HPLC analyses have been studied for the determination of human uric acid, which is the end product of purine metabolism [25,26].

Recently, some methods for selective analysis of nucleosides through the pretreatment of serum or plasma prior to HPLC analysis have been reported. The prefractionation techniques of urine nucleosides with boronate affinity gel [27] were applied to the determination of inosine and adenosine in human plasma by Pfadenhauer and Tong [28]. The boronate-treated human plasma was analyzed by reversed-phase HPLC. Because of the increased sensitivity as well as selectivity of this method, measurable amounts of adenosine were found in normal human plasma.

Serum and plasma uridine levels in mice, rats, and humans were determined by Karle et al. [29] using the same pretreatment with boronate gel and reversed-phase HPLC with isocratic elution mode of acetate buffer. Human serum uridine levels of 2-9 nmol/ml were relatively constant and were unaffected by a fasting period, suggesting the presence of a regulatory mechanism of its levels.

Using the same prefractionation techniques with boronate gel prior to reversed-phase HPLC, Zumwalt et al. [30] found six nucleosides, including pseudouridine, uridine, 1-methylinosine, 1-methylguanosine, N^2-methylguanosine, and N^2,N^2-dimethylguanosine, in 1.66 ml of pooled serum. They found that the concentrations of modified nucleosides in normal serum are lower by at least 50-100 factors than those in urine. This study is an extension of their excellent analyses of urine nucleosides which are described later. These methods are applicable only to nucleosides which contain a ribose group.

Other selective analysis procedures for plasma nucleosides and bases were examined by Taylor et al. [31]. They used an initial reversed-phase HPLC separation for 500 μl of deproteinized plasma to allow the collection of a number of effluent fractions, and then reapplied these fractions to a second reversed-phase HPLC separation. Using this method, the nucleosides and bases, hypoxanthine, thymine, inosine, and thymidine, were identified in normal and patient plasma, and their circulating levels were measured.

The normal human serum or plasma levels of nucleosides, bases, and their metabolites, and the chromatographic conditions for selective analysis discussed in this section, are listed in Table 2.

B. Saliva

So far, HPLC analysis of the endogenous compounds in saliva has seldom been performed, with a few exceptions, because interferences caused by diet may make the interpretation of data difficult.

Based on the analysis system of reversed-phase HPLC for serum samples established by Brown et al., Nakano et al. [32] investigated the low-molecular-weight UV-absorbing compounds in saliva.

Table 2 Normal Human Serum or Plasma Levels of Nucleosides, Bases, and Their Metabolites, and Chromatographic Conditions for Selective Analysis

Compounds	Samples[a]	Means ± S.D. or range (nmol/ml)	Stationary[b] phase	Mobile[b] phase	Ref.
Adenosine	HP (6)	0.07 ± 0.03	PXS 10/25 SCX (Whatman) or µBondapak C_{18} (Waters)	0.1 M $NH_4H_2PO_4$ (pH 4.5), isocratic or 0.005 M KH_2PO_4: 25% methanol (v/v), pH 7.5	43
Deoxyadenosine	HP (6)	< 0.06			
Adenosine	P (M, 5; F, 1)	0.23 ± 0.18[c]	Spherisorb ODS (Spectra Phys.)	0.05 M H_3PO_4 (pH 3.05), isocratic	28
Inosine	P (M, 5; F, 1)	0.91 ± 0.71[c]			
Uridine	S (13)	1.9–8.4	PXS 5/25 ODS (Whatman)	0.01 M Sodium acetate + acetic acid (pH 4.5), isocratic	29
Thymine	HP (14)	< 0.1	µBondapak C_{18} (Waters) LiChrosorb 10 RP18 (Merck)	0.05 M Ammonium acetate (pH 5.0), isocratic 0.025 M Ammonium acetate (pH 5.0), isocratic	31
Inosine	HP (14)	0.68 ± 0.37[d]			
Hypoxanthine	HP (13)	1.2 ± 0.72[d]			
Thymidine	HP (12)	0.43 ± 0.21[d]	Zorbax C_8 (DuPont)	0.05 M Acetic acid, isocratic	

Table 2 (Continued)

Compounds	Samples[a]	Means ± S.D. or range (nmol/ml)	Stationary[b] phase	Mobile[b] phase	Ref.
Pseudouridine	PS (3)	4.17	μBondapak C_{18} (Waters)	Step-gradient of 0.01 M $NH_4H_2PO_4$ buffer, (A) pH 5.3 with 2.5% methanol (36 min) and (B) pH 5.1 with 8.0% methanol (60 min)	30
Uridine	PS (3)	5.73			
1-m-Inosine	PS (3)	0.106			
1-m-Guanosine	PS (3)	0.036			
N^2-m-Guanosine	PS (3)	0.242			
N^2-dm-Guanosine	PS (3)	0.104			

[a]For abbreviations, see Table 1.
[b]Chromatographic conditions for analytical HPLC only are shown.
[c]Mean and S.D. values are calculated from the data reported.
[d]S.D. values are calculated from SEM values reported.

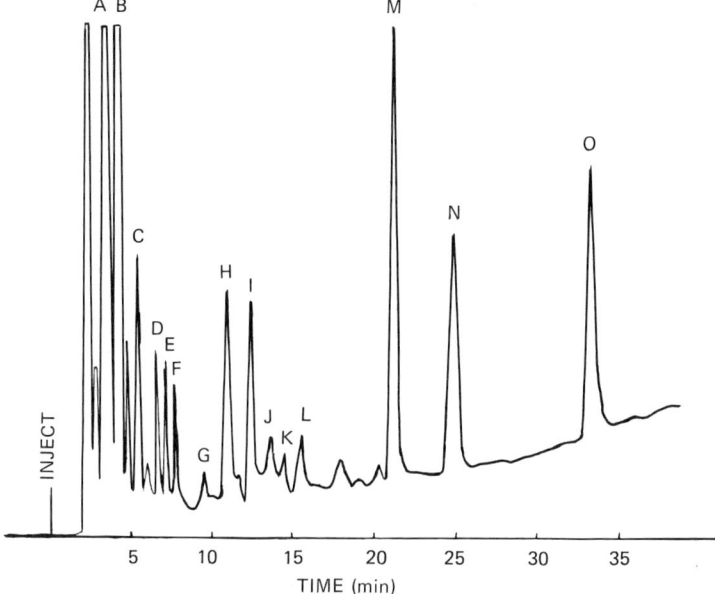

Figure 2. Chromatogram of the Saliva from a normal donor. Injection volume: 100 µl. Column: µBondapak C_{18}, 10-µm particle diameter, 3.9 × 300 mm. Eluents: low strength, 0.02 mol/liter KH_2PO_4, pH 5.7; high strength 100% methanol. Gradient: linear, 0-24% of high strength eluent in 35 min. Flow rate: 1.5 ml/min. Detection: 254 nm, 0.02 aufs. Temperature: ambient. Peaks: A = creatinine; B = uric acid; C = tyrosine; D = hypoxanthine; E = uridine; F = xanthine; G = kynurenine; H = 5-hydroxytryptophan; I = inosine; J = guanosine; K = hippuric acid; L = tryptophan; M = theobromine; N = para-xanthine/theophylline; O = caffeine.

Figure 2 shows a representative profile of UV-absorbing compounds in saliva. These compounds in saliva were very similar to the serum constituents, but their concentrations were much lower. Table 3 lists the concentrations of the saliva constituents found in fasting subjects and their relation (in percentage) to values measured in serum. Caffeine and its metabolites are reported as the concentrations observed 3 hr after caffeine ingestion.

In this study, the changes with time in concentrations of caffeine and its major metabolite, paraxanthine, in saliva and serum after administration of caffeine have been examined. The half-lives of these drugs in both saliva and serum were determined. Thus, it was demonstrated that this HPLC system was also useful in the pharmacokinetics study of some drugs.

Table 3 Normal Human Saliva Levels of Nucleosides, Bases, and Their Metabolites and Their Relation to Serum Levels

Compounds	Saliva levels (μmol/liter)	$\dfrac{[C]_{serum}}{[C]_{saliva}} \times 100$
Uric acid	44.8	16.4
Hypoxanthine	1.09	15.3
Uridine	0.214	6.75
Xanthine	1.18	45.0
Inosine	1.13	20.1
Paraxanthine	5.06	83.8
Caffeine	6.07	94.0

Source: Modified from Ref. 32.

C. Urine

1. Profile analysis

In 1967 Scott and collaborators [5] developed a high-resolution urine analyzer with microreticular anion exchangers (Dowex 1-X8) to separate the UV-absorbing constituents in physiological fluids. This system, called "the UV analyzer," separated up to 143 chromatographic peaks from urine, serum, and cerebrospinal fluid in an elution time of 40 hr [6]. Fifty of these UV-absorbing constituents in urine were identified by using gas chromatography (GC) and mass spectrometry (MS). Included are the following purine and pyrimidine metabolites: β-pseudouridine, uracil, 5-acethylamino-6-amino-3-methyluracil, 7-methylxanthine, hypoxanthine, xanthine, 3-methylxanthine, 1-methylxanthine, uric acid, orotidine, and orotic acid [33]. Also, cytosine, thymine, and adenine were tentatively identified by HPLC [6].

Using the UV analyzer, Kelley and Wyngaarden [34] examined the effect of dietary purine restriction, allopurinol, and oxipurinol on urinary excretion of UV-absorbing compounds. It was demonstrated that dietary purine restriction was related to a decreased excretion of purine and pyrimidine metabolites in urine, indicating that these compounds originate, at least partly, exogenously.

Scott and Lee [35] used the sequential columns of microreticular anion-exchange resin (Aminex A-27) and pellicular resin (Pellionex AS) for the high-resolution separation of UV-absorbing constituents in urine. This method produced a more rapid separation (8 hr) than that observed for microreticular resin alone and a higher resolution than that obtained with pellicular resin alone. However, these methods were not suitable for routine clinical analyses of physiological fluids because of a long analysis time.

Recently, Seta et al. have therefore studied a new ion-exchange chromatographic system using a macroreticular anion-exchange resin (Diaion CDR-10) for the rapid separation of UV-absorbing components in body fluids. The elution was performed with a linear acetate gradient from 0 to 6.0 M [36] or a concave ammonium perchlorate gradient from 0 to 0.25 M [37]. In these methods, about 100 UV peaks in urine were detectable in an analysis time of 120 min. Forty-eight peaks, assigned by coinjection of standards and stopped-flow scanning spectrophotometry, consist of nucleic acids, amino acid constituents, organic acids, dietary compounds, and their metabolites. Purine and pyrimidine metabolites in urine observed by these analyses were almost the same as those identified by Mrochek et al. [33].

2. Selective analysis

It has been suggested that minor or modified nucleosides and bases excreted in human urine are possible biomarkers in cancer detection. There have been a number of efforts to develop rapid, sensitive procedures for analysis of these compounds in urine using GC or HPLC.

Mrochek et al. [38] studied the analytical techniques for selective separation of minor nucleic acid components, such as pseudouridine, 1-methylinosine, N^2,N^2-dimethylguanosine, 1-methylguanosine, and 7-methylguanosine in human urine by means of the analytic UV analyzer described in the previous section, as well as identified these compounds by GC, MS, and GC-MS. The quantitative analyses were first performed by using an anion-exchange (Aminex A-27) column with temperature gradient and by rechromatography on a cation-exchange (Aminex A-7) column. However, the analysis times of 6 hr or more were very time-consuming and not applicable to routine analyses.

Uziel et al. [27] developed highly selective prefractionation procedures prior to the HPLC analysis for nucleosides in human urine. They used boronate affinity gel containing an immobilized phenylboronic acid group that specifically binds cis-diols, as in ribonucleosides, which can then be quantitatively recovered by elution with acetic acid. Individual application of several nucleosides to this affinity column gave recoveries of 100 ± 5%.

Davis et al. [39] modified slightly the synthesis of boronate gel and applied it in the selective isolation of nucleosides in urine, serum, and amniotic fluid. The reversed-phase HPLC methods developed by Gehrke et al. [40] were applied to the prefractionated samples equivalent to 25 µl of urine. Within less than 1 hr, these analyses showed excellent resolution and recovery of pseudouridine, 1-methyladenosine, 1-methylinosine, N^2-methylguanosine, adenosine, and N^2,N^2-dimethylguanosine in concentrations ranging from 0.4 to 60 µg/ml, and about 20 other nucleosides, 0.2 to 4 µg/ml in normal urine.

Further, Gehrke et al. [41] investigated the parameters of urinary nucleosides separation, such as the mobile phase flow rate, pH, methanol concentration, column temperature, and injection volume. Thus, they have developed four sets of chromatographic conditions for the reversed-phase HPLC separation, including (a) an improved isocratic method for urinary nucleosides, (b) a rapid method for pseudouridine [42], (c) a two-buffer step-gradient method, and (d) a rapid method for N^2,N^2-dimethylguanosine. Figure 3 shows the HPLC separation of nucleosides in urine analyzed by methods (a) and (c).

The selective HPLC methods for nucleosides in physiological fluids by Gehrke and his group [39,40-42,45] are characterized by the following features: (a) sensitivity at the nanogram level, (b) high chromatographic resolution and sensitivity, (c) direct measurement of nucleosides with accuracy and precision, (d) nondestructive and high selectivity of nucleosides in terms of boronate gel, and (e) comparatively rapid analysis time of about 1 hr. However, the isolation procedure for nucleosides using an open column with the boronate gel is tedious and time-consuming for routine analyses. If this procedure could be improved, their HPLC methods would be more effective for clinical use.

Another selective method for the analysis of adenosine and adenine compounds in urine, plasma, erythrocytes, and lymphocytes has been developed by Kuttesch et al. [43]. Their compounds were measured in picomole amounts using reversed-phase HPLC separation of the fluorescent $1,N^6$-etheno derivatives. In this method, adenine compounds were isolated using an anion-exchange resin column (Dowex Ag1-X2) prior to HPLC analysis.

Evans and Tieckelmann [44] developed a selective method for the analysis of neutral pyrimidine bases and nucleosides in urine. For the initial isolation of these compounds in urine, they used small, disposable, mixed-bed ion-exchange columns of both Ag 50W-X8 and Ag1-X8 prior to HPLC on a silica gel column.

The normal human urine levels of nucleosides, bases, and their metabolites, and the chromatographic conditions for selective analyses discussed in this section, are listed in Table 4.

Figure 3. (a) Isocratic reversed-phase HPLC separation of nucleosides in urine. Sample: 25 µl urine; column: µBondapak C_{18}, 4 × 600 mm; buffer: 0.01 M $NH_4H_2PO_4$, pH 5.1, with 6.0% methanol; flow rate: 1.0 ml/min; temperature: 35°C; detection: 254 nm, 0.01 aufs (upper trace), 280 nm, 0.01 aufs (lower trace). PCNR = 2-pyridone-5-carboxamide-N'-ribofuranoside. (b) Step-gradient reversed-phase HPLC separation of nucleosides in urine. Sample: 25 µl pooled ovarian cancer patient urine; buffer: 0.01 M $NH_4H_2PO_4$, (A) pH 5.3, with 2.5% methanol, (B) pH 5.1, with 8.0% methanol. Others are the same as the above. (From Ref. 41.)

Free Nucleosides in Physiological Fluids / 261

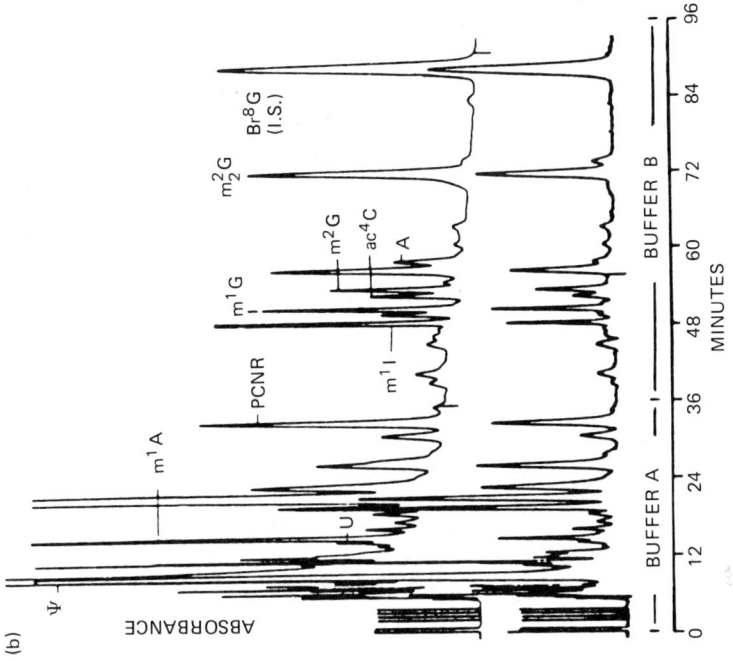

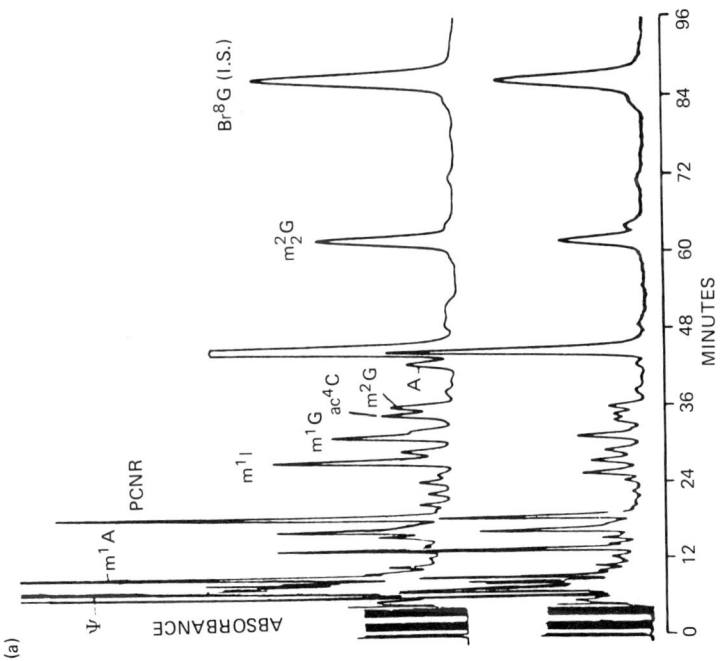

Table 4 Normal Human Urine Levels of Nucleosides, Bases, and Their Metabolites, and Chromatographic Conditions for Selective Analysis

Compounds	Samples[a]	Means ± S.D.	Stationary phase[b]	Mobile phase[b]	Ref.
Pseudouridine	U (17)	0.94 ± 0.16	Aminex A-27 (Bio-Rad)	Ammonium acetate-acetic acid (pH 4.3), 0.015-6.0 M linear gradient	38
1-m-Guanosine	U (5)	0.07 ± 0.018			
N^2-dm-Guanosine	U (17)	0.06 ± 0.014	Aminex A-7 (Bio-Rad)	0.77 M Ammonium acetate-acetic acid (pH 4.65), isocratic	
1-m-Inosine	U (17)	0.06 ± 0.009			
7-m-Guanine	U (5)	0.10 ± 0.23 (mg/kg/24 hr)			
Adenosine	U (4)	5.4 ± 4.5	PXS 10/25 SCX (Whatman) or μBondapak C_{18} (Waters)	0.1 M $NH_4H_2PO_4$ (pH 4.5), isocratic or 0.005 M KH_2PO_4: 25% methanol (v/v), pH 7.5	43
Deoxyadenosine	U (4)	< 0.01			
Adenine	U (4)	3.2 ± 2.3 (nmol/ml)			
Uracil	U (6)	12.2 ± 2.0	LiChrosorb SI-100 (E. Merck)	Methylene chloride: methanol: 1 M ammonium formate (75:22:3, v/v/v) pH 3.0, isocratic	44
Pseudouridine	U (6)	108.5 ± 36.0 (μg/mg creatinine)			

Pseodouridine	U (M, 40)	22.4 ± 2.10	µBondapak C$_{18}$ (Waters)	0.01 M NH$_4$H$_2$PO$_4$ (pH 5.10) with 6% methanol (v/v), isocratic	40
	U (F, 40)	26.7 ± 4.50			45
1-m-Adenosine	U (M, 40)	1.77 ± 0.29			
	U (F, 40)	1.76 ± 0.48			
1-m-Inosine	U (M, 40)	1.15 ± 0.27			
	U (F, 40)	1.18 ± 0.39			
N^2-m-Guanosine	U (M, 40)	0.35 ± 0.07			
	U (F, 40)	0.41 ± 0.12			
N^2-dm-Guanosine	U (M, 40)	1.20 ± 0.15			
	U (F, 40)	1.44 ± 0.38			
		(nmol/µmol creatinine)			

[a] Abbreviations used: U, urine. See Table 1 for others.
[b] Chromatographic conditions for analytical HPLC only are shown.

REFERENCES

1. G. B. Chheda, in *Handbook of Biochemistry*, 2nd ed., (H. A. Sober, ed.), Chemical Rubber Co., Cleveland, Ohio, 1970, p. G-106.
2. T. Akaike, Y. Sakurai, I. Gonoi, N. Takai, and M. Seno, Proc. 38th Annual Meeting of the Japanese Cancer Association, Tokyo, 1979, p. 1040.
3. H. Miyagi, J. Miura, Y. Takata, S. Kamitake, S. Ganno, and Y. Yamagata, *Advances in Chromatography*, Proc. 18th International Symposium, Tokyo, 1982, p. 431.
4. M. Uziel, C. Koh, and W. Cohn, Anal. Biochem., 25, 77 (1968).
5. C. D. Scott, J. E. Attrill, and N. G. Anderson, Proc. Soc. Exp. Biol. Med., 125, 181 (1967).
6. R. L. Jolley and C. D. Scott, Clin. Chem., 16, 687 (1970).
7. P. R. Brown, A. M. Krstulovic, and R. A. Hartwick, J. Clin. Chem. Clin. Biochem., 14, 282 (1976).
8. R. A. Hartwick and P. R. Brown, J. Chromatogr., 126, 679 (1976).
9. A. M. Krstulovic, P. R. Brown, and D. M. Rosie, Anal. Chem., 49, 2237 (1977).
10. A. M. Krstulovic, P. R. Brown, D. M. Rosie, and P. B. Champlin, Clin. Chem., 23, 1984 (1977).
11. A. M. Krstulovic, R. A. Hartwick, P. R. Brown, and K. Lohse, J. Chromatogr., 158, 365 (1978).
12. R. A. Hartwick, S. P. Assenza, and P. R. Brown, J. Chromatogr., 186, 647 (1979).
13. R. A. Hartwick, A. M. Krstulovic, and P. R. Brown, J. Chromatogr., 186, 659 (1979).
14. R. A. Hartwick, D. V. Haverbeke, M. McKeag, and P. R. Brown, J. Liq. Chromatogr., 2, 725 (1979).
15. S. P. Assenza and P. R. Brown, J. Chromatogr., 181, 169 (1980).
16. S. P. Assenza and P. R. Brown, J. Liq. Chromatogr., 3, 41 (1980).
17. M. Zakaria and P. R. Brown, Anal. Biochem., 120, 25 (1982).
18. A. M. Krstulovic, R. A. Hartwick, and P. R. Brown, Clin. Chim. Acta, 97, 159 (1979).
19. M. Zakaria, P. R. Brown, M. P. Farnes, and B. E. Barker, Clin. Chim. Acta, 126, 69 (1982).
20. W. E. Wung and S. B. Howell, Clin. Chem., 26, 1704 (1980).
21. A. Mcburney and T. Gibson, Clin. Chim. Acta, 102, 19 (1980).
22. R. J. Simmonds and R. A. Harkness, J. Chromatogr., 226, 369 (1981).
23. R. A. Hartwick and P. R. Brown, J. Chromatogr., 143, 383 (1977).
24. C. A. Koller, P. L. Stetson, L. D. Nichamin, and B. S. Mitchell, Biochem. Med., 24, 179 (1980).

25. E. J. Kiser, G. F. Johnson, and D. L. Witte, Clin. Chem., 24, 536 (1978).
26. L. A. Pachla and P. T. Kissinger, Clin. Chem., 25, 1847 (1979).
27. M. Uziel, L. H. Smith, and S. A. Taylor, Clin. Chem., 22, 1451 (1976).
28. E. H. Pfadenhauer and S.-D. Tong, J. Chromatogr., 162, 585 (1979).
29. J. M. Karle, L. W. Anderson, D. D. Dietrick, and R. L. Cysyk, Anal. Biochem., 109, 41 (1980).
30. R. W. Zumwalt, K. C. Kuo, D. Phan, and C. W. Gehrke, in preparation (1982).
31. G. A. Taylor, P. J. Dady, and K. R. Harrap, J. Chromatogr., 183, 421 (1980).
32. K. Nakano, S. P. Assenza, and P. R. Brown, J. Chromatogr., 233, 51 (1982).
33. J. E. Mrochek, W. C. Butts, W. T. Rainey, Jr., and C. A. Burtis, Clin. Chem., 17, 72 (1971).
34. W. N. Kelley and J. B. Wyngaarden, Clin. Chem., 16, 707 (1970).
35. C. D. Scott and N. E. Lee, J. Chromatogr., 83, 383 (1973).
36. K. Seta, M. Washitake, T. Anmo, N. Takai, and T. Okuyama, J. Chromatogr., 181, 311 (1980).
37. K. Seta, M. Washitake, I. Tanaka, N. Takai, and T. Okuyama, J. Chromatogr., 221, 215 (1980).
38. J. E. Mrochek, S. R. Dinsmore, and T. P. Waalkes, J. Nat. Cancer Inst., 53, 1553 (1974).
39. G. E. Davis, R. D. Suits, K. C. Kuo, C. W. Gehrke, T. P. Waalkes, and E. Borek, Clin. Chem., 23, 1427 (1977).
40. C. W. Gehrke, K. C. Kuo, G. E. Davis, R. D. Suits, T. P. Waalkes, and E. Borek, J. Chromatogr., 150, 455 (1978).
41. C. W. Gehrke, K. C. Kuo, and R. W. Zumwalt, J. Chromatogr., 188, 129 (1980).
42. K. C. Kuo, C. W. Gehrke, R. A. McCune, T. P. Waalkes, and E. Borek, J. Chromatogr., 145, 383 (1978).
43. J. F. Kuttesch, F. C. Schmalstieg, and J. A. Nelson, J. Liq. Chromatogr., 1, 97 (1978).
44. J. E. Evans, H. Tieckelmann, E. W. Naylor, and R. Guthrie, J. Chromatogr., 163, 29 (1979).
45. J. Speer, C. W. Gehrke, K. C. Kuo, T. P. Waalkes, and E. Borek, Cancer, 44, 2120 (1979).

14
Cyclic Nucleotides

ROBERT F. BURGOYNE Waters Associates, Milford, Massachusetts

I. Introduction 267
II. Background 269
III. Chromatographic Developments 271
   References 282

I. INTRODUCTION

A special class of compounds in nucleic acid chemistry is cyclic nucleotides. The structures of these compounds are analagous to nucleotides in that they contain a purine or pyrimidine base bonded to a sugar moiety, either β-D-ribose or β-D-2-deoxyribose, to which is bonded a phosphate group. In a nucleotide compound, the phosphate group is linked to the sugar via a phosphate ester linkage. The linkage is usually formed at the C-5 position but can also form at the C-3 position of the pentose. Cyclic nucleotides are synthesized in vivo from the corresponding 5'-triphosphate nucleotide by the action of a specific membrane-bound cyclase enzyme. Figure 1 illustrates the synthesis of adenosine 3',5'-monophosphate (cyclic AMP) from adenosine 5'-triphosphate (ATP) through the action of adenyl cyclase. A similar reaction could be developed for the guanosine, cytidine, uridine, and inosine cyclic nucleotides.

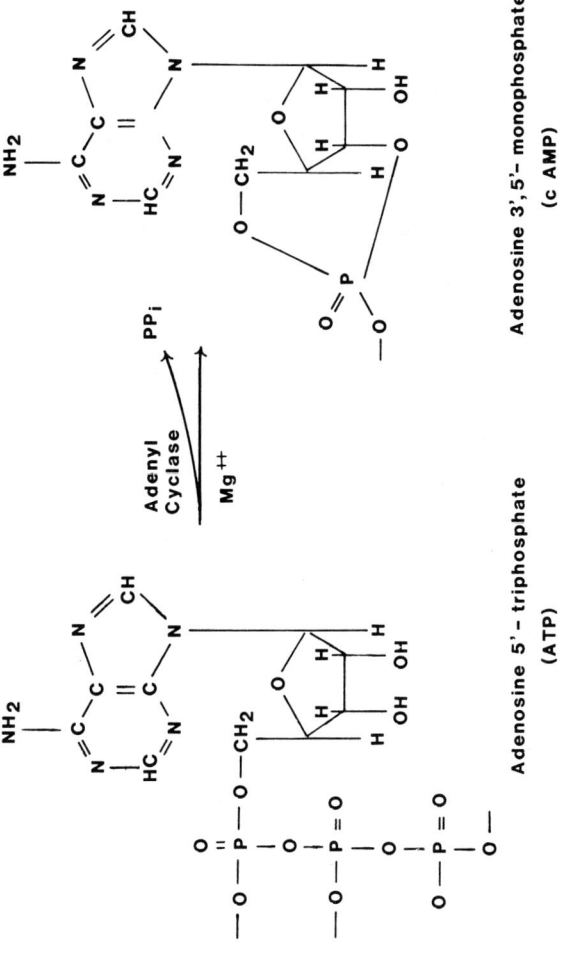

Figure 1. Adenyl-cyclase-activated synthesis of 3'-5'-cyclic AMP.

Although the isolation of all five cyclic nucleotides from biological samples has been reported [1-6], the vast bulk of biochemical research in this field has involved primarily the adenosine and guanosine forms. The first to be discovered was cyclic AMP, which was found to play a role in the mediation of hormone activity. Earl Sutherland et al. began studies in the 1950s to elucidate the mechanism of action of hormones on glycogen synthesis and breakdown [1]. His landmark studies indicated that the addition of epinephrine or glucagon resulted in the activation of the enzyme phosphorylase and that the activation could occur in cell-free homogenates. He further found that the response to the hormones decreased upon centrifugation, indicating that an essential part of the hormone response system was contained in the particulate fraction of the homogenate. Later experiments proved the existence of a two-step response system: the interaction of the hormone with a plasma membrane receptor followed by increased adenyl cyclase activity within the cell to alter the cyclic AMP concentration. The discovery of this *second messenger* mechanism has had immense impact on biochemical research and has placed the measurements of cyclic nucleotides at the forefront of analytical methods [7-8].

II. BACKGROUND

The second messenger model fueled a major expansion of scientific inquiry concerned with studying hormone action at the molecular level. None of the cyclic nucleotides has been analyzed in greater depth than cyclic AMP. Along with mediating hormone action within the cell, cyclic AMP has been implicated in neurotransmitter action. The concentration of adenyl cyclase, which catalyzes cyclic AMP synthesis, is highest in brain tissue. The later discovery of cyclic GMP and guanylate cyclase in the central nervous system has illustrated the fundamental significance of the cyclic nucleotides in neural function. Research has focused on numerous studies involving cyclic nucleotide mediation, including enzyme activity associated with dopamine [9], biogenic amines [10], calcium binding [11], amino acids [12], prostaglandins [12], drugs [12], and environmental influences [13,14].

The rapid increase in research in the field has resulted in the development of many analytical techniques for monitoring cyclic nucleotide formation and enzyme activity. The availability of rapid, accurate, and sensitive assays for cyclic nucleotides in biological samples is critical for the successful determination of physiological mechanisms. The primary difficulties to be surmounted by the methods are the extremely low levels of cyclic nucleotides contained in the biological samples and frequently sample size limitations. Physiological concentrations of cyclic nucleotides vary depending on the sample matrix. Extensive work involving cyclic AMP and cyclic GMP indicate concentrations ranging from picomolar in tissue to nanomolar in cerebrospinal fluids and plasma to micromolar in urine. In general, the concentrations of cyclic GMP in these matrices is one or more orders of magnitude less than cyclic AMP levels [15-19].

Some of the cyclic nucleotide analysis methods used employ specific interactions of the cyclic nucleotide species to enzymes such as luciferase [20] or binding to specific proteins [21-23]. Other methods available include thin-layer chromatography [24,25], gas-liquid chromatography [26], fluorescence [27,28], and spectrophotometric techniques [29,30]. There are considerable variations among these techniques with respect to their limits of detection, ease of use, time consumption, and extent of sample purification required prior to application of the method. The use of high performance liquid chromatography (HPLC) in the field of nucleotide research is a relatively recent development [34]. The technique has several advantages, including highly specific, sensitive detection devices, high speed of analysis, relative ease of use, and availability of many different microparticulate packing materials.

For a high performance liquid chromatographic method to be valid in the biochemical analysis of cyclic nucleotides, it must meet several requirements. The speed of analysis will be an important factor in the selection of an analytical method. Competitive binding and enzymatic assays are exceptionally sensitive analytical techniques. However, the methods are time-consuming and are frequently not entirely specific for the compounds of interest. For example, in radioimmunoassay techniques, nonspecific binding can reduce the sensitivity of the method when biological sample matrices are involved. Also, one must often handle hazardous materials such as radioactive tracers or compounds which degrade easily. Thin-layer chromatographic techniques yield sufficiently selective separations of cyclic nucleotide compounds but can require relatively large samples and lengthy elution times. Detection of the compounds after separation is also a problem with thin-layer procedures.

The use of high performance liquid chromatographic techniques for analysis of cyclic nucleotide compounds has advantages over other analytical techniques. A minimal amount of sample preparation is necessary prior to analysis, since potential interferences can be separated chromatographically from the compounds of interest. This reduction in preparation time allows an investigator to process samples more rapidly than when other analysis techniques are used. The separation powers of HPLC also make possible the simultaneous determinations of multiple cyclic nucleotide compounds during a single chromatographic run. This represents a distinct advantage over immunoassay techniques, which may not be totally selective in their determinations. A wide range of separation techniques is available to choose from, allowing the investigator to tailor the HPLC method for the specific compounds of interest. The mode of detection may also be selected to optimize a tailored separation system. Through the use of microvolume flow-through detectors, extremely sensitive detection and quantitation of the eluted compounds can be obtained. Detection of cyclic nucleotides is typically by ultraviolet (UV) absorbance measurements at discrete wavelengths. Detection efficiencies can be maximized by the selection of a monitoring wavelength which approximates the λ_{max}, the wavelength of maximum molar absorptivity, of a specific compound. For the cyclic nucleotides, the most efficient range of UV absorbance detection can be obtained within the 250-275 nm range [39], with 254 nm being the most commonly used wavelength for complete cyclic nucleotide profiles. Finally, HPLC methodologies can be automated to allow for unattended sample analyses. This feature of HPLC instrumentation can dramatically increase the sample throughput of the laboratory.

III. CHROMATOGRAPHIC DEVELOPMENTS

The development of high performance liquid chromatographic methods for the analysis of nucleotide compounds (Table 1) resulted from a large amount of preliminary research into the many operation parameters of the separation systems. Cohn [31] studied the mechanisms of ion-exchange chromatography, which are as applicable to high performance columns as they are to conventional ion-exchange columns. A continuous column effluent monitoring system was devised by Anderson [32], and the operating parameters of high performance liquid chromatographic systems have been studied in great depth by Kirkland and Felton [33] and by Horvath et al. [34-35]. Brown [36] combined this information and used the nucleotide analyses in biomedical studies. A 3-m × 1-mm-i.d. capillary column packed with pellicular anion-exchange material (Varian) was utilized for the separation. Flow rates up to 12 ml/hr were used, yielding sample analysis times of the order of 60 min. An ionic-strength gradient elution technique was employed for full nucleotide profiles.

One of the aims constantly addressed by high performance liquid chromatography is the acceleration of analyses to reduce sample analysis times. To this end, Brooker [37] developed a method specifically for the rapid analysis of cyclic AMP only. The same pellicular anion-exchange column was run at 80°C with a chloride-based isocratic eluent system to accelerate mass transfer of the solute within the column. The result of these changes was a specific method for cyclic AMP with a sample analysis time of 10 min. In addition, Brooker was able to reduce significantly the minimum detectable quantity of cyclic AMP to 30 pmol by using a sensitive, 8-μl, 1 cm pathlength UV absorbance detector set at 254 nm (Fig. 2). Brooker used his method in monitoring the enzymatic activity levels of adenyl cyclase. This HPLC method replaced more tedious procedures involving enzyme activation of the glycogen phosphorylase system or incubation with radioactive ATP.

Table 1 Chromatographic Conditions for Selected Cyclic Nucleotide Analyses

Compounds resolved[a]	Stationary phase	Mobile phase	Temp (°C)	Flow rate (ml/min)	Analysis time (min)	Ref. no.
3',5'-Cyclic ribonucleotides of Ade and Gua and 5'-nucleotides of Cyt, Ura, Thy, Ade, and Gua	Pellicular anion exchange (3 m × 1 mm, Varian)	A: 15 mM KH_2PO_4 B: 0.25 M KH_2PO_4, 2.2 M KCl Gradient: add B at 0.1 ml/min into 50-ml reservoir of A	Ambient	0.2	70	36
3',5'-Cyclic AMP	Pellicular anion exchange (3 m × 1 mm)	6.3 mM HCl	80	0.2	11	37
3',5'-Cyclic ribonucleotides of Ade and Ino and 5'-nucleotides of Ade, Ino, and Gua	µBondapak C_{18} (30 cm × 3.9 mm, 10 µm, Waters)	A: 50 mM $NH_4H_2PO_4$, pH 6.0 B: 95 × {50 mM $NH_4H_2PO_4$, pH 6.0} 5 × {methanol} Gradient: stepwise 0% to 100% B at 7 min	Ambient	2.0	25	38
3',5'-Cyclic ribonucleotides and 5'-triphosphate nucleotides of Cyt, Gua, and Ade	µBondapak C_{18}	90 × {0.2 M ammonium phosphate, pH 5.1} 10 × {methanol}	Ambient	1.0	18	39
3',5'-Cyclic ribonucleotides of Ade, Cyt, Ura, Ino, and Gua	Aminex A-27 (5.2 × 0.62 cm, Bio-Rad)	25 mM sodium citrate, 1 mM potassium perchlorate, 0.3 mM NaN_3, pH 7.5	70	0.7	40	40
3',5'-Cyclic ribonucleotides of Cyt, Ade, Gua, Ura, and Ino and Adenosine 5'-monophosphate	µBondapak C_{18}	A: 20 mM KH_2PO_4, pH 3.7 B: methanol/H_2O (60:40) Gradient: Linear, 0% to 25% B in 30 min	Ambient	1.5	25	41

Analytes	Column	Mobile phase	Temperature (°C)	Flow rate (mL/min)	Run time (min)	Ref.
3',5'-Cyclic ribonucleotides of Cyt, Ade, Gua, Ura, and Ino	μBondapak C_{18} Radial-Pak cartridge (8 mm × 10 cm, 10 μm, Waters)	A: 20 mM KH_2PO_4, pH 3.7 B: methanol/H_2O (60:40) Gradient: linear, 0% to 25% B in 17.3 min	Ambient	4.0	12	—
3',5'-Cyclic ribonucleotides of Ade and Gua and 5'-nucleotides of Ade, Gua, Ino, and Cyt	Radial-Pak A (C_{18}) cartridge (8 mm × 10 cm, 10 μm, Waters)	A: 65 mM KH_2PO_4, 0.8 mM tetra-butylammonium phosphate, pH 3.2 B: 95 × {65 mM KH_2PO_4, 0.9 mM TBAP, pH 3.2} 5 × {acetonitrile} Gradient: concave, 0% to 100% B in 10 min	Ambient	2.5	30	43
3',5'-Cyclic AMP and 5'-nucleotides of Ade, Gua, Cyt, and Ino	ODS-Hypersil (Shandon Southern Products)	88 × {75 mM phosphate, 1.25 mM 11-aminoundecanoic acid, pH 5.65} 12 × {methanol}	25	Not reported	15	44
3',5'-Cyclic ribonucleotides of Ade, Gua, Ura, Hyp, and Cyt	Zorbax-ODS (250 × 4.6 mm)	Isocratic 4.0 mM (NH_4) HPO_4 and 4.0 mM (NH_4) H_2PO_4 pH 3.0	35	2.5	10	45
3',5'-Cyclic ribonucleotides of Ade and Gua and their associated nucleotides, nucleosides and bases	Zorbax-ODS (250 × 4.6 mm)	Isocratic 4.0 mM (NH_4) HPO_4 and 4.0 mM (NH_4) H_2PO_4 pH 3.0	35	1.5	19	45

[a] Abbreviations: Ade = adenine, Cyt = cytosine, Gua = guanine, Ino = Inosine, Thy = thymine, Ura = uracil.

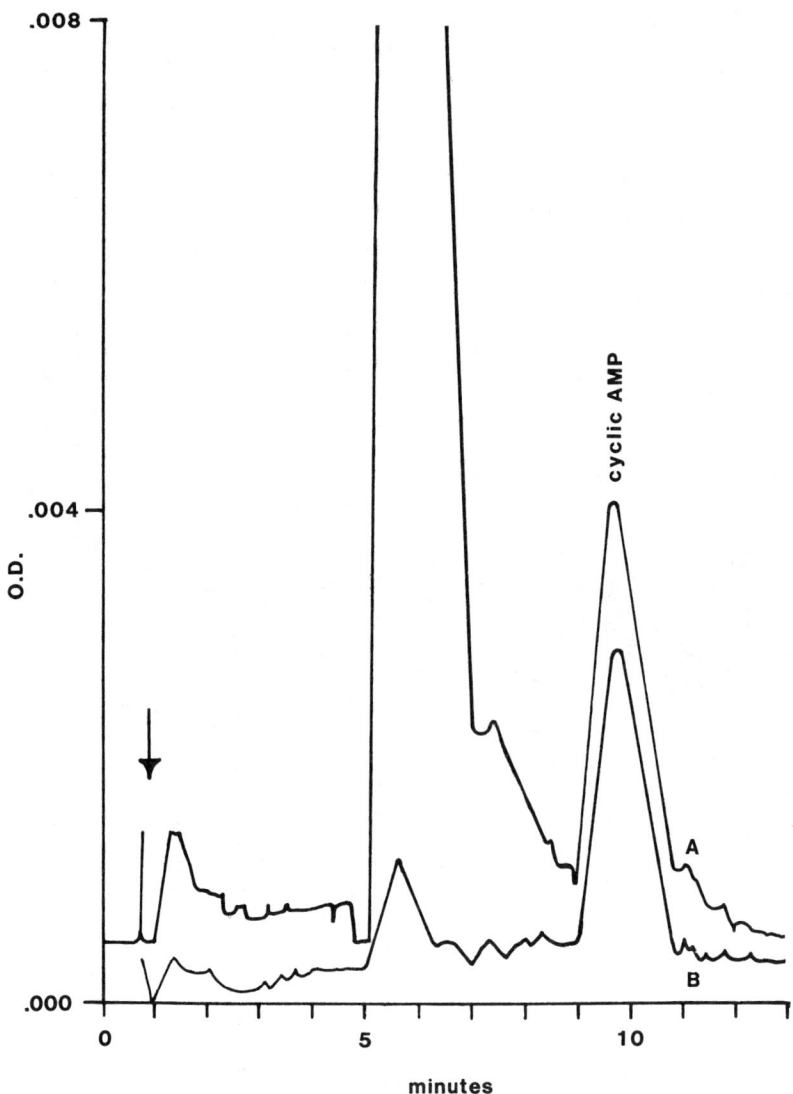

Figure 2. A. Chromatogram of cyclic AMP produced enzymatically. B. Chromatogram of 30 pmol of cyclic AMP. (Reprinted with permission from Ref. 37, p. 1109, Copyright 1970, American Chemical Society.)

The enzyme is incubated with ATP and the amount of cyclic AMP formed is determined by HPLC.

There are several problems associated with gradient elution techniques on ion-exchange columns. The detection sensitivity of the system can potentially be reduced by impurities in the buffers, which may result in extraneous peaks during the gradient or in a sharply increasing baseline. This increase is seen particularly when detectors are run at their most sensitive settings. Trace impurities in the buffers may also poison the ion-exchange packings, thereby reducing their capacities and selectivities. Finally, it is imperative that the high performance liquid chromatograph be capable of generating gradients in an accurate, reproducible fashion.

To avoid some of these problems, Anderson and Murphy [38] have developed a step-gradient system using reversed-phase (RP) columns which they applied to analysis of classes of nucleic acid compounds. The pellicular ion-exchange packings are replaced by totally porous, chemically bonded, microparticulate (5-10 μm) packings which offer significantly greater efficiencies and selectivities. Much higher rates of mass transfer are possible, and the increased surface areas generate higher sample loading capacities. Because solute-column interactions are weaker in reversed-phase systems, the operation of the system is simplified and the reequilibration of the column during solvent changeover is much more rapid. As Anderson demonstrates in his separations of nucleic acid classes, it is possible to analyze a larger number of substances of a wider range of polarity on a reversed-phase system. The chromatogram in Fig. 3 illustrates the RP-HPLC analysis of all the adenine species. The linear nucleotides are ionized in solution, resulting in relatively low affinities for the reversed-phase packings. However, there are sufficient hydrophobic interactions so that separation of the mono-, di-, and triphosphate nucleotides can be obtained. The remaining compounds, adenine, adenosine, and cyclic AMP, are retained strongly on the hydrophobic support. By increasing the solvent strength via a step-gradient procedure, these compounds can also be rapidly eluted. Anderson and Murphy have applied this liquid chromatograhic method to the identification and quantitation of ATP metabolites from the perfused cat spleen. Minimal sample preparation was required prior to sample injection onto the chromatograph. More recently, Martinez-Valdez [39] et al. have used similar chromatographic procedures in isolating, identifying, and quantitating ribonucleotide, deoxyribonucleotide, and cyclic nucleotide components of the nucleotide pool from HeLa cells.

Up to this point, the separations discussed have involved primarily cyclic AMP and cyclic GMP methodologies. However, it has been established that there are three additional naturally occurring cyclic

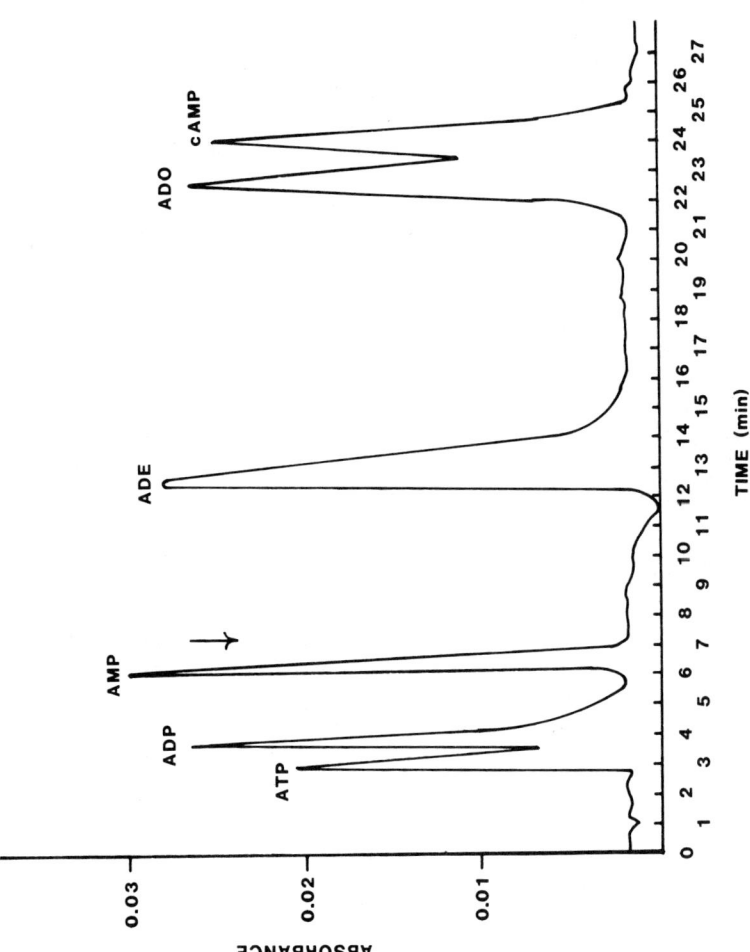

Figure 3. Separation of ATP, ADP, AMP (ca. 200 ng each in 10 μl total), and adenine, adenosine, and cyclic AMP (ca. 500 ng each in 10 μl total). A step gradient from 0.05 M $NH_4H_2PO_4$ buffer at pH 6.0 to phosphate buffer with 5% methanol was initiated at the arrow after AMP eluted. (From Ref. 38, p. 254.)

nucleotide compounds: cyclic CMP, cyclic UMP, and cyclic IMP. For some studies the chromatographic methods must be adapted for analyses of all five compounds simultaneously from biological samples. Khym [40] was the first investigator to report chromatographic separations of all the cyclic nucleotides. The chromatogram in Fig. 4 illustrates the profile generated on an Aminex A-27 anion-exchange column. This packing material differs from the pellicular packing described previously. The Aminex material is an 8% cross-linked styrene-divinylbenzene copolymer to which is bonded a quaternary ammonium functionality. The packing is less easily poisoned than the pellicular packings, and provides greater loading capacities due to its increased surface area. This method is hindered, however, by problems associated with sample complexity. Khym reports that it is imperative that structurally similar compounds be removed prior to chromatography, specifically linear nucleotides and nucleotide sugars.

An alternative procedure for complete cyclic nucleotide determinations using reversed-phase packings has been reported by Krstulovic et al. [41]. An eluent system similar to Anderson's was used, with the exception that a linear gradient replaced the previous step-gradient procedure (Fig. 5). Krstulovic also reported an in-depth treatment of chromatographic peak identifications of the cyclic nucleotide compounds. Positive identification was obtained through the use of stopped-flow ultraviolet scanning, absorbance ratio, and enzymatic peak shift techniques. Minimum detectable quantities of cyclic nucleotides by this method were reported to be 50-100 pmol. Application to rat brain extracts indicate that no sample concentration step is required prior to chromatography.

Krstulovic's procedure requires sample analysis times of approximately 30 min when column reequilibration time is included. By utilizing the advantages associated with radial compression technology [42], analysis times can be lowered significantly. Engineering innovations have resulted in reduced system backpressures, which allow the investigator to run chromatographic assays at elevated flow rates. Figure 6 illustrates the chromatography obtained using a Radial-Pak cartridge containing μBondapak C_{18} packing eluted under the same solvent conditions as in Fig. 5. Sample analysis times have been reduced by 50% by this technology, which in turn could double the sample throughput of a laboratory.

The primary problem associated with reversed-phase chromatography of nucleotides is the relative lack of affinity of the solutes, particularly the linear nucleotides, for the hydrophobic packing materials. Several investigators have reported the use of ion-pairing agents for nucleotide analysis on reversed-phase systems. Darwish and Prichard [43] have used the ion-pairing reagent tetrabutylammonium phosphate

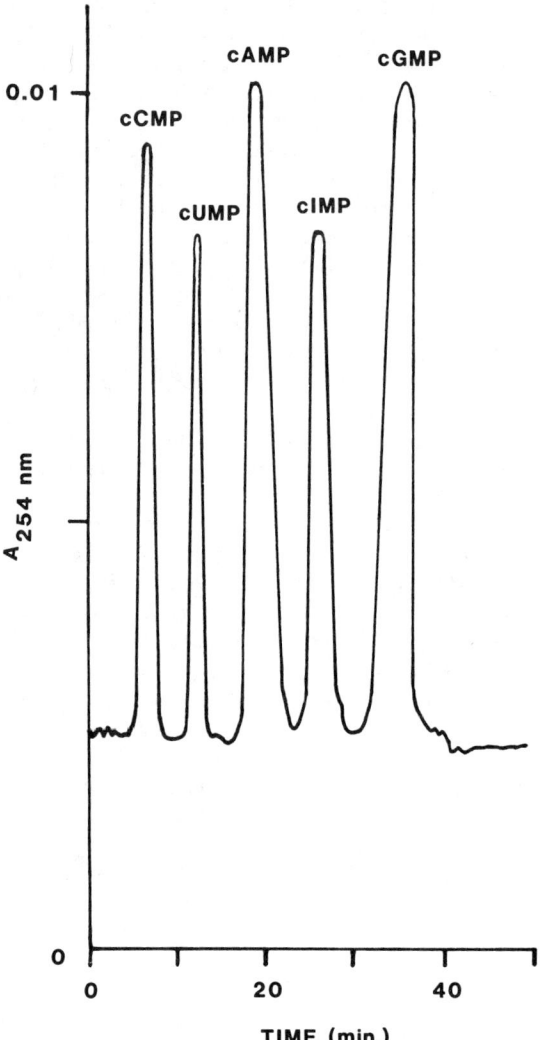

Figure 4. Separation of the 3',5'-cyclic ribonucleotides was done on a 5.2 × 0.62 cm. column of Aminex A-27 converted to the citrate-perchlorate form. The column was equilibrated for ca. 2 hr with eluent composed of 25 mM sodium citrate + 1 mM potassium perchlorate (pH 7.5), and also containing 0.3 mM NaN_3 to prevent bacterial growth; flow rate was 0.7 ml/min; column temperature, 70°C. (From Ref. 40, p. 422.)

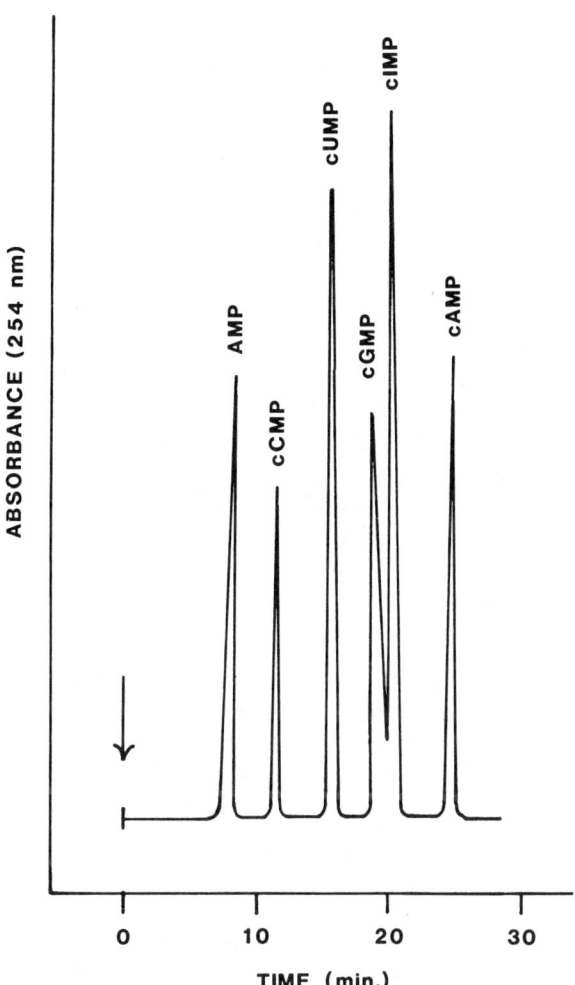

Figure 5. Separation of reference compounds detected at 254 nm. Chromatographic conditions: column: µBondapak C_{18}; low-concentration eluent: 20 mM KH_2PO_4, pH 3.7; high-concentration eluent: methanol/water (3/2 by vol); gradient: linear from 0 to 25% of the high-concentration eluent in 30 min; flow rate: 1.5 ml/min. (From Ref. 41, p. 236.)

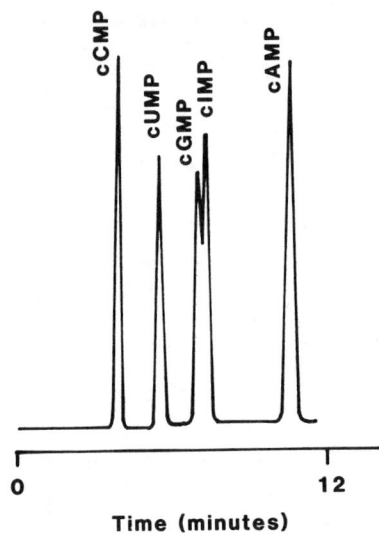

Figure 6. Separation of reference compounds on a Z-Module Radial Compression System and a Radial-Pak μBondapak C_{18} cartridge (8 mm × 10 cm, 10 μm). Low-concentration and high-concentration eluents are the same as in Fig. 5. Gradient: 0 to 25% of the high-concentration eluent in 17.3 min; flow rate: 4 ml/min.

in conjunction with a reversed-phase gradient chromatographic method for the analysis of linear and cyclic nucleotides and coenzymes (Fig. 7). A more novel method of ion pairing has been demonstrated by Knox and Jurand [44], who utilized a zwitterion-pairing agent, 11-amino undecanoic acid. In general, the use of ion-pairing chemistries increases the potential for successful reversed-phase chromatography of linear nucleotides. The formation of hydrophobic ion pairs allows for greater utilization of the immense selectivities of reversed-phase systems.

High performance liquid chromatographic methods are rapidly replacing alternative procedures for the analysis of cyclic nucleotides. HPLC offers the opportunity to select from multiple elution modes, to choose a highly sensitive and selective detection device, and to decrease sample analysis times significantly by the use of high-speed elution techniques and automated instrumentation. HPLC methods will become increasingly important in research areas associated with cyclic nucleotides, such as enzyme activity and biochemical pathway studies.

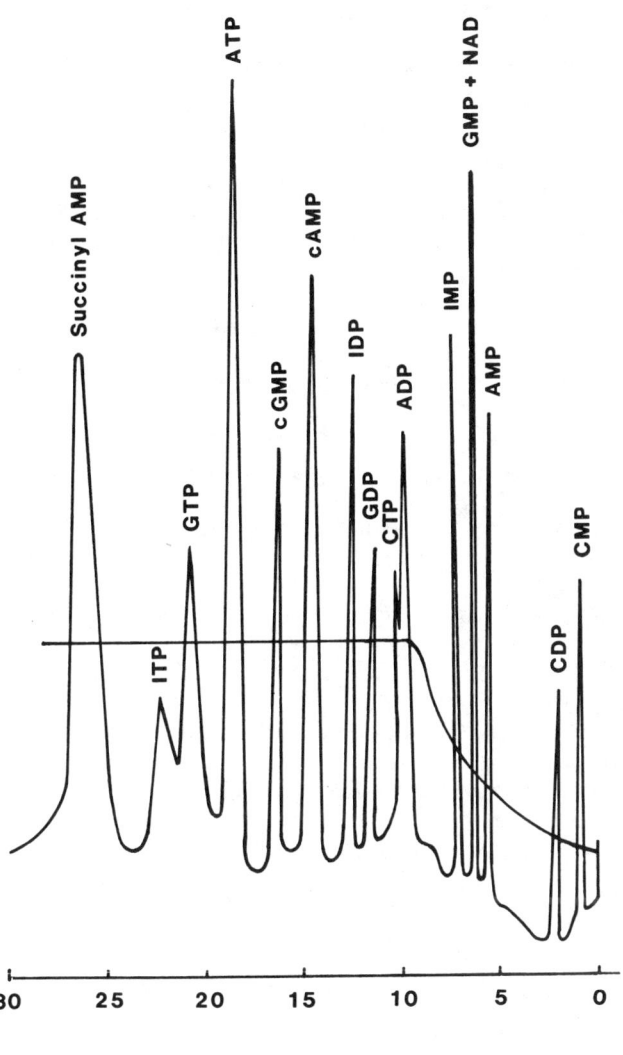

Figure 7. Gradient elution of standards including cyclic AMP, cyclic GMP, and succinyl AMP in the presence of triphosphate nucleotides on Radial-Pak A cartridge (8 mm × 10 cm, 10 μm, C_{18}). Eluent A: 65 mM KH_2PO_4 + 0.9 mM tetrabutylammonium phosphate (TBAP), pH 3.2; eluent B: 95 parts buffer A + 5 parts acetonitrile; gradient: 0 to 100% B, concave, in 10 min; flow rate: 2.5 ml/min. (From Ref. 43, p. 1521.)

REFERENCES

1. G. A. Robison, R. W. Butcher, and E. W. Sutherland, *Cyclic AMP*, Academic Press, New York, 1971.
2. J. G. Hardman and E. W. Sutherland, J. Biol. Chem., *244*, 6363 (1969).
3. A. Bloch, Biochem. Biophys. Res. Commun., *58*, 652 (1974).
4. A. Bloch, Biochem. Biophys. Res. Commun., *64*, 210 (1975).
5. J. Ishiyama, Biochem. Biophys. Res. Commun., *65*, 286 (1975).
6. S. P. Dutta, A. Mittelman, and G. B. Chhed'a, Anal. Biochem., *75*, 649 (1976).
7. R. Rodnight, *International Review of Biochemistry*, vol. 26 (K. F. Tipton, Ed.), University Park Press, Baltimore, 1979.
8. T. W. Rall, Pharmacol. Rev., *24*, 399 (1972).
9. J. W. Kababian, Y. C. Clement-Cormier, G. L. Petzold, and P. Greengard, Adv. Neurol., *9*, 1 (1975).
10. R. F. Oeda and M. Katz, Proc. Nat. Acad. Sci. U.S., *73*, 3398 (1976).
11. M. J. Berridge, Adv. Cyclic Nucleotide Res., *6*, 1 (1975).
12. H. Shimizu, H. Ichishita, and H. Odagiri, J. Biol. Chem., *249*, 5955 (1974).
13. L. Volicer and B. P. Hurter, J. Pharmacol. Exp. Ther., *200*, 298 (1977).
14. G. Biggio, B. B. Brodie, E. Costa, and A. Guidotti, Proc. Nat. Acad. Sci. U.S., *74*, 3592 (1977).
15. G. Brooker, J. F. Harper, W. L. Terasaki and R. D. Moylan, Adv. Cyclic Nucleotide Res., *10*, 1 (1979).
16. A. E. Broadus, N. I. Kaminsky, J. G. Hardman, E. W. Sutherland, and G. W. Liddle, J. Clin. Invest., *49*, 2222 (1970).
17. F. Murad, T. W. Rall, and M. Vaughan, Biochem. Biophys. Acta, *192*, 430 (1969).
18. R. W. Butcher and E. W. Sutherland, J. Biol. Chem., *237*, 1244 (1962).
19. F. Murad and C. Y. Pak, N. Engl. J. Med., *286*, 1382 (1972).
20. M. S. Ebadi, B. Weiss, and E. Costa, J. Neurochem., *18*, 183 (1971).
21. M. Honma, T. Satoh, J. Takezawa, and M. Ui, Biochem. Med., *18*, 257 (1977).
22. I. F. Harper and G. Brooker, J. Cyclic Nucleotide Res., *1*, 207 (1975).
23. G. T. Siggins, E. F. Battenberg, B. J. Hoffer, F. E. Bloom, and A. L. Steiner, Science, *179*, 585 (1973).
24. G. P. Beardsley and H. T. Abelson, Anal. Biochem., *105*, 311 (1980).
25. N. C. Bols, B. W. M. Bowen, K. G. Khor, and S. A. Boliska, Anal. Biochem., *106*, 230 (1980).
26. G. Krishna, Fed. Proc., *27*, 649 (1968).

27. W. D. Lust, G. K. Feussner, E. K. Barbehenn, and J. V. Passonneau, Anal. Biochem., *110*, 258 (1981).
28. J. G. Carter, S. J. Berger, and O. H. Lowry, Anal. Biochem., *100*, 244 (1979).
29. C. E. Dreiling and C. Mattson, Anal. Biochem., *102*, 304 (1980).
30. T. Frielle, A. A. Crimaldi, and C. J. Coffee, Anal. Biochem., *97*, 239 (1979).
31. W. E. Cohn, J. Am. Chem. Soc., *72*, 1471 (1950).
32. N. G. Anderson, Anal. Biochem., *4*, 269 (1962).
33. J. J. Kirkland and H. Fleton, J. Chromatogr. Sci., *7*, 7 (1969).
34. C. G. Horvath, B. A. Preiss, and S. R. Lipsky, Anal. Chem., *39*, 1422 (1967).
35. C. Horvath and S. Lipsky, Anal. Chem., *41*, 1227 (1969).
36. P. R. Brown, J. Chromatogr., *52*, 257 (1970).
37. G. Brooker, Anal. Chem., *42*, 1108 (1970).
38. F. S. Anderson and R. C. Murphy, J. Chromatogr., *121*, 251 (1976).
39. H. Martinez-Valdez, R. M. Kothari, H. V. Hershey, and M. W. Taylor, J. Chromatogr., *247*, 307 (1982).
40. J. X. Khym, J. Chromatogr., *151*, 421 (1978).
41. A. M. Krstulovic, R. A. Hartwick, and P. R. Brown, Clin. Chem., *25*, 235 (1979).
42. G. J. Fallick and C. W. Rausch, Am. Lab., *11*, 87 (1979).
43. A. A. Darwish and R. K. Prichard, J. Liq. Chromatogr., *4*, 1511 (1981).
44. J. H. Knox and J. Jurand, J. Chromatogr., *203*, 85 (1981).
45. S. P. Assenza and P. R. Brown, J. Chromatogr. Biomed. App. *272*, 373 (1983).

15
Enzyme Assays by HPLC

ANNE P. HALFPENNY[*] University of Rhode Island, Kingston, Rhode Island

I. Introduction 286
   A. Background 286
   B. Purines 287
   C. Pyrimidines 288
II. Kinetic Considerations 288
   A. Background 288
   B. K_m and V_{max} 289
   C. Measurement of K_m 290
   D. pH 291
   E. Temperature 293
   F. Units for enzyme activity 293
III. Analytical Methods and Procedures 294
   A. Sample preparation 294
   B. Chromatography 294
   C. Quantification: activity calculations 299
IV. Conclusions 300
   References 301

---

[*]*Current affiliation*: Memorial Sloan-Kettering Cancer Center, New York, New York.

I. INTRODUCTION

A. Background

Enzymes are protein molecules which function as biological catalysts. They are diverse in structure and size, ranging from molecular weights of approximately 10,000 to 2,000,000. Enzymes are widely distributed in all tissues, cells, and biological fluids and throughout all species, from single prokaryotic organisms to humans. Through their catalytic function, enzymes control the dynamic equilibrium in all living systems. Thus changes in enzyme activity levels will reflect and correlate with changes in the state of the organism. It is from this correlation that the interest in enzyme analysis is derived.

An enzyme assay is a measurement of the activity of an enzyme. Assays are usually performed by measuring the increase in concentration of the product and/or the decrease in concentration of the substrate and relating this measurement to the amount of enzyme present. Therefore, the information obtained is in terms of quantity of change caused, rather than an absolute quantity, such as mass. Clearly an enzyme assay requires careful consideration of the reaction conditions, as well as the means for measurement of the substrate and product. A number of methods have been employed in enzyme assays. Some of the most commonly used techniques include: spectroscopy, fluorescence, colorimetry, radioisotopic labelling followed by separation of the product by precipitation, high-voltage electrophoresis, and chromatography [1-5].

In recent years, high performance liquid chromatography (HPLC) has played a key role in multicomponent analysis of physiological fluids. In particular, HPLC has been a valuable technique for the study of nucleotides, nucleosides, and bases [6-9] in biological systems. These compounds are precursors of the nucleic acids, DNA and RNA, as well as participants in intermediary metabolism. The utility of HPLC stems not only from its power as a separation technique but also from its versatility. The modes available, such as ion exchange, reversed and normal phase, and techniques such as ion pairing and ion suppression permit separations and measurement of the most complex samples. Therefore HPLC has tremendous potential for enzyme assays and has been employed in assays for a number of diverse enzyme systems [10-14]. One outstanding feature of the technique, in terms of its use for enzyme analysis, is that measurement of the change in both substrate and product concentration can be obtained concurrently. Biological matrices are highly complex media, and the potential for interference by competing reactions is high. With both product and substrate information available, discrepancies in quantification are readily noted. The purpose of this chapter is to discuss the variables which are important in the assay of enzymes, and present examples of this technique for enzymes affecting the purine and pyrimidine metabolic pathways.

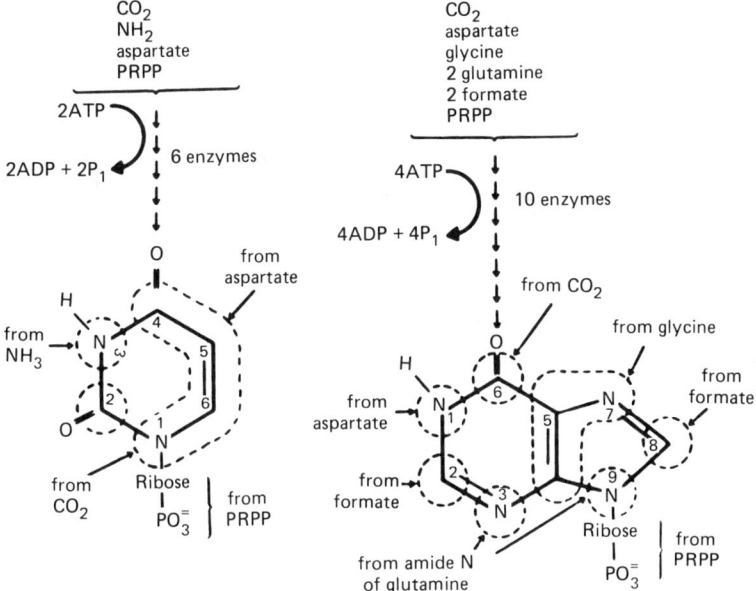

Figure 1. Left: de novo synthesis of pyrimidines. Right: de novo synthesis of purines.

B. Purines

The enzymatic formation of purine nucleotides may occur in vivo in one of two ways. The first is through the purine salvage pathway, which forms 5-inosine monophosphoric acid (IMP) by reutilizing purines from the breakdown of nucleic acids. The other means is through synthesis de novo, a series of 10 reactions starting from glutamine and 5-phosphoribosyl pyrophosphate (PRPP) and terminating in IMP (Fig. 1). Biochemically, a balance exists between salvage and de novo synthesis. Although the precise mechanism of balance between these routes remains unclear, it is considered likely that control is exerted in the first step of de novo synthesis. This step involves the transfer of the amide nitrogen of glutamine to the phosphoribosyl moiety of PRPP to form 5-phosphoribosyl-1-amine catalyzed by the enzyme PRPP amidotransferase. The complexity of these pathways and interrelationships among them has been the focus of intensive research for a number of years. Some of the interest in these reactions stems from the association of purine enzymes with a number of disease states. For example, there is mounting evidence to suggest an important role for purines in control of the immune response [15,16]. Adenosine deaminase (ADAase), purine nucleoside phosphorylase (PNPase), and hypoxanthine-guanine phosphoribosyl transferase (HGPRTase) are members of the purine salvage pathway. PNPase converts to hypoxanthine the inosine formed from the deamination of adenosine by adenosine deaminase. The hypoxanthine is subsequently converted to IMP by HGPRTase. ADAase was the first enzyme to be associated with severe combined immunodeficiency (SCID), where disturbance of both T-cell and B-cell function is observed [17,18]. In 1975 Giblett described a child with severely defective T-cell function and normal B-cell immunity, who was deficient in PNPase [19]. This information is intriguing and may suggest the possibility of multienzyme mediation of the immune response in humans. Total deficiency of another salvage enzyme, HGPRTase, was described in 1967 by Seegmiller et al. [20]. This defect was observed in a patient with Lesch-Nyhan syndrome. This syndrome was discovered in 1964 [21] and is usually characterized by excessive production of uric acid, mental retardation, and a behavioral disorder resulting in extreme aggressiveness and self-mutilization. The existence of these disease states can be confirmed by enzyme assay.

C. Pyrimidines

The de novo synthesis of pyrimidine nucleotides is somewhat different from that of purines (Fig. 1). The ring is formed from orotic acid before the attachment of the ribose 5-phosphate. Uridylic acid is then formed and phosphorylated to uridine triphosphate (UTP). UTP becomes the precursor for the other pyrimidine nucleotides. Overall, the de novo synthesis of purines is more complex than that for pyrimidines, requiring 10 enzymatic steps versus the six for the pyrimidine pathway. The pyrimidines have also been implicated in a number of disease states [22-23]. One example is orotic aciduria, which is a genetic disorder of pyrimidine biosynthesis in humans in which orotic acid accumulates in the body. This defect causes aberrations in the normal growth and development of children with the disease.

II. KINETIC CONSIDERATIONS

A. Background

The simplest model for enzyme kinetics was developed by Michaelis and Menten in 1913 [24]. The model assumes that an enzyme E will reversibly complex with the substrate S and form the enzyme-substrate intermediate ES. This complex ultimately breaks down to form the product of the reaction and the free enzyme.

$$E + S \underset{k_2}{\overset{k_1}{\rightleftharpoons}} (ES) \overset{k_3}{\rightarrow} E + P$$

The rate equation derived from this model is called the Michaelis-Menten equation:

$$V_0 = \frac{V_{max} * (S)}{K_m + (S)}$$

While this equation described the behavior of a one substrate enzyme reaction, it is also useful in determining the conditions for many enzyme assays. The utility of this expression arises from its provision of a quantatative relationship between enzymatic rate and parameters that can be measured, such as K_m and V_{max}, and those that can be controlled in an experiment, such as the substrate concentration.

B. K_m and V_{max}

V_{max} is the maximum initial velocity of the reaction. This maximum is obtained when the concentration of ES is such that all the available enzyme is complexed with substrate. Thus the maximum velocity will be directly proportional to the enzyme concentration. It is through this proportionality that enzyme activities can be used as estimates of enzyme concentration.

The Michaelis constant K_m is operationally defined as

$$K_m = \frac{k_2}{k_1}$$

According to this definition, K_m is the dissociation constant for the enzyme-substrate complex. This definition was extended for use in a general case by Briggs and Haldane [25]:

$$K_m = \frac{k_2 + k_3}{k_1}$$

In enzyme systems where k_2 is much larger than k_3, these definitions are equivalent. Therefore, K_m gives some information concerning the enzyme's affinity for a substrate. Given an enzyme with a number of possible substrates, the one with the smallest K_m is the most preferred, that with the largest K_m the least preferred.

A useful measure of K_m may be obtained by plotting the initial velocity vs. substrate concentration and applying the Michaelis-Menten equation to the resultant curve (Fig. 2). This curve is essentially the mixing of first- and zero-order dependence of rate on concentration. An important relationship is obtained by examining the initial

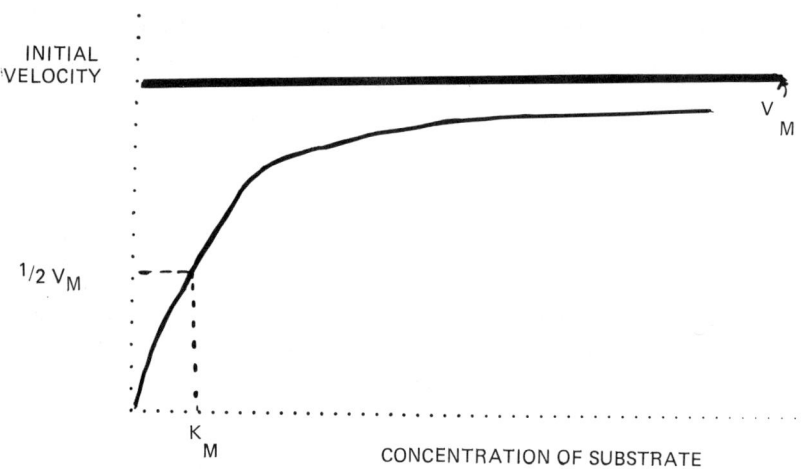

Figure 2. Plot of initial velocity vs. substrate concentration for an enzyme that follows Michaelis-Menten kinetics.

velocity at one-half its possible maximum. Under this constraint, the substrate concentration is exactly equal to K_m. Therefore, the Michaelis constant is equal to the substrate concentration at which the initial reaction velocity is half-maximal.

As the substrate concentration approaches infinity, the initial velocity approaches its maximum. In order to ensure that the assay is run under zero-order kinetics, or on the Michaelis plateau, K_m must be measured for every factor which participates in the reaction. The substrate concentration needed to achieve maximum velocity is infinite, but the approach is asymptotic. Therefore, when the substrate concentration is 10 K_m, the rate is 91% of V_{max}, when the substrate is set at 20 K_m, the velocity is approximately 95% V_{max}, and so on. Generally, the higher the substrate concentration in the incubation media, the closer the assay measurement is to the actual maximum velocity. The K_m is not a static quantity; it may vary with pH, temperature, and ionic strength. A more complete discussion of enzyme kinetics can be found in references: 22, 24-28.

C. Measurement of K_m

K_m may be estimated from a graph such as that in Fig. 2, but in order to obtain an accurate measurement, the Michaelis-Menten equation is usually transformed into a straight line and regression techniques applied. A discussion of the merits of all the available transformations of this equation is beyond the scope of this chapter; however, a number of references on the subject exist [28-33]. In our laboratory, the following transformation has been used:

$$\frac{(S)}{V_0} = \frac{(S)}{V_m} + \frac{K_m}{V_m}$$

a plot of $(S)/V_0$ vs. (S) (Fig. 3) gives a straight line with slope $1/V_m$, y intercept K_m/V_m, and x intercept $-K_m$. This plot is obtained by varying the substrate concentration and measuring the velocity. These measurements must be performed under conditions of constant pH, ionic strength, and temperature.

D. pH

Enzyme activity varies as a function of pH. This variation is due to the pKs of groups present in the enzyme molecule, the pKs of the substrate, or both. Most enzymes have a characteristic pH at which their activity is at a maximum. The pH activity relationship of many enzymes is the familiar bell-shaped curve, but this is not always the case. Figure 4 compares this relationship for the enzymes PNPase and

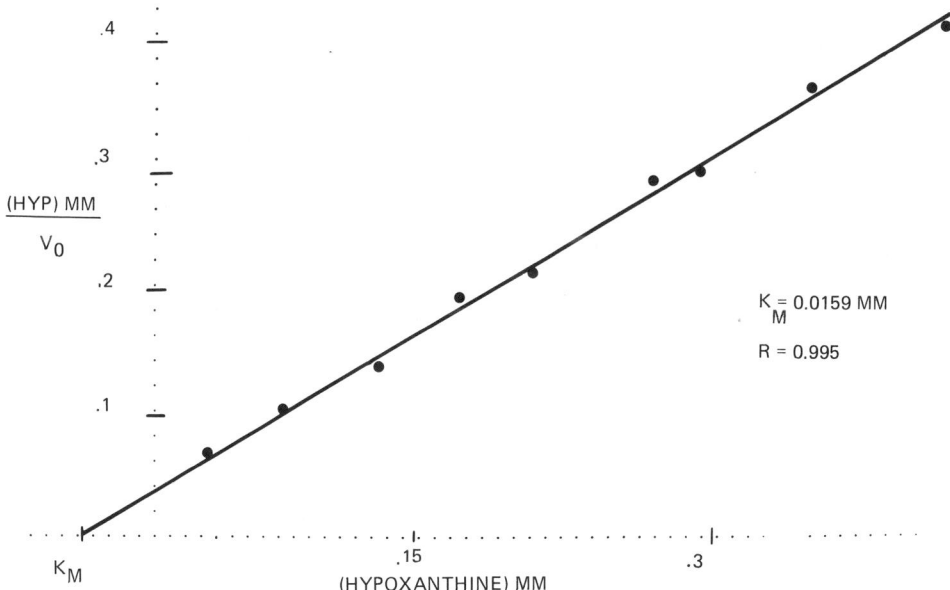

Figure 3. Linear transformation of the Michaelis-Menten equation in order to determine K_m for the enzyme HGPRTase.

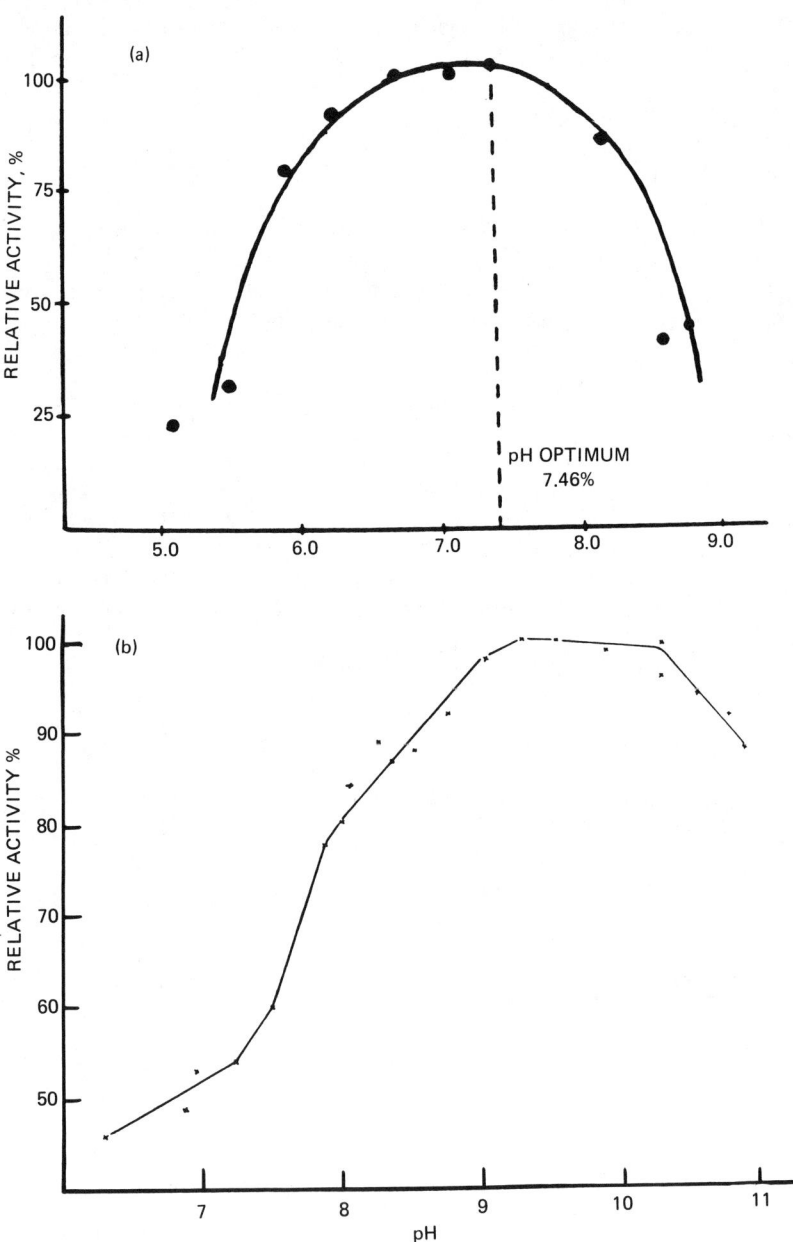

Figure 4. (a) Activity of PNPase as a function of pH. (b) Activity of HGPRTase as a function of pH. (From Ref. 42.)

HGPRTase. Many published assays for this latter enzyme are performed at pH 7.4. Although this pH is closer to physiological pH, an assay at this pH would have lower sensitivity than an assay performed at the pH optimum. When developing the conditions for an enzyme assay, the pH-activity relationship must be explored by experiment. While holding all other variables constant (including ionic strength), the pH is varied and the activity measured.

E. Temperature

The rate of many enzyme reactions is highly dependent on temperature. This dependence is due both to the thermodynamic effect of temperature on all reactions and on the effect of temperature on the physical state of the enzyme. It must be remembered that all enzymes are proteins, and therefore may be inactivated or denatured by changes in temperature. The majority of enzymes in human physiological systems are stable at 37°C, since this is the temperature at which they function in the body. Presently, there is a debate in the clinical world as to which is the best temperature to use in enzyme assays. The International Union of Biochemistry recommended 25°C in 1961, but changed this to 30°C in 1964. Unfortunately, many laboratories had been using 37°C, and continued to do so. The benefit of performing the assay at 37°C is that the increased rate allows higher sensitivity and to some extent increased accuracy. Proponents of the lower, 25°C temperature argue that the long-term stability of the enzyme is reduced at higher temperature. Regardless of the temperature chosen, it is imperative to ensure minimum temperature fluctuation. Ideally the temperature should be controlled to ±0.05 of the set point, since a 1.0 degree change in temperature can introduce a 10% variation in rate.

F. Units for enzyme activity

The activity of an enzyme is the rate at which a quantity of substrate is consumed or product formed. While this is a seemingly simple statement, the number of ways in which this number is presented is staggering. The Commission on Enzymes recommended that the unit of enzyme activity be defined as the quantity of enzyme which will catalyze the reaction of 1 μmol of substrate per minute. This is the international unit (U). The enzyme concentration is to be expressed as quantity per liter or milliliter, whichever is most convenient. A new international unit for enzyme activity is the katal. The katal is the amount of enzyme that converts 1 mol per second. Thus, 6e + 7 units = 1 katal. Presently there are large discrepancies in results for enzyme activities from different laboratories. Many of these differences are the result of different assay conditions and methodologies; it is hoped that the eventual adoption of a universal unit will help to correct these differences.

III. ANALYTICAL METHODS AND PROCEDURES

A. Sample preparation

The goal of sample preparation is the removal of all protein material before injection into the HPLC column. Most methods for protein removal can also be used for reaction termination in an enzyme assay. One of the most common methods for protein removal is precipitation with acid, followed by neutralization prior to injection. Perchloric or trichloracetic acids followed by neutralization with potassium hydroxide, or an amine-Freon solution, will extract nucleotides from cells and therefore can be used to terminate a reaction and clean up the sample for injection [34-38]. In our laboratory, cold, concentrated HCL has been useful for precipitation of protein from red blood cell lysates [39]. Phosphoric acid has also been used [40]. Another method for protein removal is immersion of the reaction vessel in boiling H_2O followed by centrifugation or filtration [41,42]. The heat precipitates the protein, which is removed in the next step. This method is applicable only after it has been determined that the enzyme is not stable at high temperatures. If HGPRTase reactions are deproteinated or terminated in this manner, spuriously high results will be obtained. Membrane cones have been used to deproteinize serum and plasma samples [43]. However, this technique is useful only if the reaction has already been terminated by other means, such as with the addition of an inhibitor of the enzyme.

It has been suggested that is is not always necessary to terminate the reaction [44]. The authors give no measure of precision for their assay, and we have observed that this practice may cause substantial variance in the measurement. If one wishes to examine time behavior of one reaction, it is best to withdraw aliquots, terminate the reaction in the individual aliquots, and inject the sample.

B. Chromatography

The minimum goal of chromatography methods development for enzyme assays is rapid separation of all participants in the reaction. It is necessary to obtain separation of all substrates, products, and cofactors. Additionally, it is good practice to examine the matrix for any additional enzymes which might participate with the substrate or product of the reaction monitored. For example, inosine and adenosine are the product and substrate, respectively, of ADAase. Thus, in an assay for ADAase it is desirable to monitor the reaction under conditions which will also separate AMP and adenine, in the event that other enzymes are operative in the matrix. In many cases the separation for an enzyme reaction mixture is somewhat simpler than for other biological fluids. Generally, the substrate is in large excess, and the amount of actual sample (blood, tissue, etc.) is very small. Therefore

some of the constraints present on other separations are not present.
The mode of chromatography chosen depends on the compounds present
in the sample, but there is considerable flexability in choice of separation conditions.

In one of the earliest enzyme assays by HPLC, Brooker measured
the activity of adenyl cyclase [45]. The column was a 3-m capillary
pellicular anion-exchange column with a pH 2.2 HCl mobile phase. The
separation was performed isocratically, with the substrate cyclic AMP
eluting in approximately 10 min. Pennington then applied essentially
the same conditions to measurement of cyclic AMP phosphodiesterase.
In this paper, he showed separations of adenosine, AMP, cyclic AMP,
and cyclic IMP. Therefore he was able to detect the interference from
5-nucleotidase which converted the AMP formed in his assay to its
base adenosine [46].

More recently, phosphodiesterase activities in crude tissue extracts
have been assayed using strong anion exchange (Partisil-Sax) in both
the isocratic and step-gradient modes [47]. A separation of AMP, cyclic AMP, GMP, and cyclic GMP was obtained in 10 min. Other assays
developed using anion exchange include one for guanylate cyclase [48],
one for ADAase [49], and recently one for pyrimidine 5-nucleotidase
in erythrocyte lysates [50]. Reversed-phase methods have also been
applied to enzyme assays. Hartwick et al. [41] developed an assay
for ADAase in erythrocyte lysates. The reaction conditions were optimized for sensitivity and speed. Due to the presence of endogenous
PNPase in red cells, it was necessary to measure adenosine, inosine,
and hypoxanthine. Separations were performed on either a µBondapak
C_{18} or a Partisil 10 ODS-3; both had particle size of 10 µm. Separation
was accomplished isocratically in 7 min (Fig. 5). Acid and alkaline
phosphatase catalyze the formation of adenosine from AMP. A reversed-phase assay for these enzymes was developed by Krstulovic et
al. [51]. After ensuring that ADAase was not operative under the
conditions used, and that 5-nucleotidase was inhibited by Ni, the assay was applied to sera from subjects with no known disease and those
with cirrhosis of the liver (Fig. 6).

An assay for the measurement of nicotinate phosphoribosyltransferase activity was developed using reversed-phase. The substrates ATP
and nicotinate and the products nicotinate mononucleotide and ADP
were resolved on a µBondapak C_{18} at pH 8, with a 25 mM $(NH_4)_3PO_4$
mobile phase [52].

Optimized assays for both PNPase and HGPRTase have been developed [42,39]. The method for PNPase is applicable to either the forward or the reverse reaction. For the forward reaction, xanthine
oxidase (XO) is added to the incubation mixture to prevent large
quantities of hypoxanthine from accumulating. XO will covert the
hypoxanthine to uric acid. The decrease of insoine and the increase
of hypoxanthine, and uric acid with time is shown in Fig. 7. Both

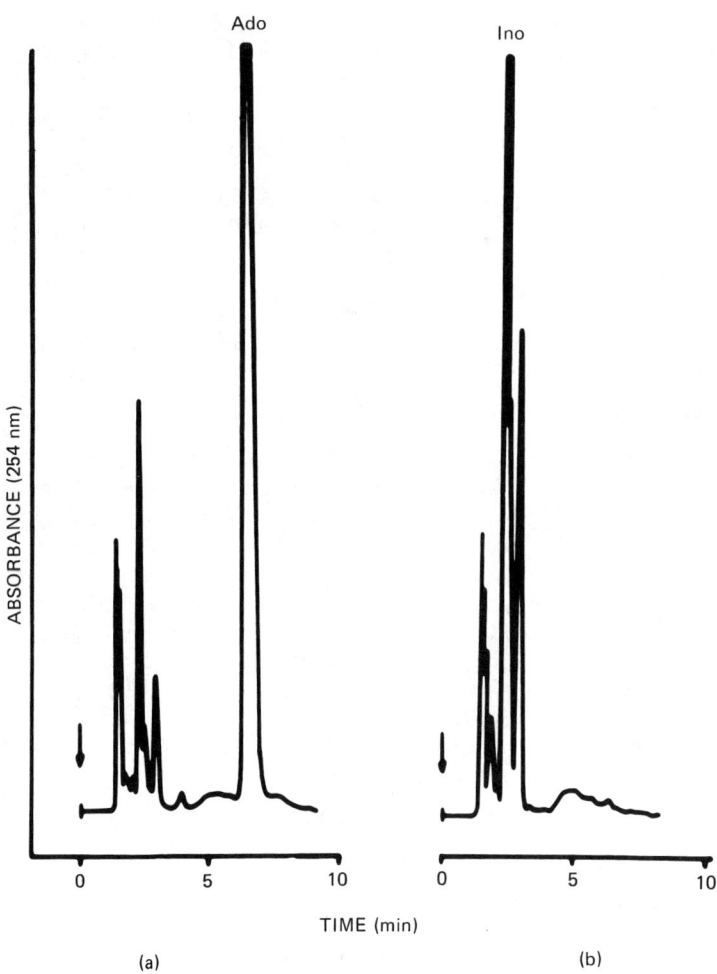

Figure 5. Separation of the participants in the reaction for ADAase.

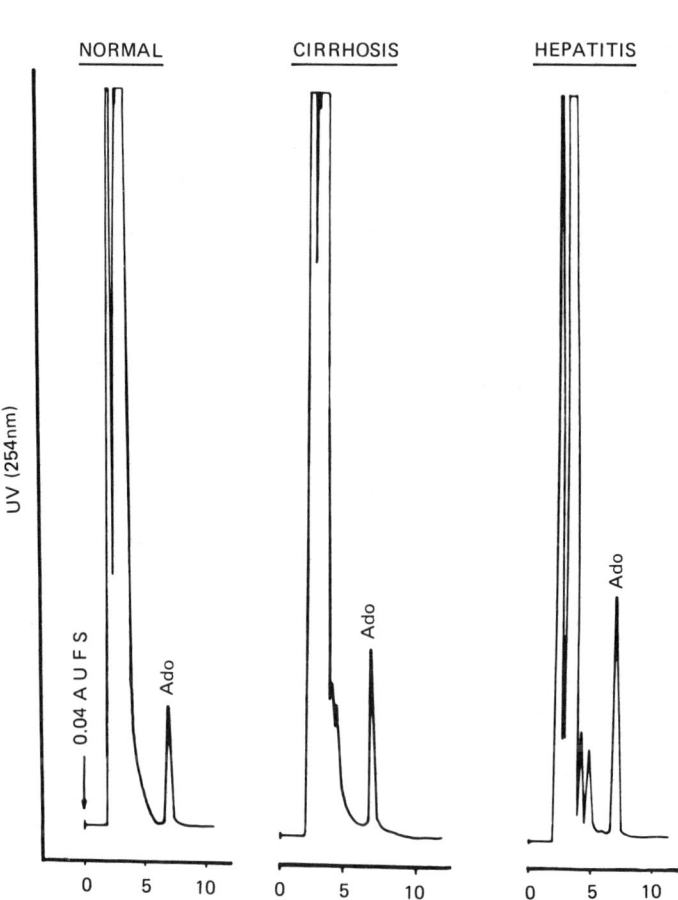

Figure 6. Comparison of acid phosphatase activity in normal serum, serum from a patient with cirrhosis, and serum from a patient with hepatitis. (From Ref. 51.)

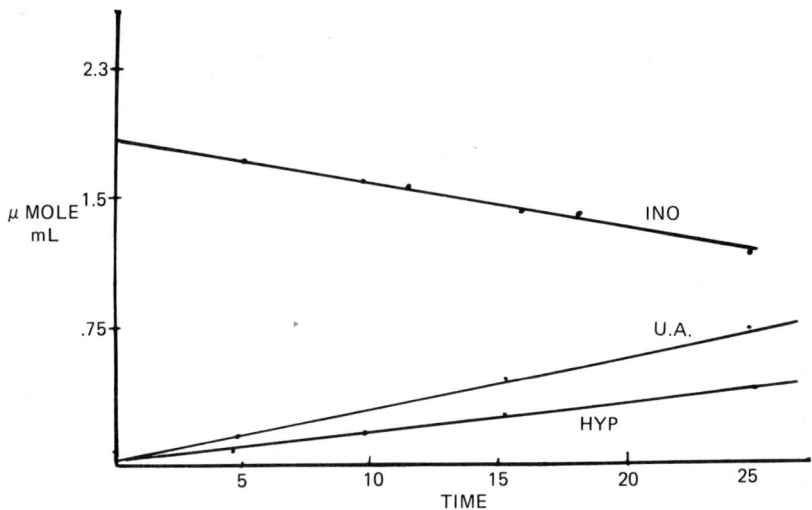

Figure 7. Plot of the decrease of inosine and increase of hypoxanthine, and uric acid with time during the course of reaction with erythrocytic PNPase. (From Ref. 42.)

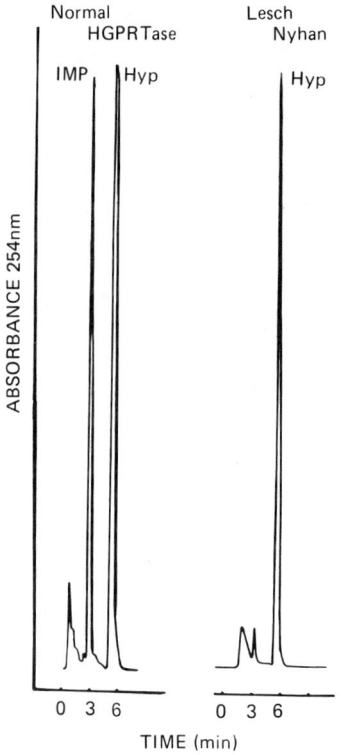

of these assays are run isocratically and on 5-μm C_{18} columns (Partisil 5-ODS). Figure 8 shows a chromatogram obtained by assay of normal and Lesch-Nyhan samples for HGPRTase. These individual assays have been used as the basis for a simultaneous assay of HGPRTase and PNPase. The separation of the reactants, IMP, inosine, and hypoxanthine, permits the assay of both of these enzymes, under zero-order conditions, in one analysis.

Ion pairing has been employed in the assay of an enzyme of clinical importance, creatine kinase [53]. By adapting the Oliver assay for HPLC, separation of AMP, ADP, and ATP was accomplished in 16 min using a reversed-phase C_{18} column and a mobile phase consisting of 88% 0.1 M KH_2PO_4, 0.25 M butylammonium hydrogen sulfate, and 12% methanol.

Rossomando et al. demonstrated the utility of HPLC with fluorimetric detection for a number of enzyme assays [44]. In their study, they used the fluorescent ATP analog, formycin 5-triphosphate, and its mono-, di-, and cyclic monophosphate forms to measure enzyme activities in the organism, *Dictyostelium discoideum*. Fluorescence measurement of these derivatives permitted detection limits of subpicomoles to be obtained.

Recent developments in the separation of proteins [54-56] by HPLC have opened up the area of monitoring isoenzymes [57]. Isoenzymes are structurally different proteins which catalyze the same chemical reaction. For example, creatine kinase has at least three isoenzyme forms, BB, MM, and MB. The separations of the isoenzymes were obtained by Schlabach and Regnier [58]. After separation, the isoenzymes were assayed by post-column reaction with the appropriate substrates and cofactors as well as two immobilized coupling enzymes. The authors also applied this system to the measurement of alkaline phosphatase (AP) activities. Detection of liver and intestinal AP was accomplished by post-column reaction with the substrate p-nitrophenyl phosphate. This type of analysis has great utility in medical and clinical disciplines, where the presence of certain isoenzymes indicates a disease state.

In addition, Sloan has shown the usefulness of NPLC in studies of enzyme kinetics and in determining enzyme reaction mechanisms (59).

C. Quantification: activity calculations

The enzymatic activities can be calculated as follows. The response factor (area per micromole) of the product(s), substrate(s), and any intermediates is calculated. The activity is calculated as the increase of product or the decrease of substrate per time. In order to obtain

---

Figure 8. Chromatogram of sample obtained by assay of HGPRTase from a Lesch-Nyhan patient and a normal control.

Table 1 Activity Calculations

$$\text{Fraction converted} = \frac{(\text{area of product} \times \text{RF of product})}{(\text{area of substrate} \times \text{RF substrate} + \text{area product} \times \text{RF product})}$$

where

$$\text{RF} = \text{response factor (area/micromole)}$$

$$\text{Activity (U/mL RBCs)} = \frac{\text{fraction converted} \times \text{micromoles of substrate}}{\text{time (min)} \times \text{volume of red blood cells}}$$

$$\text{U} = \text{international unit}$$

maximum accuracy, the percent conversion of substrate can be used. Since the amount of substrate is known, this method can minimize errors due to pipetting and injection without the use of an internal standard. However, if an internal standard is to be used, it should be added after reaction termination. Alternatively, experiments must be performed to ensure that the compound chosen for standardization does not activate, inhibit, or participate with the enzyme in any way. The activity is calculated from the percent as shown in Table 1.

IV. CONCLUSIONS

HPLC offers some unique advantages over many traditional methods of enzyme assays. Some of these advantages are the following:

1. Multiple enzymes can be assayed simultaneously.
2. Both the forward and reverse reactions can be monitored.
3. Matrix effects can be detected rapidly.
4. The behavior of the enzyme in the presence of inhibitors can be readily studied.

For purine and pyrimidine compounds, HPLC with ultraviolet detection provides good sensitivity. This sensitivity is greatly enhanced if the compound exhibits fluorescence or can be derivitized to do so. The sensitivity of the anlaysis, coupled with the relative ease of the method, makes this technique highly attractive for enzyme assays. Additionally, the increasing utility of HPLC in protein separations will provide a means of simultaneous separation and assay of enzymes and their isoenzyme forms.

REFERENCES

1. B. Emmerson, C. J. Thompson, and D. Wallace, Ann. Inter., 76, 285 (1972).
2. P. Cartier and M. Hamet, Clin. Chim. Acta, 20, 205 (1968).
3. A. Giacomeilo and C. Salerno, Anal. Biochem., 79, 263 (1977).
4. G. Guilbault, *Enzymatic Methods of Analysis*, Pergamon Press, Elmsford, N.Y., 1970.
5. H. Bergmeyer, *Methods of Enzymatic Analysis*, vol 1-4, Academic Press, New York, 1974.
6. M. Zakaria and P. R. Brown, J. Chromatogr., 226, 267 (1981).
7. P. R. Brown, A. Krstulovic, and R. A. Hartwick, in *Advances in Chromatography*, vol. 18 (J. C. Giddings, E. Gruska, J. Cazes and P. R. Brown, Eds.), Marcel Dekker, New York, 1980.
8. H. A. Scobie and P. R. Brown, in *Advances and Perspectives in Chromatography*, vol. 3 (C. Horvath, Ed.), Academic Press, New York, 1983.
9. P. R. Brown and A. M. Krustulovic, Anal. Biochem., 99, 1 (1979).
10. K. Omichi and S. Tokujiikenaka, J. Chromatogr., 230, 415 (1982).
11. G. C. Davis and P. T. Kissinger, Anal. Chem., 51, 1960 (1982).
12. L. Cantoni, R. Ruggierie, Dal Fuime, and M. Rizzardine, J. Chromatogr., 229, 311 (1982).
13. P. Pietta, A. Calatroni, and M. Pace, J. Chromatogr., 241, 409 (1982).
14. S. L. Pahdja, J. Albert, and T. W. Reid, J. Chromatogr., 225, 27 (1981).
15. B. Ullman, L. Gudas, S. Clept, and D. W. Martin, Proc. Natl. Acad. Sci. U.S., 76, 1074 (1979).
16. R. M. Goldblum, F. C. Schmalstieg, J. A. Nelson, and G. C. Mills, Birth Defects: Original Article Series, XIV (6A), 73 (1978).
17. E. Giblett, J. Anderson, F. Cohen, B. Pollarn, and H. Meuwissen, Lancet, 2, 1067 (1972).
18. J. Dissing and B. Knudsen, Lancet, 2, 1316 (1972).
19. E. Giblett, J. Amman, R. Sandman, D. Ware, and L. K. Diamond, Lancet, 1, 1010 (1975).
20. J. E. Seegmiller, F. M. Rosenbloom, and W. N. Kelley, Science, 155, 1696 (1967).
21. M. Lesch and W. L. Nyhan, Am. J. Med., 36, 561 (1964).
22. A. L. Lehninger, *Biochemistry*, Worth, New York, 1975.
23. A. H. Van Gennip, J. Grift, P. K. De Bree, B. J. Zegers, J. W. Stoop, and S. K. Wadman, Clin. Chim. Acta, 93, 419 (1979).
24. L. Michaelis and M. L. Menten, Biochem. Z., 49, 333 (1913).
25. G. E. Briggs and J. B. S. Haldane, Biochem. J., 19, 338 (1925).
26. W. Ferdinand, *The Enzyme Molecule*, John Wiley & Sons, New York, 1976.
27. L. Styrer, *Biochemistry*, W. H. Freeman, San Francisco, 1975.

28. S. Ainsworth, *Steady-State Enzyme Kinetics*, University Park Press, 1977.
29. W. W. Cleland, Advan. Enzymol., *20*, 1 (1967).
30. A. Cornish-Bowen and R. Eisenthal, Biochem. J., *139*, 721 (1974).
31. R. Eisenthal and A. Cornish-Bowden, Biochem. J., *149*, 775 (1975).
33. R. C. Kohberger, Anal. Biochem., *101*, 1 (1980).
34. G. Kraus and H. Reinboth, Anal. Biochem., *78*, 1 (1977).
35. B. Bakay, E. Nissinen, and L. Sweetman, Anal. Biochem., *86*, 65 (1978).
36. S. Chen, D. M. Rosie, and P. R. Brown, J. Chromatogr. Sci., *15*, 218 (1977).
37. A. M. Krstulovic and C. Matzura, J. Chromatogr., *176*, 217 (1979).
38. J. X. Khym, Clin. Chem., *21*, 1245 (1975).
39. A. P. Halfpenny and P. R. Brown, Abstract 285, Pittsburgh Conference on Analytical and Applied Spectroscopy, 1981.
40. L. W. Yu and R. S. Fager, Anal. Chem., *54*, 1904 (1982).
41. R. H. Hartwick, A. Jeffries, A. M. Krstulovic, and P. R. Brown, J. Chromatogr. Sci., *16*, 427 (1978).
42. A. P. Halfpenny and P. R. Brown, J. Chromatogr., *199*, 275 (1980).
43. R. A. Hartwick, A. M. Krstulovic, and P. R. Brown, J. Chromatogr., *186*, 737 (1979).
44. E. F. Rossamando, J. Jahngen, and J. F. Eccleston, Anal. Biochem., *116*, 80 (1981).
45. G. Brooker, Anal. Chem., *42*, 1108 (1970).
46. S. N. Pennington, Anal. Chem., *43*, 1707 (1971).
47. D. S. Hsu and S. Chen, J. Chromatogr., *247*, 369 (1982).
48. S. Mehta, D. Dunlap, J. Lapeyre, and A. Maziel, J. Liq. Chromatogr., *3*, 1517 (1980).
49. J. Uberti, J. J. Lightbody, and R. M. Johnson, Anal. Biochem., *80*, 1 (1977).
50. T. Sakai, S. Yanagihara, and K. Ushio, J. Chromatogr., *239*, 717 (1982).
51. A. M. Krstulovic, R. H. Hartwick, and P. R. Brown, J. Chromatogr., *163*, 19 (1979).
52. L. Hanna and D. L. Sloan, Anal. Biochem., *103*, 230 (1980).
53. N. D. Danielson and J. A. Huth, J. Chromatogr., *221*, 39 (1980).
54. F. E. Regnier, Anal. Biochem., *126*, 1 (1982).
55. F. E. Regnier and K. M. Godding, Anal. Biochem., *103*, 1 (1980).
56. J. D. Pearson, N. T. Lin, and F. E. Regnier, Anal. Biochem., *124*, 217 (1982).
57. D. N. Vacik and E. C. Toren, J. Chromatogr., *228*, 1 (1981).
58. T. D. Schlabach and F. E. Regnier, J. Chromatogr., *158*, 349 (1978).
59. D. L. Sloan in *Advances in Chromatography*, vol. 23 (J. C. Giddings et al., Eds.), Marcel Dekker, New York, 1983).

16

Nucleotide Coenzymes

MARY E. DWYER University of Rhode Island, Kingston, Rhode Island

I. Introduction 303
II. Chemistry and Structures 304
III. HPLC of Nucleotide Coenzymes 308
IV. Conclusion 313
   References 313

I. INTRODUCTION

Since the discovery of the first nucleotide coenzyme, nicotinamide adenine dinucleotide (NAD)*, in 1905 by Harden and Young [1] and its subsequent isolation and purification in 1936 [2,3], it has been found that nucleotide coenzymes can catalyze a wide variety of biochemical reactions. These reactions include oxidation-reduction processes and the synthesis of such macromolecules as nucleic acids, polysaccharides, and glycerides.

Thus the study of nucleotide coenzymes is of great significance in the field of enzymology, and rapid and sensitive analysis of these compounds is of great value to the researcher. In this chapter some of the important background information on the chemistry and structures of these coenzymes will be highlighted and the advances made in the separation and the detection of these compounds by liquid chromatography will be discussed.

---

*Wherever NAD or NADP appears in the text, it is assumed that the species is positively charged (NAD^+ and $NADP^+$).

II. CHEMISTRY AND STRUCTURES

Numerous enzymes, catalyzing a broad spectrum of biochemical reactions, require the presence of a nonprotein prosthetic group, more or less tightly linked to the protein, for efficient performance of catalytic function. Since these prosthetic groups are frequently participants in the overall reactions, they are also termed cofactors or coenzymes. A cofactor may be either a metal ion (the divalent cations Mg^{2+}, Zn^{2+}, Mn^{2+} being most common) or a complex organic molecule called a prosthetic group. Coenzymes are classified as organic cofactors, and some may be considered biosynthetically related to the vitamins. For example, the coenzyme NAD, important for cellular energy metabolism, incorporates the vitamin niacin into its chemical makeup. Other coenzymes, ATP for example, are synthesized de novo by most mammalian cells. In addition, coenzymes may be regarded as cosubstrates undergoing a chemical transformation during the enzyme reaction (e.g., NAD reduced to NADH). Reversal of this enzyme requires a separate enzyme, possibly in a different cellular location.

The nucleotide coenzymes will be described starting with the nucleoside 5'-polyphosphates (ATP, ADP, etc.). The oxidation-reduction coenzymes will be described next; these compounds are biochemically active by virtue of oxidation-reduction processes which take place in the non-nucleotidic portion of the molecule. The structures of these coenzymes are shown in Fig. 1 and their functions are summarized in Table 1.

The nucleoside 5'-polyphosphates which are most widespread in nature are adenosine diphosphate (ADP) and adenosine triphosphate (ATP). The fundamental role of the hydrolysis of ATP as the driving force for biochemical processes was first clearly recognized by Lipmann [4]. Biochemical processes known to be dependent on ATP hydrolysis include muscle contraction, photosynthesis, bioluminescence, and the biosynthesis of protein, nucleic acids, complex carbohydrates, and lipids, among many others.

Other nucleoside di- and triphosphates, namely, those of uridine, cytidine, and guanosine, play similar roles in phosphorylation reactions, coenzyme synthesis, and nucleic acid synthesis. For example, UTP is necessary for the synthesis of glycogen and for the conjugation of bile pigment. GTP is required during the RNA-directed biosynthesis of protein [5]. The coenzymatic function of cytidine nucleotide derivatives was first recognized by Kennedy and Weiss [6,7], who discovered that CDP-choline and CDP-ethanolamine are essential intermediates in the biosynthesis of lecithins, an important class of lipids. There are, of course, many other functions for each of the nucleoside di- and triphosphates.

The pyridine nucleotides were the first coenzymes to be recognized. These include nicotinamide adenine dinucleotide (NAD), also called diphosphopyridine nucleotide (DPN) or coenzyme I (Co I) and

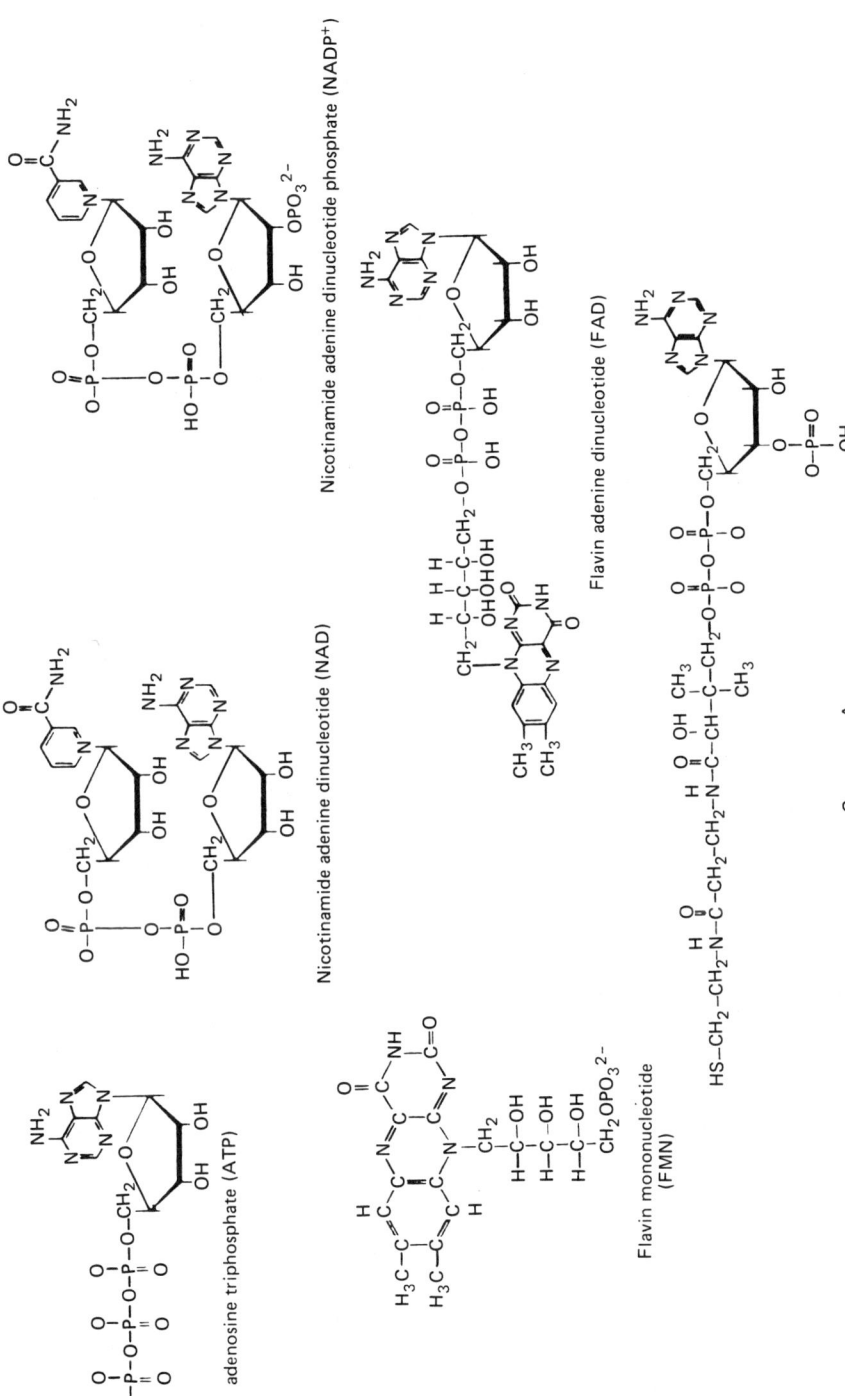

Figure 1. Structure of nucleotide coenzymes: (a) ATP, (b) NAD, (c) NADP, (d) FMN, FAD, (e) Coenzyme A.

Table 1 Summary of Nucleotide Coenzymes and Their Functions

Coenzyme	Biochemical Function
Nucleoside 5'-polyphosphates	Phosphorylation reactions
	Coenzyme synthesis
	Nucleic acid synthesis
Nicotinamide adenine dinucleotide	Oxidation-reduction
Nicotinamide adenine dinucleotide phosphate	Oxidation-reduction
Flavin mononucleotide	Oxidation-reduction
Flavin adenine dinucleotide	Oxidation-reduction
Coenzyme A	Acyl transfer
	Fatty acid synthesis
	Glyceride synthesis

nicotinamide dinucleotide phosphate (NADP), which is less frequently called triphosphopyridine nucleotide (TPN), or coenzyme II (Co II). NAD and NADP are derived biosynthetically in the mammal, from nicotinamide (niacin) or nicotinic acid. A nutritional deficiency of the precursors results in pellagra (in the human), a disease characterised by dermatitis, diarrhea, and dementia.

The pyridine coenzymes, in association with appropriate protein apoenzymes, are cofactors for a wide variety of dehydrogenation reactions [8]. One common dehydrogenation reaction which is catalyzed by NAD^+ is the oxidation of alcohols to aldehydes or ketones:

$$CH_3CH_2OH + NAD^+ \rightarrow CH_3CHO + NADH + H^+ \tag{1}$$

The most important sources of alcohol dehydrogenase are yeast [9] and liver [10].

Aldehydes (or their hydrates) are dehydrogenated to carboxylic acids by enzymes which require either NAD or NADP as cofactors [11,12]:

$$H_2O + RCHO + NADP^+ \rightarrow RCOOH + NADPH + H^+ \tag{2}$$

The flavin coenzymes, flavin adenine dinucleotide (FAD) and riboflavin mononucleotide (FMN), are oxidation-reduction coenzymes by virtue of the reversible oxidation and reduction of the flavin moiety. The number of known flavoproteins utilizing either FMN or FAD as coenzyme is large, and such enzymes are universally distributed in nature. Typical substrates dehydrogenated by flavoproteins include pyridine nucleotides, α-amino acids, α-hydroxy acids, aldehydes, and substances containing saturated carbon-carbon bonds that are converted to olefins. For example, FAD will oxidize an α-amino acid to an imine as shown in Eq. (3):

$$NH_2CHRCOOH + 2FAD \rightarrow 2FADH + NH=CRCOOH \tag{3}$$

In contrast to the pyridine and flavin coenzymes, which function in oxidation-reduction reactions, coenzyme A (CoASH) is the most prominent acyl group transfer coenzyme in living systems. Its universal occurrence, the large number of enzymatic reactions in which it is involved, and the variety of reaction types in which its derivatives are concerned emphasize the importance of this coenzyme. The most important CoASH derivative is undoubtedly acetyl-CoA, or activated acetate.

Coenzyme A was originally identified as a heat-stable cofactor required for certain biological acetylations [13,14], and was subsequently implicated as a cofactor for the incorporation of acetate into acetoacetate and citrate [15,16]. The formation of citrate from oxalacetate and S-acetyl CoA is illustrated in Eq. (4):

$$\begin{array}{c} CO \cdot COOH \\ | \\ CH_2COOH \end{array} + CoASCCH_3 \longrightarrow CoASH + \begin{array}{c} CH_2 \cdot COOH \\ | \\ COH \\ | \\ CH_2 \cdot COOH \end{array} \tag{4}$$

The structure of coenzyme A (see Fig. 1) was elucidated in the laboratories of Lipmann, Snell, and Baddiley [17]. This molecule is quite complex structurally and contains a multiplicity of possible functional groups.

From this brief summary it can be seen that the nucleotide coenzymes are of fundamental importance in biological oxidations, which constitute the core of the energy-producing mechanisms of the body. Therefore the detection and quantitation of these compounds are of vital importance in biochemistry and more specifically enzymology.

III. HPLC of Nucleotide Coenzymes

In the past, nucleoside 5'-polyphosphates have been assayed in several different ways [18-22], including an enzymatic method [23], a fluorimetric procedure [24], column chromatography, and high performance liquid chromatography (HPLC) [25]. At first ion-exchange HPLC using a linear gradient was utilized for separating and quantitating the nucleotides of human blood [26-28]. Rao et al. [29] demonstrated that nucleoside 5'-polyphosphates can be separated rapidly and quantitated accurately by reversed-phase chromatography (RPLC) using isocratic elution. The time taken for each analysis could be further reduced by using radially compressed columns. Their method is fast, (approximately 24 min with steel columns; 10 min with radially

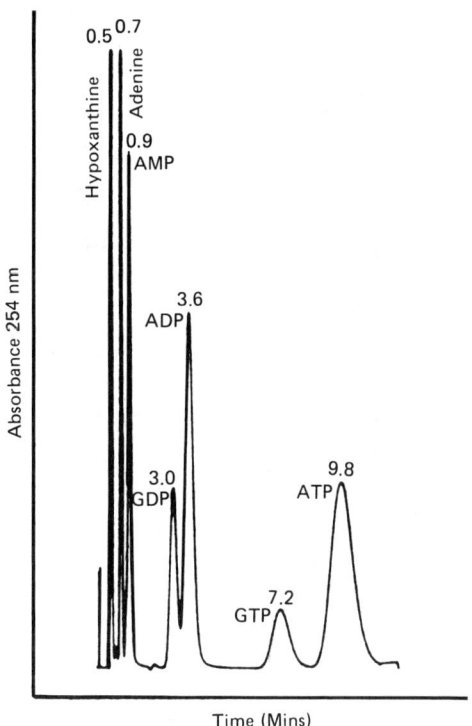

Figure 2. Separation of various nucleotide standards from a mixture by isocratic elution using a μBondapak C_{18} column. Complete separation of all the nucleotides (including cyclic AMP from AMP) at ambient temperatures was achieved in approximately 24 min. Chromatographic conditions: column, 30 cm × 4 mm; packing, μBondapak C_{18}; sample 10 μl. (From Ref. 29.)

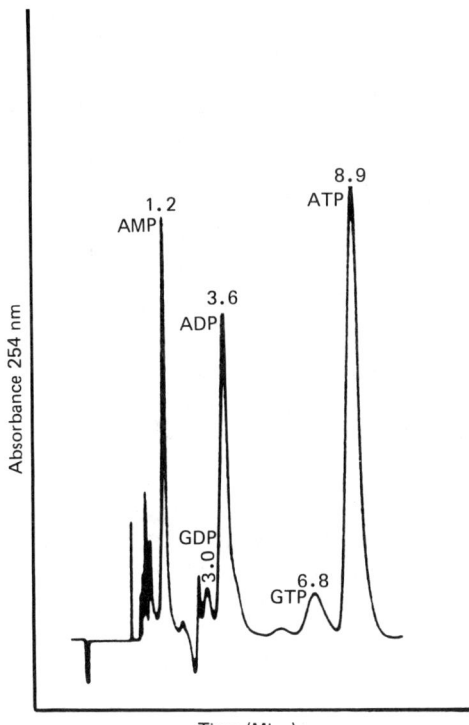

Figure 3. Separation of platelet nucleotides by isocratic elution using a radially compressed column. Identification of individual peaks was done by comparison of retention times of standards added to the platelet extract and assayed under identical conditions. Also, standard additions of known nucleotides to platelet extracts and their analysis served as a supplementary confirmation for peak identification. Retention times for AMP, ADP, and ATP were 1.2, 3.6, and 8.9 min. Chromatographic conditions as in Fig. 2. (From Ref. 29.)

compressed columns), efficient and provides excellent separation of platelet nucleotides (see Figs. 2 and 3). The values obtained using this method agree with results published by others using linear gradient elution to achieve complete separation of nucleotides [27].

Nicotinamide adenine dinucleotide (NAD) is an important cofactor in a number of oxidoreductase enzyme systems. The detection and quantitation of its reduced form (NADH) is the basis for a number of methods which determine both substrates and enzyme activity. Chromatographic methods including paper, thin-layer, and open columns have helped isolate some NADH breakdown and reaction products [30-34].

These methods, however, are too slow for the analysis of transient species and lack the efficiency needed to separate the complex mixtures formed when NADH degrades.

Margolis et al. developed an analytical RPLC procedure for rapid assessment of purity of NADH [35]. They found that preparative chromatography on DEAE-cellulose gives NADH that is free from adenine nucleotides as well as other impurities commonly present in NADH.

Miksic and Brown [36] utilized reversed-phase and anion-exchange modes of HPLC to separate breakdown products and impurities in solutions of NADH, to follow the course of breakdown reactions, and to characterize some products of the reactions (see Fig. 4). In 1978 Miksic and Brown [37] used RPLC to isolate acid breakdown products of NADH and products produced when NADH breakdown is catalyzed by glyceraldehyde 3-phosphate dehydrogenase. They found that formation of the major acid products fit a three-step, first-order mechanism curve, thus reporting the first HPLC kinetic study on NADH breakdown.

Kissinger and his group have demonstrated the potential of electrochemistry in the detection of enzymatically generated NADH [38,39]. Liquid chromatography with electrochemical detection (LCEC) provides selectivity and sensitivity. Davis et al. [40], utilizing RPLC and

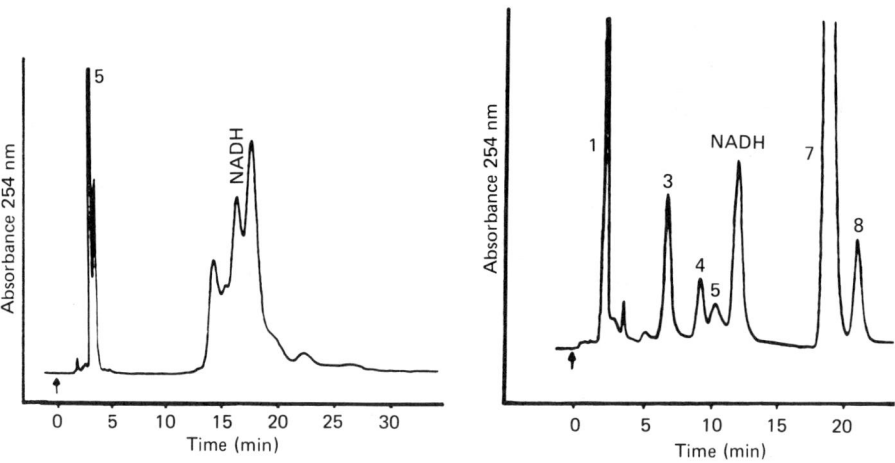

Figure 4. Left: The separation of NADH and acidic breakdown products by ion-exchange HLPC (mM NADH stored in 0.05 F KH_2PO_4, PH 5.2, 1 day). Right: The separation of NADH and acidic breakdown products by reversed-phase HPLC (mM NADH, stored in 0.05 F sodium acetate, pH 5.41, for 19 hr). (From Ref. 36.)

alcohol dehydrogenase (ADH), used LCEC to analyze enzymatically generated NADH for the detection of blood alcohol.

In 1979, Margolis and Schaffer [41] developed a RPLC system for the qualitative and quantitative evaluation of NADP and NADPH in addition to NAD and NADH. The phosphorylated coenzymes undergo reactions similar to NAD and NADH [42-44]. The separation required less than 1 hr for completion.

Heard and Tritz [45] have achieved separation of the six pyridine compounds which comprise the pyridine nucleotide cycle as well as NADP. The pyridine nucleotide cycle consists of the intermediates nicotinic acid, nicotinic acid mononucleotide, nicotinic acid adenine dinucleotide, nicotinamide adenine dinucleotide, nicotinamide mononucleotide, and nicotinamide. They found that optimum separation is achieved using ion pairing in RPLC with a C_{18} stationary phase. Their separation and quantitative analysis of these compounds were achieved in a matter of minutes.

Riboflavin is present in the body as either riboflavin phosphate (FMN) or flavin adenine dinucleotide (FAD). Previous methods of separation include ion-exchange chromatography [46], selective adsorption to Sephadex or Bio-Gel resins [47] and FMN- or FAD- specific immobilized enzymes [48,49]. Light et al. in 1980 [50] reported a rapid, convenient, and quantitative separation of FAD and FMN using RPLC. The method allows the separation of oxidized from reduced forms of the coenzymes when their stability does not preclude it. The method also facilitates the separation and purification of radioactive analogs at the FMN and FAD levels.

Speek et al. [51] developed a RPLC method specifically for the analysis of FAD in whole blood, with the ability to separate FAD, FMN, and riboflavin from each other and from interfering compounds. A reliable and sensitive HPLC method for the detection of FAD was obtained by selecting the optimal pH of the mobile phase and optimal adjustment of the fluorescence detector. This method is suitable for large-scale routine analysis of FAD in whole blood.

Pietta and coworkers [52] have described a method of RPLC for the direct determination of FAD, FMN, and riboflavin in blood plasma using a fixed-wavelength detector (254 nm) and isocratic elution (see Fig. 5). The sensitivity of the method was estimated to be 0.5 µg/ml and was adequate for following the hydrolysis of FAD and FMN.

In 1983, Nielson and coworkers [53] described HPLC techniques which can be used for analytical and preparative separations of isomeric phosphoric acid esters of riboflavin and riboflavin analogs. Good analytical separation could be achieved by RPLC both with and without ion pairing.

Different methods have been used to measure coenzyme A (CoASH) in biological extracts. At present, indirect enzyme reactions are generally used [54-56]. Paper chromatography is employed, although

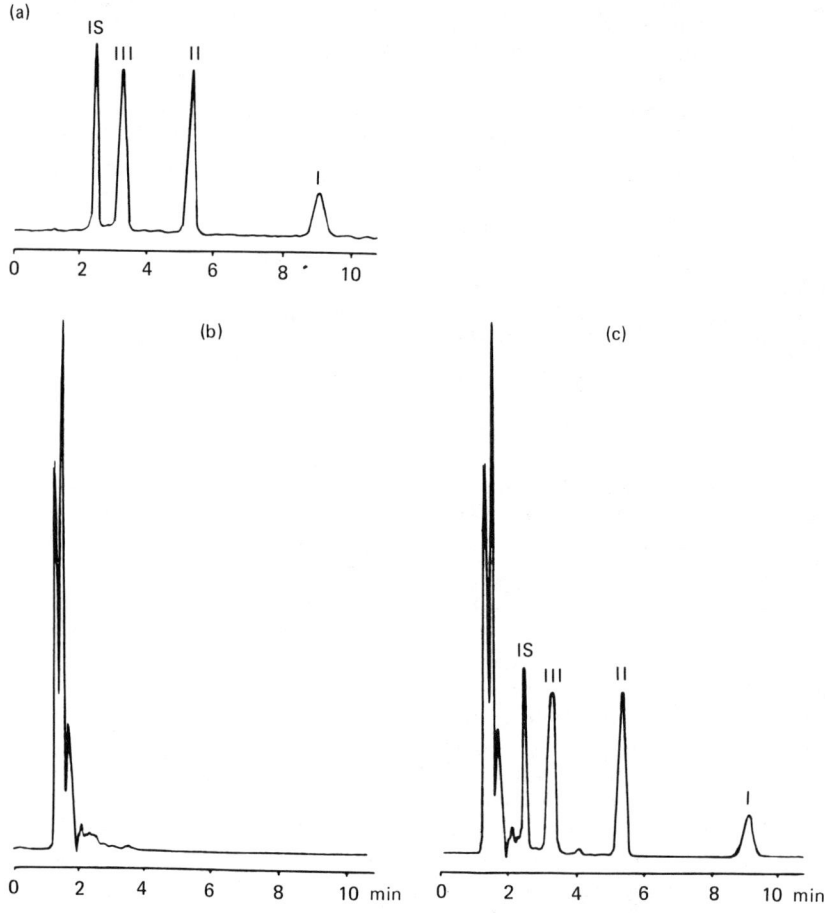

Figure 5. Chromatograms of (a) a standard mixture of internal standard, nicotinamide (IS, 45 ng), FAD (III, 65 ng), FMN (II, 65 ng), and riboflavin (I, 25 ng); (b) control plasma; (c) plasma fortified with internal standard (IS, 45 ng), FAD (III, 65 ng), FMN (II, 65 ng) and riboflavin (I, 25 ng). (From Ref. 52.)

this is a time-consuming procedure [57]. However, a HPLC system was developed with baseline separation of CoASH from dephospho-CoASH and acetyl-CoASH using isocratic elution [58]. The chromatographic separation was achieved in a reversed-phase system with the eluate monitored with an ultraviolet detector at 254 nm with the limit of detection at around 5 pmol.

Numerous other separations of nucleotide coenzymes have been achieved. From the research and results that have been reported, it is quite evident that high performance liquid chromatogrpahy provides a rapid, sensitive, and selective means for separation and quantitation of these coenzymes.

IV. CONCLUSION

The use of high performance liquid chromatography has provided rapid, specific, and reproducible methods of analysis of nucleotide coenzymes. The recent advances in state-of-the-art instrumentation, column technology, and data acquisition and storage, as well as advantages which include flexibility, efficiency, and sensitivity, greatly enhance the continuing development of HPLC methods for nucleotide coenzyme research.

Future work in the fields of biochemistry and more specifically enzymology will no doubt utilize these advances and advantages to make HPLC the method of choice for nucleotide coenzyme analysis.

REFERENCES

1. A. Harden and W. J. Young, Proc. Roy. Soc. B, *78*, 369 (1906).
2. H. von Euler, H. Albers, and F. Schlenk, Z. Physiol. Chem., *240*, 113 (1936).
3. O. Warburg and W. Christian, Biochem. Z., *287*, 291 (1936).
4. F. Lipmann, Adv. Enzymol., *1*, 99 (1941).
5. E. B. Keller and P. C. Zamecnik, J. Biol. Chem., *221*, 45 (1956).
6. E. P. Kennedy and S. B. Weiss, J. Am. Chem. Soc., *77*, 250 (1955).
7. E. P. Kennedy and S. B. Weiss, J. Biol. Chem., *222*, 193 (1956).
8. M. Dixon and E. C. Webb, *Enzymes*, Academic Press, New York, 1958.
9. R. J. P. Williams, F. L. Hoch, and B. L. Vallee, J. Biol. Chem., *232*, 465 (1958).
10. A. D. Merrit and G. M. Tomkins, J. Biol. Chem., *234*, 2778 (1959).
11. J. E. Seegmiller, J. Biol. Chem., *201*, 629 (1953).
12. T. E. King and V. H. Cheldelin, J. Biol. Chem., *220*, 177 (1956).
13. F. Lipmann, J. Biol. Chem., *160*, 173 (1945).
14. D. Nachmansohn and M. Berman, J. Biol. Chem., *165*, 551 (1946).
15. M. Soodak and F. Lipmann, J. Biol. Chem., *175*, 999 (1948).
16. J. R. Stein and S. Ochoa, J. Biol. Chem., *179*, 491 (1949).
17. J. Baddiley, Adv. Enzymol., *16*, 1 (1955).

18. P. Fantl and H. A. Ward, J. Biochem., *64*, 747 (1956).
19. G. V. R. Born, J. Biochem., *68*, 695 (1958).
20. H. Holmson, Scand. J. Clin. Lab. Invest., *17*, 230 (1965).
21. U. Goetz, M. Da Prada, and A. Pletscher, J. Pharmacol. Exp. Ther., *178*, 210 (1971).
22. H. Holmson, I. Holmsen, and A. Bernharden, Anal. Biochem., *17*, 456 (1966).
23. D. C. B. Mills and D. P. Thomas, Nature (London), *222*, 991 (1969).
24. J. L. Gordon and A. H. Drummond, J. Biochem., *138*, 165 (1974).
25. C. G. Horvath, B. Preiss, and S. R. Lipsky, Anal. Chem., *39*, 1922 (1967).
26. E. H. Scholar, P. R. Brown, R. E. Parks, and P. Calabresi, Blood, *41*, 927 (1973).
27. G. H. R. Rao, A. A. Jachimowicz, and C. J. Witkop, J. Lab. Clin. Med., *84*, 839 (1974).
28. L. D. D'Souza and H. I. Glueck, Thromb. Haemostas., *38*, 990 (1977).
29. G. H. R. Rao, J. D. Peller, and J. G. White, J. Chromatogr., *226*, 466 (1981).
30. T. Sakai, T. Uchida, and I. Chibata, J. Chromatogr., *66*, 111 (1972).
31. G. Disabato, Biochim. Biophys. Acta, *167*, 646 (1968).
32. P. Strandjord and K. J. Clayson, J. Lab. Clin. Med., *67*, 144 (1966).
33. E. Silverstein, Anal. Biochem., *12*, 199 (1965).
34. F. Pastore, J. Biol., Chem., *236*, 2314 (1961).
35. S. A. Margolis, B. F. Howell, and R. Schaffer, Clin. Chem., *22*(8), 1322 (1976).
36. J. R. Miksic and P. R. Brown, J. Chromatogr., *142*, 641 (1977).
37. J. R. Miksic and P. R. Brown, Biochemistry, *17*, 2234 (1978).
38. W. R. Heineman and P. T. Kissinger, Anal. Chem., *50*, 166R (1978).
39. P. T. Kissinger, Anal. Chem., *48*, 17R (1976).
40. G. C. Davis, K. L. Holland, and P. T. Kissinger, J. Liq. Chromatogr., *2*(5), 663 (1979).
41. S. A. Margolis and R. Schaffer, J. Liq. Chromatogr., *2*(6), 837 (1979).
42. O. H. Lowry, J. U. Passonneau, and M. K. Rock, J. Biol. Chem., *236*, 2756 (1961).
43. T. J. Williams, T. K. Lee, and R. B. Dunlap, Arch. Biochem. Biophys., *181*, 569 (1977).
44. J. Everse, E. C. Zoll, L. Kahan, and N. O. Kaplan, Bioorg. Chem., *1*, 207 (1971).

45. J. T. Heard and G. J. Tritz, J. Liq. Chromatogr., 4(2), 325 (1981).
46. H. W. Shmukler, J. Chromatogr. Sci., 10, 137 (1972).
47. T. Jones, R. Spencer, and C. Walsh, Biochemistry, 17, 4011 (1978).
48. S. G. Mayhew and M. J. J. Strating, Eur. J. Biochem., 59, 539 (1975).
49. V. Massey and L. D. Mendelsohn, Anal. Biochem., 95, 156 (1979).
50. D. R. Light, C. Walsh, and M. Marletta, Anal. Biochem., 109, 87 (1980).
51. A. J. Speek, F. Van Schaik, J. Schrijver, and W. H. P. Schreur, J. Chromatogr., 228, 311 (1982).
52. P. Pietta, A. Calatroni, and A. Rava, J. Chromatogr., 229, 445 (1982).
53. P. Nielson, P. Rauschenbach, and A. Bacher, Anal. Biochem., 130, 359 (1983).
54. P. K. Tubbs and P. B. Garland, Methods Enzymol., 13, 535 (1969).
55. G. Michal and H. U. Bergmeyer (Eds.), *Methods of Enzymatic Analysis*, vol. 4, Academic Press, New York, 1974.
56. S. Skrede and J. Bremer, Eur. J. Biochem., 14, 465 (1970).
57. C. M. Smith, J. Nutr., 108, 863 (1978).
58. O. C. Ingebretsen and M. Farstad, J. Chromatogr., 202, 439 (1980).

17
HPLC Analyses of Pyrimidine and Purine Antimetabolite Drugs

JOY R. MIKSIC Biodecision Laboratories, Inc., Pittsburgh, Pennsylvania

I. Introduction 317
   A. The purine and pyrimidine drugs 317
   B. The pyrimidine antimetabolites 318
   C. The purine antimetabolites 321
II. Analytical Considerations 321
   A. General 321
   B. Sample preparation for HPLC 323
   C. HPLC columns and conditions 332
III. Summary 334
IV. Abbreviations 334
   References 334

I. INTRODUCTION

A. The purine and pyrimidine drugs

The purine and pyrimidine drugs are analogs of naturally occurring nucleotide bases and nucleosides. The drugs are formed by substitution, rearrangement, or addition to the purine or pyrimidine base or to the sugar moiety (Table 1). These drugs are important as antimetabolites, substances which interfere in the use or function of some

endogenous metabolite. They can become incorporated into DNA or
RNA and interfere in normal cell growth. The drugs are generally
active as antineoplastics, although a number of them have other uses:
antiviral, antibacterial, immunosuppressive, or uricostatic. In addition to numerous drugs on the market, many compounds are currently
under investigation, and many of the established antimetabolites are
being modified to enhance solubility, stability, or absorption properties. The methylxanthines (theophylline, theobromine, and caffeine)
are structurally related drugs which function as bronchiodialators,
central nervous system (CNS) stimulants, and diuretics, but do not
act as antimetabolites. Since the brochiodialators represent a very
important class of related drugs, but have very different biochemical
and analytical characteristics from the other purine and pyrimidine
drugs, the reader is referred to other sources for review [1-4]. This
chapter will focus on the high performance liquid chromatographic
methods which can be used for the analyses of those purine and pyrimidine drugs which act as antimetabolites.

B. The pyrimidine antimetabolites

The pyrimidine bases which, after structural modification, are most
useful clinically are uracil and cytosine (see Table 1a). The uracil
drugs include the widely used 5-fluorouracil (5-FU, fluracil), a precurser of 5-FU, ftorafur, and drugs which have the same active metabolites as 5-FU, including 1-hexylcarbamoyl-5-fluorouracil and 5-
fluoro-2'-deoxyuridine (floxuridine). These 5-FU drugs are therapeutic in the treatment of solid tumors such as breast, ovary, and
gastrointestinal cancers. Another structural analog of uracil, idoxuridine, is used topically as an antiviral agent to treat *Herpes simplex*.

A frequently used cytosine analog, cytosine arabinoside (Ara-C,
cytarabine) differs from most antimetabolites because the nucleoside,
cytidine, is chemically modified on the sugar moiety rather than the
base. Ara-C is one of the most effective agents in treating acute
myelocytic or lymphocytic leukemia. Ara-C is metabolized primarily
through deamination to form inactive uracil arabinoside [5], but its
therapeutic effect seems to be correlated with incorporation into DNA
of another of its metabolites, arabinosyl cytosine triphosphate (ARA-
CTP) [64]. Other cytosine analogs are 5-fluorocytosine (5-FC, flucytosine), used in combination with amphotericin B to treat severe
mycotic infection, and 5-azacytidine, used clinically for acute leukemia. Serious instability problems with 5-azacytidine have led to the
investigation of a similar drug, 5-aza-2'-deoxycytidine [6] and stabilization of azacytidine by addition of bisulfite ion on the pyrimidine
ring [7].

Table 1a The Pyrimidine Drugs

Drug	Abbreviations	Structure	Function
Uracil analogs			
Fluorouracil	5-FU		Antineoplastic
Ftorafur	FT		5-FU precurser
Idoxuridine			Antiviral
Cytosine analogs			
Cytosine arabinoside	Ara-C		Antileukemic
5-Fluorocytosine	5-FC		Antifungal
5-Azacytidine	Aza-C		Antileukemic

Table 1b. The Purine Drugs

Drug	Abbreviations	Structure	Function
6-Mercapto-purine	6-MP		Antileukemic
6-Thioguanine	6-TG		Antileukemic
Azathioprine			Immunosuppressive
Allopurinol			Xanthine oxidase inhibitor (gout)
Acyclovir			Antiviral
Adenine arabinoside	Ara-A		Antiviral

C. The purine antimetabolites

Of the purine analogs, 6-mercaptopurine (6-MP) and allopurinol are clinically the most important (see Table 1b). 6-MP has been widely used since the late 1950s to treat acute leukemia. 6-Thioguanine (6-TG), a purine analog and metabolic product of 6-MP, has similar function and mechanism. Both 6-TG and 6-MP require conversion to thioguanine deoxyribonucleotides, which are then incorporated into DNA to exert their effects [8]. Azathioprine, another thiopurine, is used for immunosuppression in organ transplants. Azathioprine is cleaved to form 6-MP and methylnitrosoimidazole, with the 6-MP probably responsible for immunosuppression [9,10]. Allopurinol, a structural analog of hypoxanthine, and its primary active metabolite oxypurinol, are xanthine oxidase inhibitors used in treating gout and hyperuricemia associated with leukemia and lymphomas. Recently, acyclovir, a guanine analog, has been shown to be an effective antiviral drug, strongly active against the *Herpes* viruses. Adenine arabinoside (Ara-A), another nucleoside modified on the sugar, is an effective antiviral agent, but solubility and dosage problems have limited its usefulness.

II. ANALYTICAL CONSIDERATIONS

A. General

1. Analytical problems

The purine and pyrimidine antimetabolites present unique problems to the analyst, whether for clinical, pharmacological, or pharmaceutical applications. One of the most serious problems is isolation of all of the species of interest from a biological matrix which contains many similar components. The pharmacologist may also need to measure multiple metabolic products, and the clinician must be concerned with co-administration of similar drugs. Studying the pharmacology or biochemistry of the antimetabolites requires sensitive detection for the parent compound, the metabolites, and the natural products which are affected. In pharmaceutical quality control, the measurement of synthetic by-products is crucial because related purine and pyrimidine analogs may be toxic. To detect formulation instability, extremely accurate quantitative methods are needed for monitoring small losses in the active ingredient.

2. Purification

Purification of the species of interest from a biological matrix or dosage form and isolation for specific quantitation can be difficult because of the high water solubility of the antimetabolites. Characteristically, drugs are more lipophilic than many endogenous serum components and can be selectively extracted into an organic solvent after pH adjustment. The antimetabolites, however, frequently have to be purified by precipitation of proteins, ultrafiltration, or other nonextractive methods. The large number of similar endogenous substances, nucleosides, and bases complicate isolation of the species of interest.

3. Chromatography

Chromatographic separation is usually needed for specific analysis of the antimetabolites. Separation by high performance liquid chromatography (HPLC) is generally preferred for isolation of the antimetabolites if the sensitivity of detection is sufficient. In order to achieve separation, the pH of the mobile phase can be adjusted to take advantage of the differences in the active groups. Other factors such as stationary phase and organic solvent in the mobile phase can be manipulated to optimize resolution. The HPLC system is usually kept at room temperature; however, a decrease in temperature can be used with a thermally unstable analyte or an increase in temperature can be utilized to improve resolution. For further structural confirmation, the HPLC effluent can be collected and the drug can be reanalyzed. Although gas chromatography represents a very sensitive analytical technique, it has major drawbacks for isolating the antimetabolites. Because the antimetabolites require high temperatures to vaporize, thermal instability can be a concern. Active groups, especially amines and hydroxyls, cause on-column adsorption and nonsymmetrical peaks. These problems necessitate derivatization, which is often undesirable because it can lead to long preparation times and masking of metabolites.

Tables 2a and 2b summarize selected procedures for analyzing the antimetabolites by HPLC published through June 1982. Approximate sensitivities and analysis times are compared within each major antimetabolite group. The use of different types of HPLC columns is more specifically defined in the following paragraphs.

4. Detection

There are a number of good methods of detecting the purine and pyrimidine antimetabolites. The antimetabolite drugs are strong absorbers of ultraviolet light. Also, spectra usually change enough from compound to compound, and with pH changes, that some identification can be made spectrophotometrically. However, sensitivity is usually in the mid to high microgram/milliliter range. When coupled with a HPLC separations, ultraviolet (UV) abosrbance measurements can extend sensitivity to the low nanogram/milliliter region for the antimetabolites. Fluorescence methods are utilized for static measurement as well as for on-line HPLC detection [8] with 10- to 50-fold gains in sensitivity over UV methods. The use of radioactive labels continues to be an inherently sensitive technique used very commonly in animal metabolism studies. When used in combination with HPLC, tritium labels permit detection in the low picogram range [11]. Gas chromatographic (GC) detectors such as the nitrogen-sensitive, electron-capture, and selective-ion mass spectrometer are useful in measuring into the low nanogram/milliliter drug region. In many cases HPLC with UV or fluorescence detection can achieve significantly greater sensitivities than GC methods.

B. Sample preparation for HPLC

1. Extraction

Selective extraction is the method of choice for purifying and concentrating analytes, but frequently results in poor yields for the antimetabolites. The antimetabolites are extracted from biological samples adjusted to pH 4-6. The organic solvent must be polar; usually ehtyl acetate or ethyl acetate in combination with another organic solvent is used. Chloroform [12-14] and ether/n-propanolol solutions [15] have also been found useful. The extraction of 6-MP results in recoveries ranging from 18% for a GC method requiring multiple extraction steps [16] to 80% for a single extraction method [17]. Azathioprine, being more lipophilic, has reported yields up to 98% [18]. Extractive yields for 5-FU commonly are lower, 58 [19] to 80% [15,20], and FT can be extracted up to 90% [19].

2. Precipitation

The majority of HPLC methods reported for biological samples involve precipitation of proteins from the aqueous matrix. Precipitants which have been used are perchloroacetic acid [8,11,18,21-24], trichloroacetic acid [25-29], alcohol [28,30], and ammonium sulfate/barium hydroxide [31]. The precipitation process can often be made more efficient if the sample is cooled or left to stand for a few hours. Although the protein cleanup is nearly 100% [32], deproteination does not isolate the antimetabolite from endogenous nucleosides and bases, metabolites, or coadministered drugs as some extractive techniques do. Therefore, chromatography must be utilized to eliminate these interferences. Concentration after use of a precipitation method is difficult because of the aqueous matrix. Precipitation methods are frequently inadequate for measurements required in metabolism studies or clinical assays where tissue concentrations are low.

3. Ultrafiltration

An alternative method of protein removal is ultrafiltration through membranes, such as Amicon CF25 or GE2500 membranes [20,30,33,34]. As with precipitation methods, ultrafiltration separates only the protein fraction, and therefore suffers from the same types of incomplete

Table 2a Pyrimidine Drugs—Selected Chromatographic Conditions

Compound measured[a]	Stationary phase[b]	Mobile phase	Sensitivity (ng/ml)	Analysis time (min)	Ref.
5-Fluorouracil drugs—clinical					
5-FU and FT	μBondapak C_{18} (Waters)	3-6% $CH_3OH/C_2H_3O_2^-$, pH 4	<10 (FU) <3500 (FT)	30	20
5-FU	μBondapak C_{18}	PO_4^{3-}, pH 3	25	8	41
5-FU	Aminex A-25 SAX (BioRad)	$C_2H_3O_2^-$, pH 4.5	100	20	15
FT	μBondapak C_{18}	10% CH_3OH/H_2O	100	8	44
5-FU Drugs—pharmacokinetic					
5-FU	Partisil PXS ODS-2 (Whatman)	PO_4^{3-}, pH 5.5	100	11	43
FT	Zorbax SIL (Dupont)	4% $C_2H_5OH/C_2H_4Cl_2$	25	8	12
Ft, FU, and metabolites	μBondapak C_{18}	5% $CH_3OH/C_2H_3O_2^-$, pH 4	20 (FU) 100 (FT, met.)	20	42 19
FT and metabolites	μBondapak C_{18}	H_2O	—	36	33
1-Hexylcarbamoyl	μBondapak C_{18}	30 % CH_3CN/H_2O	10 (serum)	16	45
5-FU and metabolites		15% $CH_3CN/35$% THF/H_2O	250 (tissue)		45
		32% CH_3OH/H_2O	1	17	14

5-Deoxy-5-FU	LiChrosorb RP-18 5 µm (Altex)	1.5% CH$_3$CN/1.5% CH$_3$OH/H$_2$O	50	15	61
5-F-2'DeoxyUMP + nucleotides	Four separate chromatographic conditions		10 fmol/10^6 cells	30-70	11

5-FU Drugs—chromatography and pharmaceutical

5-FU and impurity 5-F-2'-deoxyuridine	LiChrosorb RP-18 10 µm	1% CH$_3$CN/H$_2$O	—	20	62
5-FU Nucleosides and nucleotides	RSIL C$_{18}$HL, 18% bonded (RSL)	Ion pair: tetrabutylammonium hydroxide in 5% CH$_3$OH/PO$_4^{3-}$ pH 5	100	12	40

Idoxuridine—pharmaceutical

Idoxuridine from impurities	µBondapak C$_{18}$	13% CH$_3$OH/H$_2$O	—	25	49
	10 different C$_{18}$ columns	4% CH$_3$OH or 5% CH$_3$CN or 2% THF in H$_2$O	—	—	63

Cytosine arabinoside drugs—clinical

Ara-C and Ara-U	Spherisorb ODS-5 (Latek, GFR)	PO$_4^{-3}$, pH 7	50	10-15	37
Ara-C	Partisil PXS 10/25 SCX (Whatman)	CHO$_2^-$, pH 4.8	50	15	58

Table 2a (Continued)

Compound measured[a]	Stationary phase[b]	Mobile phase	Sensitivity (ng/ml)	Analysis time (min)	Ref.
Ara-C Drugs—pharmacokinetic					
Ara-C from metabolites	Aminex A-27SAX (BioRad)	Citrate, tetraborate pH 9.3 at 65°C	~5 ng (on column)	20-70	30
	μBondapak C_{18}	PO_4^{3-}, pH 5.6	~0.4 ng (on column)	15	30
Ara-C-5'-tri-phosphate	ABX (DuPont)	Gradient: PO_4^{3-}, pH 3-4.4	—	50	64
Ara-C, Ara-U	Aminex A-7 SCX	PO_4^{3-}, pH 4.3	—	12	57
Ara-C—Chromatography					
2-F-2-deoxyarabino-furanosyl nucleosides	Partisil ODS-1, ODS-3 5, 10 μm Whatman	4% CH_3OH in PO_4^{3-}, pH 5.3, to 20% CH_3OH/ same buffer	—	25	see 60, 65
Ara-C and Ara-U and analogs	2 Nucleosil 10 C_{18} 10 μm	PO_4^{3-}, pH 2.0	2 (Ara-C) 100 (Ara-U)	30	50

Azacytidine drugs—pharmaceutical

5-Azacytidine from degradation products	Aminex A-6 SCX	NH_4CHO_2, pH 4.6	—	8	51
	Zorbax C-8	PO_4^{3-}, pH 6.5	—	5	7

Flucytocine—clinical

5-FC	Varian pellicular cation exchange	$(NH_4)_3PO_4$, pH 2.5 (74°C)	5000	40	38
	μBondapak C_{18}	PO_4^{3-}, pH 7	5000	10	25

[a] For abbreviations see Tables 1a and 1b.
[b] Manufacturer listed only first time referenced.

Table 2b Purine Drugs—Selected Chromatographic Conditions

Compound measured[a]	Stationary phase[b]	Mobile phase	Sensitivity (ng/ml)	Analysis time (min)	Ref.
6-Mercaptopurine drugs—clinical					
6-MP, azathioprine	μBondapak C_{18}	15% $CH_3OH/HC_2H_3O_2$ (0.02%)	0.5	12	18
6-MP	Two Spherisorb 10-ODS	PO_4^{3-}, pH 6.35	3	15	29
6-MP drugs—pharmacokinetic					
6-MP	LiChrosorb RP8-18 10 μm	1% $CH_3OH/0.5\%$ $CH_3CN/60$ mg DTE/PO_4^{3-}, pH 4.0	5	8	46
Azathioprine	μBondapak C_{18}	11% $CH_3CN/NaC_2H_3O_2$, pH 4	40	10	46
6-TG nucleotide metabolites of 6-MP and 6-TG	Partisil 10 SAX (Whatman)	Gradient: PO_4^{3-}, pH 3.5, to PO_4^{3-}/Cl^- pH 4.5	~1 ng	30	8
6-MP	μBondapak C_{18}	Ion pair: heptane sulfonic acid in 10% $CH_3OH/1\%$ $C_2H_3O_2^-$	200	10	26
6-MP and all metabolites	M71 cation exchange 10 μm (Beckman)	CHO_2^-, pH 4.6	~0.030 (on column)	140	58
6-MP nucleotide metabolites	Pellionex 10 SAX (Whatman)	Gradient: $PO_4^{3-}/C_2H_3O_2^-$, pH 5, to PO_4^{3-}, pH 3.9 (70°C)	—	100	52

Compound	Column	Mobile phase	Amount		Ref.
6-TG and metabolites	µBondapak C_{18}	10% $CH_3OH/C_2H_3O_2^-$, pH 3.5 25% $CH_3OH/C_2H_3O_2^-$, pH 3.5 (for 6-MeTG)	150 (6-TG)	10	21
6-MP drugs—pharmaceutical					
6-MP and azathioprine	ODS hypersil, 5 µm (Shandon-Southern Instruments)	20% $CH_3OH/79.5\%\ PO_4^{3-}/0.5\%\ C_2H_3O_2^-$, pH 4.5	0.5 (on column)	5	39
Allopurinol drugs—clinical					
Allopurinol, oxipurinol	SAS Hypersil 5 µm	0.1 M citric acid/0.2 M PO_4^{3-}	100	10	36
Allopurinol, oxipurinol, 5-FU purine metabolites	µBondapak C_{18}	PO_4^{3-}, pH 4.5	15	13	22
Allopurinol, oxipurinol	Aminex A-27 12-15 µm	$C_2H_3O_2^-$, pH 8.7 (71°C)	75	15	53
Allopurinol, oxipurinol	Aminex A-28	$C_2H_3O_2^-$, pH 6.45	500	30	54
Allopurinol drugs—pharmacokinetic, metabolic					
Purine and pyrimidines related to gout	µBondapak C_{18}	Gradient PO_4^{3-}, pH 5.8 to CH_3OH (100%)	—	30	48

Table 2b (Continued)

Compound measured[a]	Stationary phase[b]	Mobile phase	Sensitivity (ng/ml)	Analysis time (min)	Ref.
Allopurinol drugs—pharmacokinetic, metabolic (cont.)					
Allopurinol, oxipurinol	Spherisorb ODS 5 μm (LDC)	PO_4^{3-}, pH 6.0	100	30	27
Xanthine oxidase products	μBondapak C_{18}	0.05 M PO_4^{3-}, pH 4.5	25 (on column)	15	47
Acyclovir—clinical					
Acyclovir	Zorbax ODS, 4.6 μm	Ion pair: heptane sulfonic acid 0.005 M; $C_2H_3O_2^-$, pH 6.5 (35 or 50°C)	200 (plasma) 2000 (urine)	15 (serum) 35 (urine)	31
Adenine arabinoside—pharmacological					
Ara-ATP (from nucleotides)	Partisil 10-SAX (Whatman)	Gradient: 5 mM PO_4^{3-}, pH 2.8, to 750 mM PO_4^{3-}, pH 3.7	~3000/10⁹ cells	40	23

Ara-A—pharmaceutical

Ara-A-5'-formate	Zipax SCX (DuPont)	PO_4^{3-}/NO_3^-, pH 4.5	—	14	56

Ara-A—chromatography

Adenine and hypoxanthine arabinosides	Aminex A-28 SAX	0.005-0.2 M tetraborate in $C_2H_3O_2^-$, pH 6.4	—	30	34
Arabino, ribodeoxy-ribonucleosides	Aminex A-6 SCX	0.1 M boric acid in NH_4^+, pH 7.4	—	20-40	60

[a] For abbreviations see Tables 1a and 1b.
[b] Manufacturer listed only first time referenced.

purification and concentration problems. In addition, protein-bound drug will be retained unless steps are taken to "debind" it for total drug measurement. Both precipitation and ultrafiltration methods can be used prior to more extensive purification by extraction, adsorption, or chromatographic techniques [28,35].

4. Direct injection

Direct injection has been used for HPLC analyses successfully with the antimetabolites [36-38] as with some other drugs. Generally, direct injection of biological samples is not a preferred method because it limits the sample volume, can cause high pressures on the analytical column, or requires frequent change of a guard column. As with the protein removal methods, interferences must be resolved chromatographically. It is expected that column switching techniques will be developed to provide better methods for both concentration of the analyte and isolation from proteins. These techniques should avoid introduction of excessive biological material onto the analytical column while retaining the advantages of chromatographic purification and automated operation.

C. HPLC columns and conditions

1. Reversed phase

Reversed-phase chromatography is the most practical of the separation modes if the drug is lipophilic enough to be retained and separated. Reversed-phase methods most frequently use octadecylsilane (C_{18}) stationary phases bonded to 10-μm particle support. Some recent separations have been reported on the more efficient 5-μm columns [27,36,39,40] and on octylsilane (C_8) stationary phases [7]. Chromatography requires between 5 [7] and 30 min [20,27,33], depending on the complexity of the separation. Mobile phases are composed of a small amount of organic solvent (0 to 20% acetonitrile or methanol) in an aqueous buffer of pH 3-7. Partitioning would often be improved at higher pH where the amount of ionization is suppressed, but unfortunately the silica-based columns cannot withstand alkaline conditions. The 5-FU drugs are well separated by reversed-phase chromatography almost exclusively using C_{18} columns [19,20,33,40-45]. The 6-MP drugs [18,21,29,39,46] and the allopurinol drugs [22,27,36,47,48] are also chromatographed primarily by reversed phase. Other antimetabolite separations reported by reversed-phase chromatography include idoxuridine [49], 5-aza-2'-deoxycytidine [6], 5-FC [25], and Ara-C [30,50].

2. Ion-paired reversed phase

Ion-pairing reagents in combination with reversed-phase chromatography can be used with both negative and positive ionic species. Acetate ion can be used for separating 6-MP and azathioprine from impurities and degradants in dosage forms [39]. Plasma assays for 6-MP [26] and acyclovir [31] employ the anionic pairing reagent heptanesulfonic acid. The mechanism of adsorption and effectiveness of tetrabutylammonium hydroxide (TBAOH) for use with 5-fluorouracil, and the deoxyribo- and ribonucleosides and nucleotides has been investigated [40]. TBAOH extends the retention of the nucleotides, while the bases and nucleosides are unaffected. The tetrabutylammonium cation in the bromide salt form has also been used for 5-fluoro-2'-deoxyuridine monophosphate analysis [11].

3. Ion exchange

Ion-exchange chromatography has been used for HPLC analyses of the antimetabolites since the early 1970s. When reversed-phase columns with greater resolution were developed in the late 1970s, they became favored in assay development. However, a number of excellent ion-exchange methods demonstrate specificity and versatility for separating the drugs from complex biological matrices. Analysis times are generally longer for ion exchange, ranging from 8 min for a single drug determination [51] to about 2.5 hr for the analysis of all purine and 6-thiopurine bases and nucleosides [52].

Ionic strength and pH of the mobile phase are the most powerful factors in ion-exchange separations. Mobile phases chosen for cation exchange are frequently ammonium formate or potassium phosphate. For anion exchange, the most common anions used are phosphate, acetate, and chloride. Ion-exchange resins are usually operated between pH values 3 and 6; however, styrene-divinyl benzene resins can be used into the alkaline pH range [30,53].

The nucleotides are easily separated from the drugs, nucleosides, and bases using anion exchange; however, the strong interactions can create undesirably long retention times. Run times can be shortened by using pH gradients or preliminary purification steps to remove strong anions. Anion-exchange methods have been reported for the analysis of ara-ATP [23], ara-C and its metabolites [30], 5-FU [15], allopurinol/oxypurinol [53,54], and 6-MP and its nucleotide metabolites [8,55]. Anion exchange has been used for species varying only in the sugar moiety (ribonucleosides, deoxyribonucleosides, and arabinosides) by taking advantage of borate complexes which selectively hinder hydroxyl interaction with ion-exchange sites [30,34].

Another class of separations of ribonucleotides from deoxynucleotides developed for the 5-FU metabolites uses periodate oxidation of the ribonucleotides, followed by separate analysis of the ribonucleotides and deoxynucleotides on anion exchange [11].

Cation exchange is used primarily in the ion-exclusion mode. The strong anions are not retained, while the nitrogenous bases in their neutral or cationic form are separated. Some examples of cation-exchange separations are described for the 5'-O-formate ester of Ara-A [56], Ara-C [57,58], 6-MP and its metabolites [52,59], and Aza-C [51]. Borate complexes have been used to improve separation [60] with the advantage of shorter run times over similar anion-exchange separations.

III. SUMMARY

The antimetaoblite drugs represent a rapidly expanding area of pharmaceutical growth. High performance liquid chromatography has provided solutions to many of the analytical problems faced in development and biochemical monitoring of these products. High water solubility and similar endogenous biological components makes isolation and measurement of the antimetabolites especially difficult. Identification of potential interferences must be of primary concern for selection of the analytical method chosen.

IV. ABBREVIATIONS

Ara-A   = adenine arabinoside
Ara-ATP = adenine arabinoside 5'-triphosphate
Ara-C   = cytosine arabinoside
Ara-U   = uracil arabinoside
Aza-C   = azacytidine
5-FC   = 5-fluorocytosine
5-FU   = 5-fluorouracil
6-MP   = 6-mercaptopurine
6-TG   = 6-thioguanine

REFERENCES

1. M. Weinberger and C. Chidsey, Clin. Chem., 21, 834 (1975).
2. R. F. Adams, F. L. Vandemark, and G. J. Schmidt, Clin. Chem., 22, 1903 (1976).
3. J. J. Orcutt, P. P. Kozak, Jr., S. A. Gullman, and L. H. Cummins, Clin. Chem., 23, 599 (1977).

4. W. Sadee and G. C. M. Beelen, Drug Level Monitoring. Analytical Techniques, Metabolism, and Pharmacokinetics, John Wiley & Sons, New York, 1980, pp. 453-459.
5. G. W. Camiener and C. G. Smith, Biochem. Pharmacol. 14, 1405 (1965).
6. K-T. Lin, R. L. Momparler, and G. E. Rivard, J. Pharm. Sci., 70, 1228 (1981).
7. D. C. Chatterji and J. F. Gallelli, J. Pharm. Sci., 68, 822 (1979).
8. D. M. Tidd and S. Dedhar, J. Chromatogr., 145, 237 (1978).
9. P. de Miranda, L. M. Beacham, III, T. H. Creagh, and G. B. Elion, J. Pharmacol. Exp. Ther., 187, 588 (1973).
10. G. B. Elion and G. H. Hitchings, in Handbook of Experimental Pharmacology, XXXVIII/2, (A. C. Sartorelli and D. G. Johns, Eds.), Springer Verlag, Berlin, 1975, p. 404.
11. R. Dreyer and E. Cadman, J. Chromatogr., 219, 273 (1981).
12. T. Marunaka, Y. Umeno, K. Yoshida, M. Nagamachi, Y. Minami, and S. Fujii, J. Pharm. Sci., 69, 1296 (1980).
13. H. Isomura, S. Higuchi, and S. Kawamura, J. Chromatogr. Biomed. Appl., 224, 423 (1981).
14. O. Nakajima, Y. Yoshida, T. Isoda, Y. Takemasa, Y. Imamura, and Y. Koyama, J. Chromatogr. Biomed. Appl., 225, 91 (1981).
15. J. L. Cohen and R. E. Brown, J. Chromatogr., 151, 237 (1978).
16. J. M. Rosenfeld, Y. Y. Taguchi, B. L. Hillcoat, and M. Kawal, Anal. Chem., 49, 725 (1977).
17. D. G. Bailey, T. W. Wilson, and G. E. Johnson, J. Chromatogr., 111, 305 (1975).
18. S-N. Lin, K. Jessup, M. Floyd, T-P. F. Wang, C. T. Van Buren, R. M. Caprioli, and B. D. Kahan, Transplantation, 29, 290 (1980).
19. A. T. Wu, J. L. Au, and W. Sadee, Cancer Res., 38, 210 (1978).
20. C. L. Hornbeck, R. A. Floyd, J. C. Griffiths, and J. E. Byfield, J. Pharm. Sci., 70, 1163 (1981).
21. P. A. Andrews, M. J. Egorin, M. E. May, and N. R. Bachur, J. Chromatogr. Biomed. Appl., 227, 83 (1982).
22. W. E. Wung and S. B. Howell, Clin. Chem., 26, 1704 (1980).
23. L. M. Rose and R. W. Brockman, J. Chromatogr., 133, 335 (1977).
24. R. L. Dedrick, D. D. Forrester, and D. H. W. Ho, Biochem. Pharmacol., 21, 1 (1972).
25. J. O. Miners, T. Foenander, and D. J. Birkett, Clin. Chem., 26, 117 (1980).
26. J. L. Day, L. Tterlikkis, R. Niemann, A. Mobley, and C. Spikes, J. Pharm. Sci., 67, 1027 (1978).
27. W. G. Kramer and S. Feldman, J. Chromatogr., 162, 94 (1979).

28. J. Boutagy and D. J. Harvey, J. Chromatogr. Biomed. Appl., 146, 283 (1978).
29. R. A. De Abreau, J. M. Van Baal, T. J. Schouten, E. D. A. M. Schretlen, and C. H. M. M. De Bruyn, J. Chromatogr. Biomed. Appl., 227, 526 (1982).
30. M. G. Pallavincini and J. A. Mazrimas, J. Chromatogr. Biomed. Appl., 183, 449 (1980).
31. G. Land and A. Bye, J. Chromatogr. Biomed. Appl., 224, 51 (1981).
32. J. Blanchard, J. Chromatogr. Biomed. Appl., 226, 455 (1981).
33. J. A. Benvenuto, K. Lu, and T. L. Loo, J. Chromatogr., 134, 219 (1977).
34. H. G. Schneider and A. J. Glazko, J. Chromatogr., 139, 370 (1977).
35. D. B. Lakings, R. H. Adamson, and R. B. Diasio, J. Chromatogr. Biomed. Appl., 146, 512 (1978).
36. H. Breithaupt and G. Goebel, J. Chromatogr. Biomed. Appl., 226, 237 (1981).
37. H. Breithaupt and J. Schick, J. Chromatogr. Biomed. Appl., 225, 99 (1981).
38. A. D. Blair, A. W. Forrey, B. T. Meijsen, and R. E. Cutler, J. Pharm. Sci., 64, 1334 (1975).
39. A. F. Fell, S. M. Plag and J. M. Neil, J. Chromatogr., 186, 691 (1979).
40. C. F. Gelijkens and A. P. De Leenheer, J. Chromatogr., 194, 305 (1980).
41. N. Christophidis, G. Mihaly, F. Vajda, and W. Louis, Clin. Chem., 25, 83 (1979).
42. J. L. Au, A. T. Wu, M. A. Friedman, and W. Sadee, Cancer Treatment Rep., 63, 343 (1979).
43. T. A. Phillips, A. Howell, R. J. Grieve, and P. G. Welling, J. Pharm. Sci., 69, 1428 (1980).
44. N. Hobara and A. Watanabe, J. Chromatogr. Biomed. Appl., 146, 518, (1978).
45. A. Kono, M. Tanaka, S. Eguchi, Y. Hara, and Y. Matsushima, J. Chromatogr. Biomed. Appl., 163, 109 (1979).
46. T. L. Ding and L. Z. Benet, J. Chromatogr. Biomed. Appl., 136, 281 (1979).
47. G. J. Putterman, B. Shaikh, M. R. Hallmark, C. G. Sawyer, C. V. Hixson, and F. Perini, Anal. Biochem., 98, 18 (1979).
48. A. McBurney and T. Gibson, Clin. Chim. Acta, 102, 19 (1980).
49. G. R. Carr, J. Chromatogr., 157, 171 (1978).
50. P. Linssen, A. Drenthe-Schonk, H. Wessels, and C. Haanen, J. Chromatogr. Biomed. Appl., 223, 371 (1981).

51. K. K. Chan, D. D. Giannini, J. A. Staroscik, and W. Sadee, J. Pharm. Sci., *68*, 807 (1979).
52. H-J. Breter, Anal. Biochem., *80*, 9 (1977).
53. M. Brown and A. Bye, J. Chromatogr. Biomed. Appl., *143*, 195 (1977).
54. R. Endele and G. Lettenbaur, J. Chromatogr., *115*, 228 (1975).
55. T. P. Zimmerman, L-C. Chu, C. J. L. Bugge, D. J. Nelson, G. M. Lyon, and G. B. Elion, Cancer Res., *34*, 221 (1974).
56. A. J. Repta, B. J. Rawson, R. D. Shaffer, K. B. Sloan, N. Bodor, and T. Higuchi, J. Pharm. Sci., *64*, 392 (1975).
57. S. H. Wan, D. H. Huffman, D. L. Azarnoff, B. Hoogstraten, and W. E. Larsen, Cancer Res., *34*, 392 (1974).
58. R. W. Bury and P. J. Keary, J. Chromatogr. Biomed. Appl., *146*, 350 (1978).
59. H-J. Breter and R. K. Zahn, J. Chromatogr., *137*, 61 (1977).
60. B. C. Pal, J. Chromatogr., *148*, 545 (1978).
61. J. P. Sommadossi, and J. P. Cano, J. Chromatogr. Biomed. Appl., *225*, 516 (1981).
62. J. L. Day, J. Maybaum, and W. Sadee, J. Chromatogr., *206*, 407 (1981).
63. S. H. Hansen and M. Thomsen, J. Chromatogr., *209*, 77 (1981).
64. Y. M. Rustum and H. D. Preisler, Cancer Res., *39*, 42 (1979).
65. A. Feinberg, J. Chromatogr., *210*, 527 (1981).

18

Central Nervous System Drugs (Methylxanthine Drugs)

KATSUYUKI NAKANO PL Medical Data Center, Tondabayashi, Osaka, Japan

I. Introduction 339
II. Specific Applications 341
   A. HPLC analysis 341
   B. Drug metabolism and pharmacokinetics 357
   References 361

I. INTRODUCTION

The methylxanthine drugs, such as caffeine (1,3,7-trimethylxanthine), theophylline (1,3-dimethylxanthine), and theobromine (3,7-dimethylxanthine), are stimulators of the central nervous system (CNS). Caffeine is the most potent stimulator of the CNS, especially of the medullary respiratory center, and the most useful drug in the prevention of apnea in premature infants.

Theophylline is a mild diuretic agent, a moderately active CNS and myocardial stimulant, a powerful bronchodilator which relaxes the smooth muscle fibers of bronchi, and a useful antiasthmatic drug. Toxicity has been observed, involving nausea, vomiting, headache, convulsions, and even death. The clinical response to theophylline relates well to its concentration in blood, with maximal therapeutic activity occurring in the 10-20 mg/liter range and toxic symptoms appearing with increasing frequency above 20 mg/liter. Therefore, it is clinically important to monitor precisely this narrow range of therapeutic levels in blood.

A great variety of techniques for the determination of methylxanthine drugs in biological fluids has been published, ranging from spectrophotometry, fluorometry, gas chromatography (GC), and thin-layer chromatography (TLC) to comparatively recent methods such as radioimmunoassay (RIA) and enzyme immunoassay (EIA). During the past decade, a number of publications have emphasized the advantage of high performance liquid chromatography (HPLC) with several modes including ion exchange, adsorption, reversed phase, and ion pair, with respect to sensitivity, specificity, and the possibility of using microsamples.

In general, there are four methods of sample preparation prior to the HPLC analysis. These include the use of solvent extraction, deproteinization, ultrafiltration, and direct sample injection. Solvent extraction is the most common procedure for isolating the drugs from the biological fluids. This technique is time-consuming, but it has the advantages of minimizing potential interferences by endogenous compounds and concentrating the drug.

Deproteinization (protein precipitation) is also a commonly used method for sample pretreatment. This procedure is relatively easy to perform and tends to minimize the amount of protein injected onto the column. However, the technique has some limitations, such as the dilution of sample (in the use of acetonitrile or methanol), the loss of a part of protein-bound drugs (in the use of TCA or PCA), and the interferences caused by endogenous compounds and other drugs. Recently ultrafiltration or membrane filtration has become a commonly used method which is easy to operate. However, this method has some of the problems encountered in deproteinization.

Direct sample injection is not widely accepted. This procedure requires the use of a guard column to protect the analytical column from proteins. After approximately 30 injections the guard column must be replaced.

The analysis time with the HPLC method is dependent on the objective of the analysis. For example, is the objective the determination of the parent drug only or simultaneous determination of the parent drug with drug metabolites? The analysis of urine samples is generally longer than for serum or other physiological fluids because of the large number of endogenous compounds present. The detection limit of drugs by HPLC is usually 1-40 mg/liter, and as low as 0.2 mg/liter in some cases. The required sample volume is usually 100 µl to 1 ml in most cases of the ion exchange and adsorption modes, but 10 to 100 µl in the reversed-phase modes.

A number of interferences in the HPLC assay for methylxanthines, especially theophylline, have been reported. They include endogenous compounds, the drug metabolites, and other drugs, such as cephalosporin antibiotics, sulfamethoxazole, trisulfapyrimidine, acetazolamide, methicillin, ampicillin, and sulfisoxazole. In addition, interference caused by the deproteinization with acetonitrile has been observed in the theophylline assay using reversed-phase HPLC with sodium acetate-acetonitrile as a mobile phase. The noticeable interferences will be discussed later.

In this chapter, the author will discuss a number of HPLC procedures for the determination of methylxanthines and their applications to the studies of drug metabolism and pharmacokinetics.

II. SPECIFIC APPLICATIONS

A. HPLC analysis

1. Ion exchange

Ion-exchange chromatography was one of the first separation modes used for the determination of theophylline. Thompson et al. [1] developed a cation-exchange HPLC method with ammonium phosphate elution for the assay of theophylline and its metabolites in serum and urine. This method was modified slightly for use with pediatric serum samples by Weinberger and Chidsey [2]. In the procedure, a sample containing 8-chlorotheophylline as an internal standard was directly analyzed with the use of a guard column. An analysis time of more than 30 min was required for each sample injection to avoid the interferences of endogenous compounds.

In 1976, Peng et al. [3] and Jusko and Poliszczuk [4] used an extraction procedure with ethyl acetate or chloroform-isopropanol, respectively, prior to cation-exchange HPLC analysis with an eluent of 0.66% acetic acid. Although the extraction procedure from serum or saliva samples required a rather long time, this HPLC analysis was performed within 15 min and was very sensitive for the theophylline assay.

In the HPLC analysis of caffeine, a cation-exchange resin was also used to determine caffeine in coffee and the methylxanthines including caffeine, theophylline, and theobromine in serum. Walton et al. [5] examined the effects of pH, solvent composition, and counterion in their separation.

The chromatographic conditions for the determination of methylxanthine drugs and their metabolites using ion-exchange HPLC analyses are listed in Table 1.

Table 1 Chromatographic Conditions for the Determination of Methylxanthine Drugs and Their Metabolites Using Ion Exchange

Compounds resolved[a]	Stationary phase	Mobile phase
UA, (X, 1MU), (3MX, 13MU), 1MX, HX, THP and other xanthines	Aminex A-5 (13 μm, Bio-Rad)	0.45 M $NH_4H_2PO_4$, pH 3.65
UA, X, 3MX, HX, THP 8C1T	Aminex A-5 (13 μm, Bio-Rad)	0.45 M $NH_4H_2PO_4$, pH 3.65
THP, THB, CAF	Partisil SCX (10 μm, Reeve Angel)	0.66% Acetic acid in water
THB, THP, βHET, CAF, other xanthines and drugs	Zipax SCX (DuPont)	0.66% Acetic acid in water
THP, THB, CAF, DEC, and other drugs	Aminex 50W-X4 (20-30 μm, Bio-Rad)	0.05 M Na_2HPO_4, pH 7.5

[a] Abbreviations used: UA, uric acid; X, xanthine; 1MU, 1-methyluric acid; 3MX, 3-methylxanthine; 13MU, 1,3-dimethyluric acid; 1MX, 1-methylxanthine; HX, hypoxanthine; THP, theophylline; 8C1T, 8-chlorotheophylline; THB, theobromine; CAF, caffeine; βHET, β-hydroxyethyltheophylline; DEC, diethylcaffeine. The compounds in parentheses are not resolved.
[b] U, urine; SR, serum; P, plasma; B, whole blood; SL, saliva; SP, spinal fluid.

2. Adsorption chromatography

Adsorption (liquid-solid) chromatography has also been applied to the assay of theophylline. Manion et al. [6] developed a method to determine the theophylline level in biological fluids by HPLC using a Durapak OPN column. The elution solvents were 14% isopropanol in hexane followed by a linear gradient. The samples were extracted by chloroform-isopropanol containing oxazepam as an internal standard.

Later, various methods for the determination of theophylline were developed using a variety of silica gel packings and nonpolar solvents. In these methods, sample analytes were extracted with appropriate organic solvents.

The procedure developed by Weddle and Mason [7] was fast enough (21 min from receipt of blood to reporting theophylline level) to be used

Flow rate (ml/min)	temp. (°C)	Detector	Sample preparation[b]	Ref.
0.17	55	UV, 280 nm	Dowex 2-X8(U)/ direct (SR)	1
0.4	55	UV, 254 nm	Direct (SR, P)	2
1	50	UV, 275 nm	Ethyl acetate (P, SL)	3
—	amb.	UV, 280 nm	TCA/chloroform-isopropanol (SR)	4
0.2	68	UV, 270 nm	Direct/TCA (B, SR)	5

for emergency determinations. In this method, the sample, extracted with 40% 2-butanol in n-hexane, was injected directly onto the HPLC system using an eluent of 1% distilled water, 5% methanol, and 25% 2-butanol in n-hexane. No evaporation with subsequent reconstitution in the eluent was required. The same idea, that is, a one-step extraction procedure was adopted for the rapid and selective assay of theophylline in serum and saliva by means of adsorption chromatography [8].

The determination of theophylline and dyphylline levels in serum was performed by Maijub et al. [9] using adsorption chromatography with Partisil 10 packings and an eluent of chloroform/n-heptane/methanol. These compounds were confirmed by MS for the separated fractions.

Sved and Wilson [10] reported the simultaneous assay method of methylxanthine metabolites of caffeine in plasma using LiChrosorb Si-60 packings and 2.2% buffered methanol in dichloromethane as a mobile phase. In this procedure, theobromine, theophylline, paraxanthine (1,7-dimethylxanthine), prednisolone (internal standard), 3-methylxanthine, 7-methylxanthine, and 1-methylxanthine were separated within 12 min. However, this method was not applicable to urine extracts because of the rapid deterioration caused by monomethylxanthines.

The method for simultaneous determination of caffeine, theophylline, and theobromine from the serum, saliva, and spinal fluid of neonates was developed by Tin et al. [11]. Only 50 µl of samples were extracted with chloroform/isopropanol and analyzed by adsorption chromatography with LiChrosorb Si-10 and chloroform/isopropanol/acetic acid as a mobile phase.

The assay of caffeine in serum using straight-phase HPLC was reported by Van Der Meer and Haas [12]. The extractants from 0.2 ml of serum sample using dichloromethane with carbamazepine as an internal standard were analyzed by an HPLC system with a Partisil 5 column and an eluent of 20% tetrahydrofuran in dichloromethane.

Recently, Van Aerde et al. [13] reported a method applicable in the routine monitoring of therapeutic levels of theophylline in patients as well as for pharmacokinetic purposes. The theophylline levels as low as 0.2 mg/liter can be measured with acceptable precision in only 0.1 ml of plasma sample. In these adsorption chromatographic methods, a nonpolar mobile phase solvent is used. This solvent system is rather costly compared with the aqueous eluents in reversed phase. However, in the procedure with adsorption chromatography, the interference of paraxanthine with a retention time identical to that of theophylline in most reversed-phase systems was easily removed [8,10,12,13]. The chromatographic conditions for these analyses are listed in Table 2.

3. Reversed-phase HPLC

For the analysis of less polar biochemicals, reversed-phase HPLC (RPLC), in which usually the nonpolar C_{18} hydrocarbon chemically bonded to the silica surface is used, has the advantages of very high resolution and the use of an aqueous mobile phase. Thus, there has been an explosive use of RPLC in a variety of biochemical and biomedical fields.

Franconi et al. [14] developed the RPLC method for the determination of theophylline in plasma using µBondapak C_{18} as the packing material and sodium acetate/acetonitrile as the mobile phase. Plasma (1.0 ml) was mixed with 0.1 ml of borate buffer and filtered using ultrafiltration membranes prior to chromatographic analysis. They found that the RPLC method offered high recovery and accuracy for the theophylline assay compared with GC and spectroscopic methods.

Adams et al. [15] also developed a RPLC method for the assay of theophylline in 50 µl of serum using an ODS-Sil-X-I packing and acetic acid/water/acetonitrile mobile phase. The sample was prepared by extraction of the theophylline using equal volumes of chloroform and isopropanol. 8-chlorotheophylline was used as an internal standard.

In the RPLC determination of theophylline in serum and plasma, no interference was noted in the ultraviolet (UV) detection when Na or Li heparin or ethylenediaminetetraacetate (EDTA) were used as anticoagulants, but citrated plasma was found to be unsatisfactory because of a highly absorbing band which eluted with the same retention as theophylline (Nelson et al.) [16]. In this study, the interference due to dyphylline, which is also used in the treatment of asthma, was observed when the serum from patients treated with the drug was analyzed. This problem was overcome in the other RPLC methods in which the mobile phase and/or analysis temperature and detector were modified using reversed-phase packings similar to those used in the above method [17-20].

Orcutt et al. [21] reported a microscale method for determining theophylline in 30 μl of serum mixed with an equal volume of solution containing β-hydroxyethyltheophylline as an internal standard. The interferences by several xanthine derivatives including caffeine, phenobarbital, theobromine, and theophylline metabolites were excluded in this method. However, they observed an interferant presumed to be a metabolite of caffeine. This metabolite, paraxanthine, is found in samples containing high concentrations of caffeine. Paraxanthine is difficult to detect using ratios of peak height obtained at two different wavelengths because of the similarity of its UV spectrum to that of theophylline.

Thus, Miksic and Hodes [22] developed an RPLC method for theophylline assay using a mobile phase of methanol/tetrahydrofuran/sodium acetate buffer to eliminate interferences caused by ampicillin and paraxanthine. Rodriguez et al. [23] also succeeded in the separation of theophylline and paraxanthine using potassium phosphate/formamide/methanol as an eluent.

In the RPLC methods, the retention time of theophylline ranged from 1.5 to 15 min depending on many factors, such as column size, concentration of organic solvent, flow rate, temperature, and sample preparation procedure. Very recently, Kabra and Marton [24] reported the RPLC analysis of serum theophylline in less than 70 sec using a short ODS column (125 × 4.6 mm I.D.), a high flow rate (4.5 ml/min), and a high temperature (50°C).

Theophylline is commonly detected with a UV detector. The maximum wavelength of theophylline is near 273 nm, but 254 or 280 nm is often used to avoid the interference of some drugs. Electrochemical detectors were recently used to analyze theophylline with RPLC, and this detection offered greater sensitivity and higher selectivity for theophylline than UV detection [19,25].

For the analysis of caffeine and its metabolites, some RPLC methods described so far are applicable, but caffeine is a less polar compound than theophylline and has a longer retention time in the RPLC system. Therefore, Blanchard et al. [26] developed a specific RPLC procedure

Table 2 Chromatographic Conditions for the Determination of Methylxanthine Drugs and Their Metabolites Using Adsorption HPLC

Compounds resolved[a]	Stationary phase	Mobile phase
THP, CAF, 7MX, 1MX OXA, THB	Durapak OPN (36-75 μm, Waters)	14-22% Isopropanol in hexane, 1%/min linear gradient
THP	Micropak Si 10 (10 μm, Varian Aerograph)	Chloroform/isopropanol/acetic acid (84/15/1, v/v)
CAF, THP, THB, βHPT	Zorbax SIL (6-8 μm, DuPont)	6% Ethanol:94% chloroform/heptane/acetic acid (300/200/0.4, v/v)
(THP, 3MX), CAF, 1MU THB, 13MU	μ-Porasil (10 μm, Waters)	1% Water:5% methanol: 25% 2-butanol in hexane
PHE, THP, βHPT, THB, DYP	Partisil 10 (10 μm, Vydac)	Chloroform/n-heptane/methanol (39/56/5, v/v)
THB, THP, PX, PRE, 3MX, 7MX, 1MX	LiChrosorb Si-60 (5 μm, BDH Chem.)	2.2% (Methanol:ammonium formate:formic acid, 100:0.02:0.017, v/v) in dichloromethane
(THP, 1MX), (3MX, CAF), X, THB, 13MU, other xanthines and drugs	Partisil 5 (5 μm, Whatman)	Hexane/isopropanol/water (80/19/1, v/v)
CAF, THP, THB	LiChrosorb Si-60 (5 μm, Altex)	Chloroform/isopropanol/acetic acid (96/2/2, v/v)
CAR, CAF, PX	Partisil 5 (5 μm, Chrompack)	20% Tetrahydrofuran in chloromethane
CAF, THP, THB, PX	Spherisorb ODS (5%) (10 μm, Spectra Physics)	Chloroform/n-heptane/ethanol/acetic acid/water (600/400/32/0.8/1.5, v/v)

Flow rate (ml/min)	temp. (°C)	Detector	Sample preparation[b]	Ref.
1.8	37	UV, 270 nM	Chloroform:isopropanol (SR, P)/ Amberlite XAD (U)	6
0.67	amb.	UV, 273 nm	Chloroform:isopropanol (P)	36
0.4	amb.	UV, 254 nm	Chloroform:isopropanol (SR)	50
3.33	25	UV, 280 nm	2-Butanol: n-hexane (P)	7
3	amb.	UV, 254 nm	Chloroform:isopropanol (SR)	9
2	amb.	UV, 280 nm	Chloroform:isopropanol (P)	10
2	amb.	UV, 280 nm	t-Pentanol: chloroform: 0.1 % HCl (SR)	51
2	amb.	UV, 275 nm	Chloroform:isopropanol (SR, SL, SP)	11
1.5	amb.	UV, 272 nm	Dichloromethane (SR)	12
1	amb.	UV, 270 nm	Chloroform:isopropanol (SR, SL)	8

Table 2 Continued)

Compounds resolved[a]	Stationary phase	Mobile phase
CAF, 3IB, THB, THP, PX	RSil (5 μm, RSL)	Chloroform/dioxane/formic acid (95.5/4.5/0.01, v/v)

[a] Abbreviations used: βHPT, β-hydroxypropyltheophylline; 7MX, 7-methylxanthine; OXA, oxazepam; PHE, phenobarbital; DYP, dyphylline; PX, paraxanthine; PRE, prednisolone; CAR, carbamazepine; 3IB, 3-isobutyl-1-methylxanthine. See Table 1 for others. The compounds in parentheses are not resolved.
[b] For abbreviations, see Table 1.

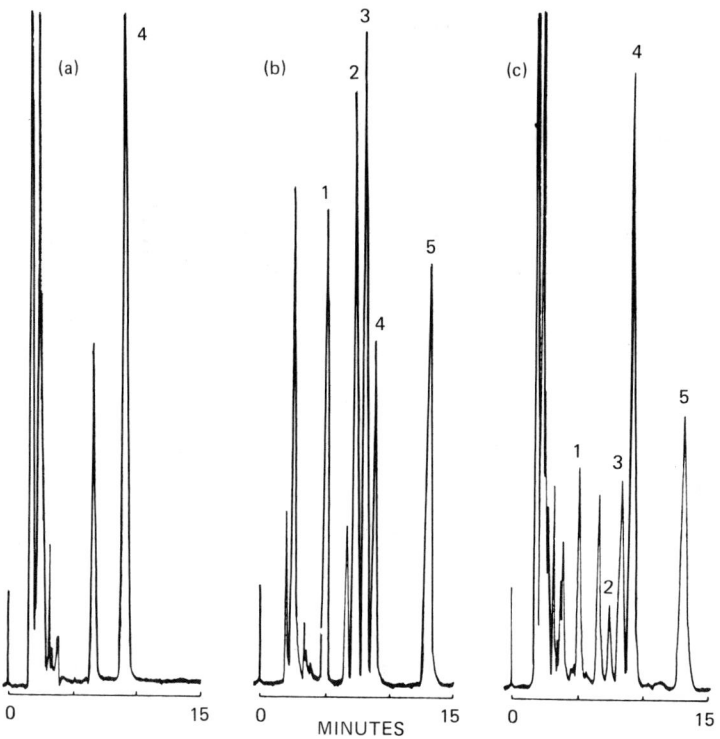

Flow rate (ml/min)	temp. (°C)	Detector	Sample preparation[b]	Ref.
1.4	amb.	UV, 273 nm	Chloroform:isopropanol (P)	13

for the assay of caffeine in plasma using a mobile phase containing a higher concentration of acetonitrile and separated the caffeine from other xanthine derivatives within 15 min.

Recently, Tse and Szeto [27] determined caffeine and its N-demethylated metabolites in dog plasma by a RPLC method. The use of a mobile phase consisting of sodium acetate, methanol, acetonitrile, and tetrahydrofuran separated theobromine, paraxanthine, theophylline, β-hydroxyethyltheophylline (internal standard), and caffeine, and made possible the quantitation of plasma levels of their xanthines as shown in Figure 1. The chromatographic conditions for RPLC analyses are listed in Table 3.

4. *Ion-pair RPLC*

Ion-pair (ion-association) RPLC methods have expanded the applicability of RPLC to highly polar and ionic compounds. Ion-pair chromatography exploits drastic changes in retention of ionic compounds in the presence

Figure 1 Chromatograms of (a) control dog plasma with β-hydroxyethyltheophylline (internal standard); (b) plasma spiked with (1) theobromine at 16 μg/ml, (2) paraxanthine, (3) theophylline, (5) caffeine, each at 32 μg/ml, and (4) β-hydroxyethyltheophylline at 20 μg/ml; and (c) plasma obtained from a dog 1 hr after the morning dose on day 8 during a 100-mg caffeine, twice-daily regimen. (From Ref. 27.)

Table 3 Chromatographic Conditions for the Determination of Methylxanthine Drugs and Their Metabolites Using Reversed-Phase HPLC.

Compounds resolved[a]	Stationary phase	Mobile phase
THP	µBondapak C_{18} (10 µm, Waters)	0.01 M Sodium acetate (pH 4.0)/acetonitrile (90/10, v/v)
THB, THP, 8ClT, CAF	ODS-Sil-X-I (13 µm, Perkin-Elmer)	1% Acetic acid/water/acetonitrile (1/48/1, v/v), pH 4.5
1MU, 3MX, 13MU, THB, THP, βHET, PHE, CAF, 8ClT	µBondapak C_{18}	0.01 M Sodium acetate (pH 4.0)/acetonitrile (93/7, v/v)
(DYP, THP), 8ClT	Partisil 10-ODS (10 µm, Whatman)	0.01 M KH_2PO_4/acetonitrile (90/10, v/v)
THB, THP, 8ClT, CAF	µBondapak C_{18}	1% Propionic acid (pH 5.0)/methanol (80/20, v/v)
THP	µBondapak C_{18}	0.01 M NaH_2PO_4/methanol (4/1, v/v)
THP, βHPT, and other xanthines	µBondapak C_{18}	0.02 M Sodium acetate (pH 4.0)/acetonitrile (90/10, v/v)
THB, THP, 8ClT, CAF	Spherisorb ODS (10 µm, Spectra Physics)	0.01 M Sodium acetate (pH 4.0)/methanol (75/25, v/v)
UA, (HX, X), 1MU, 3MX, 1MX, 13MU, THB, THP, DYP	µBondapak C_{18}	0.05 M KH_2PO_4 (pH 4.7)/methanol (88/12, (v/v)
THB, THP, PRO, CAF, 8ClT	µBondapak C_{18}	30% Methanol in 0.02 M KCl, pH 2
1MU, (8ClT, 13 MU), 3MX, THB, THP, DYP, CAF	ODS HC Sil-X-1 (Perkin-Elmer) µBondapak C_{18}	Water/acetonitrile (90/10, v/v) (94/6, v/v)
THP, 8ClT	Partisil-10 ODS	0.025 M KH_2PO_4 (pH 2.5)/methanol (65/35, v/v)

Flow rate (ml/min)	temp. (°C)	Detector	Sample preparation[b]	Ref.
2	amb.	UV, 254 nm	Ultrafiltration (P)	14
1.5	55	UV, 273 nm	Chloroform:isopropanol (SR)	15
2	amb.	UV, 254 nm	Acetonitrile:sodium acetate (P, SR, SL)	21
1.33	amb.	UV, 280 nm	Acetonitrile (SR, P)	16
1.7	amb.	UV, 280 nm	Chloroform (SR, P)	52
0.8	amb.	UV, 280 nm	Perchloric acid (SR)	53
1.8	amb.	UV, 254 nm	Chloroform:isopropanol (SR, P)	54
—	amb.	UV, 280 nm	Chloroform (SR, B)	55
1.1	amb.	UV, 254 nm	Ultrafiltration (SR, U, SL)	17
1.6	amb.	UV, 254/280 nm	TCA (SR)	56
2	amb.	UV, 275 nm	Acetonitrile (P)	18
3	amb.			
1.5	amb.	UV, 254 nm	Direct (SR)	57

Table 3 (Continued)

Compounds resolved[a]	Stationary phase	Mobile phase
THB, (THP, SMT), CAF	Partisil-10 ODS	0.01 M Sodium acetate (pH 4.0)/acetonitrile (93.5/6.5, v/v)
THB, CPP, THP, βHET, AMP, CPZ	μBondapak C_{18}	0.02 M Sodium acetate (pH 3.5)/methanol (85/15, v/v)
THB, AMP, THP, HEP, βHET, CAF, 8ClT	μBondapak C_{18}	0.01 M Sodium acetate (pH 4.0)/acetonitrile (93/7, v/v)
THB, THP, CAF, 8ClT	LiChrosorb C8 (10 μm, Altex)	Sodium acetate-acetic acid (pH 4.0)/ethanol (92/8, v/v)
(PX, AMP), THP, βHET	Partisil PXS 5/25 (5 μm, Whatman)	7% Methanol, 1% tetrahydrofuran in 0.01 M sodium acetate (pH 5.0)
PX, THP, βHET, AMP	RCM radial pak A (10 μm, Waters)	6% Methanol, 1.2% tetrahydrofuran in 0.01 M sodium acetate (pH 5.0)
THB, PAR, THP, βHET, CAF, and other xanthines	ODS-Hypersil (5 μm, Shandon So. Prod.)	0.02 M Sodium acetate (pH 4.0)/acetonitrile (92/8, v/v)
3MU, 7MU, 1MU, 7MX, 3MX, (37MU, 1MX), 13MU, THB, 17MU, (PX, THP), 13 7 MU, CAF	μBondapak C_{18}	1.5-7.5% acetonitrile in 0.5% acetic acid in 14 min by concave gradient and maintained for 15 min
(X, HX), 1MU, (3MX, 7MX), 1MX, THB, (THP, PX), PRO, CAF	Spherisorb ODS (10 μm, Altex)	0.01 M Sodium acetate (pH 4.0)/acetonitrile (85/15, v/v)
THB, THP, DYP, βHET, CAF	μBondapak C_{18}	0.01 M Sodium acetate (pH 4.0)/acetonitrile (94/6, v/v)

Flow rate (ml/min)	temp. (°C)	Detector	Sample preparation[b]	Ref.
3	amb.	UV	TCA (SR)	58
2	amb.	UV, 273/254 nm	Methanol (SR)	59
2.0	amb.	UV, 254 nm	Direct (SR, P, SL)	60
1	amb.	ECD/UV, 254 nm	Standards	25
1.5	amb.	UV, 280/254 nm	Acetonitrile (SR, SL)	22
3	amb.	UV, 280/254 nm	Acetonitrile (SR, SL)	22
1.5	amb.	UV, 273 nm	Chloroform:isopropanol (P)	61
2.0	amb.	UV, 280 nm	Chloroform:isopropanol (U, P)	45
1	amb.	UV, 273 nm	Acetonitrile (P)	26
2	40	UV, 274 nm	TCA (P)	20

Table 3 (Continued)

Compounds resolved[a]	Stationary phase	Mobile phase
3MX, AC, (SAL, THB), THP, 8ClT, (CAF, MC), APH, MEP	Spherisorb ODS (10 μm, Spectra Physics)	Water/methanol/0.01 M acetic acid/acetonitrile (788/180/16/16, v/v)
PX, THP	LiChrosorb RP-18 (7 μm, Merck)	0.05 M KH_2PO_4/formamide/methanol (66/11.5/22, v/v), pH 5.8
THB, PX, THP, βHET, CAF	μBondapak C_{18}	0.005 M Sodium acetate (pH 5) methanol/acetonitrile/tetrahydrofuran (92.5/3/2.8/1.7, v/v)
3MX, THB, PA, AC, (THP, DYP), βHET, NAPA, SAL, CAF, NPPA, 8ClT, ASA	μBondapak C_{18}	0.1 M KH_2PO_4 (pH 4.0)/acetonitrile (90.25/9.75, v/v)
THP, βHET, other xanthines and drugs	C18 RP bonded-phase (5 μm, Perkin-Elmer)	0.02 M KH_2PO_4 (pH 3.6)/acetonitrile (90.5/9.5, v/v)

[a]Abbreviations used: PRO, proxyphylline; SMT, sulphamethoxazole; CPP, cephapirin; AMP, ampicilline; CPZ, cephazolin; HEP, heparin; PAR, paracetamol; 3MU, 3-methyluric acid; 7MU, 7-methyluric acid; 37MU, 3,7-dimethyluric acid; 17MU, 1,7-dimethyluric acid; 137MU, 1,3,7-trimethyluric acid; AC, acetaminophen; SAL, salicylate; MC, mephensine carbamate; APH, acetophenetidine; MEP, mephesine; PA, procainamide; NAPA, N-acetylprocainamide; NPPA-N-propionylprocainamide; ASA, acetylsalicylate. See Tables 1 and 2 for others. The compounds in parentheses are not resolved.

[b]For abbreviations, see Table 1.

Flow rate (ml/min)	temp. (°C)	Detector	Sample preparation[b]	Ref.
1.52	amb.	UV, 273 nm	Chloroform:isopropanol (SR, P)	62
0.8	amb.	UV, 254/280 nm	Sodium acetate:acetonitrile (SL, P, U)	23
1.5	amb.	UV, 280 nm	TCA (P)	27
2	25	UV, 254 nm	Chloroform:isopropanol (SR)	63
4.5	50	UV, 273 nm	acetonitrile:ethyl acetate (SR)	24

of ion-pair reagents that have a substantial hydrophobic moiety to interact with the reversed-phase packings. This method was applied to the theophylline determination by Farrish and Wargin [28], eliminating the interferences of paraxanthine and dyphylline (Figure 2). They used tetrabutylammonium chloride as an ion-pair reagent.

The ion-pair method for the simultaneous determination of theophylline and its major metabolites in urine was also developed using tetrabutylammonium hydrogen sulfate as an ion-pair reagent and methanol gradient elution [29]. In the urinary assay, this method is superior to the RPLC methods of Desiraju et al. [17] and Grygiel et al. [30], because the urine sample contains so many background substances.

This ion-pair method was recently improved by Tang-Liu and Riegelman [31], using a slightly complicated solvent program for the simultaneous determination of 17 xanthine derivatives (Fig. 3). In this study,

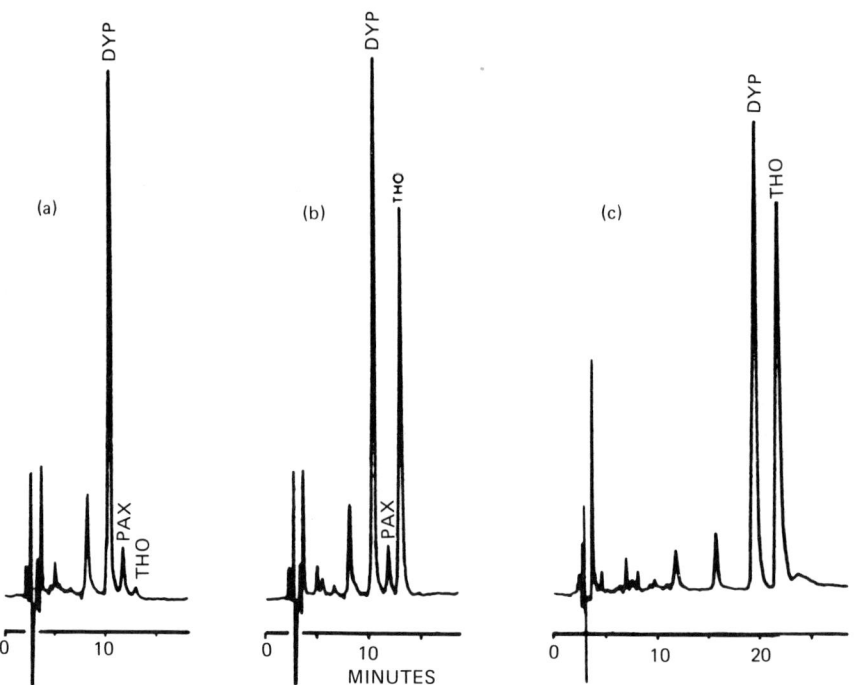

Figure 2. Chromatograms of human plasma sample before (a) and 2 hr after (b) administration of theophylline; (c) sample 2 hr after theophylline but without tetrabutylammonium chloride. DYP, dyphylline; PAX, paraxanthine; THO, theophylline. (From Ref. 28.)

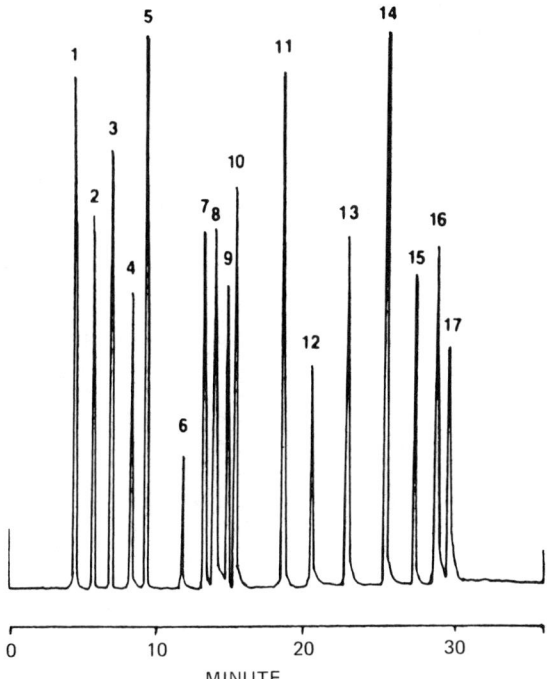

Figure 3 Chromatogram of a spiked standard mixture: (1) xanthine, (2) uric acid, (3) 3-MU, (4) 7-MX, (5) 3-MX, (6) 1-MX, (7) 3,7-MX, (8) 3,7-MU, (9) 7-MU, (10) 1-MU, (11) 1,3-MU, (12) 1,7-MX, (13) 1,3-MX, (15) 1,7-MU, (16) 1,3,7-MU, (17) 1,3,7-MX, each at 15 µg/ml; and (14) internal standard (β-hydroxyethyltheophylline). (From Ref. 31.)

13 methylxanthines were found in the urine of a volunteer. The urine sample was taken after concomitant administration of theophylline and caffeine.

A rapid procedure for the therapeutic monitoring of theophylline, acetaminophen, or ethosuximide was reported using the ion-pair technique. Quattrone and Putnam [32] used volatile, ion-pairing buffers such as triethylamine/acetic acid or N-ethylmorpholine/acetic acid in acetonitrile/water. The chromatographic conditions for these analyses are listed in Table 4.

B. Drug metabolism and pharmacokinetics

1. Theophylline

It is known that roughly 15% of theophylline is eliminated unchanged by renal excretion, and the remainder is metabolized in the liver to

Table 4 Chromatographic Conditions for the Determination of Methylxanthine Drugs and Their Metabolites Using Ion-Pair RPLC

Compounds resolved[a]	Stationary phase	Mobile phase[b]
DYP, PX, THP	Spherisorb ODS (5 μm, Laboratory Data Control)	0.05 Sodium acetate (pH 4.2):0.01 M TBAC/methanol (90/10, v/v)
(UA, 1MU, 3MU, HX, X), (1MX, 3MX), THB, AC, THP, βHET, ETH, CAF, 8C1T	μBondapak C_{18}	1.5 ml Acetic acid + 2.75 liters of water containing acetonitrile (250 ml) and triethylamine (2.0 oml)
X, UA, 3MU, 7MX, 3MX, 1MX, THB, 37MU, 7MU, 1MU, 13MU, PX, THP, 17MU, 137MU, CAF	Ultrasphere ODS (5 μm, Altex)	A: 0.01 M Sodium acetate, 0.005 M TBAHS, pH 4.9 B: A + 50% methanol, pH 4.8, 0 to 45% solvent B, 7 steps gradient

[a]Abbreviations used: ETH, ethosuximide. See Tables 1, 2, and 3 for others. The compounds in parentheses are not resolved.
[b]TBAC, tetra-n-butylammonium chloride; TBAHS, tetrabutylammonium hydrogen sulfate.
[c]For abbreviations, see Table 1.

1,3-dimethyluric acid, 3-methylxanthine, and 1-methyluric acid as major metabolites; however, the renal clearance of theophylline is low (for review see Refs. 33 and 34). Theophylline in plasma is reversibly bound to circulating plasma proteins. Studies have demonstrated that, on the average, 53 to 65% of the drug is bound in the plasma of healthy adults, and 29 to 37% in patients with hepatic cirrhosis [33].

Studies on pharmacokinetics of theophylline have shown that the theophylline half-life varies from 2.9 to 12.8 hr (mean 5.8 hr) in normal and asthmatic adults, and from 1.4 to 7.9 hr (mean 3.69 hr) in children. The half-life is prolonged in patients with congestive heart failure, liver disfunction, obesity, viral upper respiratory infections, pneumonia, and fever, while smokers show a short half-life. Several dietary factors effect theophylline half-life [33,34].

Flow rate (ml/min)	temp. (°C)	Detector	Sample preparation^c	Ref.
1.0	50	UV, 280 nm	Chloroform:isopropanol (P)	28
2.5	amb.	UV, 204 nm	Acetonitrile:ethyl acetate (SR)	32
1.5	25	UV, 280 nm	Ethyl acetate: chloroform:isopropanol (U)	31

Van Gennip et al. [35] studied the urinary excretion of methylxanthines after administration of theophylline using RPLC, TLC, and GC-MS. They found trace levels of 1-methylxanthine in addition to the three major metabolites described above in asthmatic patients, and only a small amount of 1-methyluric acid and trace amounts of 3-methylxanthine in a patient with severe combined immune deficiency disease (SCID). The presence of 1-methylxanthine as a minor metabolite of theophylline was also detected by Grygiel et al. [30]. They proposed that the secondary biotransformation of 1-methylxanthine to 1-methyluric acid was mediated by xanthine oxidase.

Several HPLC methods have been applied to the pharmacokinetic studies of theophylline [33]. Sitar et al. [36] studied the determination and pharmacokinetics of plasma theophylline in infants and children using adsorption chromatography. Their method in pharmacokinetic studies was found to be more specific and sensitive than spectroscopic methods.

Ishizaki et al. [37] compared two assay methods, RPLC by Adams et al. [15] and EIA, on the study of plasma levels and pharmacokinetics of theophylline in blood samples from asthmatic patients and normal adult males. The standard curves and plasma levels measured with the two methods correlated well with each other within a concentration range of 2.5 to 20 mg/liter; however, EIA gave lower values than RPLC, indicating slightly lower sensitivity of EIA. The calculated half-life and clearance values of plasma theophylline were also found to be different using the two methods, and the values were dissimilar to those obtained by spectroscopic methods.

2. Caffeine

The renal excretion of unchanged caffeine is negligible (0.5-1.5%) in most species, and hence the rapid decline in plasma caffeine is due to metabolism and diffusion (or transport) into tissue mass. The half-life of caffeine in plasma ranges from 2.3 to 4.5 hr (mean 3.1 hr) in adults (for review see Ref. 38).

Although the urinary metabolites of caffeine is slightly different among the species studied [38], caffeine is metabolized mainly in the liver, undergoing demethylation and oxidation. The methylxanthines, such as paraxanthine, theobromine, theophylline, 1-methylxanthine, and 3-methylxanthine, and the methyluric acids, such as 1,3,7-trimethyldihydrouric acid, 1,3-dimethyluric acid, and 1-methyluric acid, were found as the major metabolites. In addition there have been found minor metabolites, such as 7-methylxanthine, 1,3,7-trimethyluric acid, 3-methyluric acid, 1,7-dimethyluric acid, and 7-methyluric acid. Using studies such as these, routes of metabolism of caffeine have been proposed [39-42].

An HPLC system has also been used in the study of metabolism of caffeine. Sved et al. [43] studied the human metabolism and pharmacokinetic conversion of caffeine to theophylline using a modification of the adsorption chromatographic method of Sitar et al. [36]. In this study, the plasma concentration of caffeine and theophylline during 24 hr after administration of 300 mg of caffeine was examined. The same group [44] identified by HPLC and MS the dimethylxanthine metabolites of caffeine, that is, paraxanthine, theobromine, and theophylline, in human plasma following a single oral dose of caffeine.

Aldridge et al. [45] assayed caffeine and 13 of its metabolites in urine of newborns as a function of age using a RPLC method. They found that caffeine remained the predominant component (half-life, 4 days in plasma) for the first 3 months, but its percentage decreased gradually to the adult value of less than 2% by the age of 7-9 months. Caffeine and nine of its metabolites were assayed by the same RPLC method in serum and urine of the female beagle dog before and after phenobarbital and β-naphthoflavone pretreatment [46].

The pharmacokinetic interactions of caffeine with theophylline were examined using a RPLC method developed by Valia et al. [20] (see Ref. [47]). To two beagle dogs, a dose of 100 mg of aminophylline (theophylline ethylenediamine) was administered intravenously, 3 weeks before and immediately after repeated oral doses of caffeine. They also studied the pharmacokinetics of theobromine in plasma after oral administration of caffeine to two beagle dogs using the same RPLC method [48].

Recently, Nakano et al. [49] examined the ultraviolet-absorbing low-molecular-weight compounds in saliva by means of a RPLC method. In addition they determined the kinetics of caffeine and paraxanthine in saliva and serum after the oral administration of caffeine.

REFERENCES

1. R. D. Thompson, H. T. Nagasawa, and J. W. Jenne, J. Lab. Clin. Med., *84*, 584 (1974).
2. M. Weinberger and C. Chidsey, Clin. Chem., *21*, 834 (1975).
3. G. W. Peng, V. Smith, A. Peng, and W. L. Chiou, Res. Commun. Chem. Pathol. Pharmacol., *15*, 341 (1976).
4. W. J. Jusko and A. Poliszczuk, Am. J. Hosp. Pharm., *33*, 1193 (1976).
5. H. F. Walton, G. A. Eiceman, and J. L. Otto, J. Chromatogr., *180*, 145 (1979).
6. C. V. Manion, D. W. Shoeman, and D. L. Azarnoff, J. Chromatogr., *101*, 169 (1974).
7. O. H. Weddle and W. D. Mason, J. Pharm. Sci., *65*, 865 (1976).
8. J. H. G. Jonkman, R. Schoenmaker, J. E. Greving, and R. A. De Zeeuw, Pharm. Weekbl., *115*, 557 (1980).
9. A. G. Maijub, D. T. Stafford, and R. T. Chamberlain, J. Chromatogr. Sci., *14*, 521 (1976).
10. S. Sved and D. L. Wilson, Res. Commun. Chem. Pathol. Pharmacol., *17*, 319 (1977).
11. A. A. Tin, S. M. Somani, H. S. Bada, and N. N. Khanna, J. Anal. Toxicol., *3*, 26 (1979).
12. C. Van Der Meer and R. E. Haas, J. Chromatogr., *182*, 121 (1980).
13. P. Van Aerde, E. Moerman, R. Van Severen, and P. Braeckman, J. Chromatogr., *222*, 467 (1981).
14. L. C. Franconi, G. L. Hawk, B. J. Sandmann, and W. G. Haney, Anal. Chem., *48*, 372 (1976).
15. R. F. Adams, F. L. Vandemark, and G. J. Schmidt, Clin. Chem., *22*, 1903 (1976).
16. J. W. Nelson, A. L. Cordry, C. G. Aron, and R. A. Bartell, Clin. Chem., *23*, 124 (1977).

17. R. K. Desiraju, E. T. Sugita, and R. L. Mayock, J. Chromatogr. Sci., 15, 563 (1977).
18. G. W. Peng, M. A. F. Gadalla, and W. L. Chiou, Clin. Chem., 24, 357 (1978).
19. E. C. Lewis and D. C. Johnson, Clin. Chem., 24, 1711 (1978).
20. K. H. Valia, C. A. Hartman, N. Kucharczyk, and R. D. Sofia, J. Chromatogr., 221, 170 (1980).
21. J. J. Orcutt, P. P. Kozak, Jr., S. A. Gillman, and L. H. Cummins, Clin. Chem., 23, 599 (1977).
22. J. R. Miksic and B. Hodes, J. Pharm. Sci., 68, 1200 (1979).
23. F. Rodriguez, P. Rouzaud, P. Marty, and P. Puig, Therapie, 36, 659 (1981).
24. P. M. Kabra and L. J. Marton, Clin. Chem., 28, 687 (1982).
25. M. S. Greenberg and W. J. Mayer, J. Chromatogr., 169, 321 (1979).
26. J. Blanchard, J. D. Mohammadi, and K. A. Conrad, Clin. Chem., 26, 1351 (1980).
27. F. L. S. Tse and D. W. Szeto, J. Chromatogr., 226, 231 (1981).
28. H. H. Farrish and W. A. Wargin, Clin. Chem., 26, 524 (1980).
29. K. T. Muir, J. H. G. Jonkman, D.-S. Tang, M. Kunitani, and S. Riegelman, J. Chromatogr., 221, 85 (1980).
30. J. J. Grygiel, L. M. H. Wing, J. Farkas, and D. J. Birkett, Clin. Pharmacol. Ther., 26, 660 (1979).
31. D. D.-S. Tang-Liu and S. Riegelman, J. Chromatogr. Sci., 20, 155 (1982).
32. A. J. Quattrone and R. S. Putnam, Clin. Chem., 27, 129 (1981).
33. R. I. Ogilvie, Clin. Pharmacokinet., 3, 267 (1978).
34. R. G. Van Dellen, Mayo Clin. Proc., 54, 733 (1979).
35. A. H. Van Gennip, J. Grift, E. J. Van Bree-Blom, D. Ketting, and S. K. Wadman, J. Chromatogr., 163, 351 (1979).
36. D. S. Sitar, K. M. Piafsky, R. E. Rangno, and R. I. Ogilvie, Clin. Chem., 21, 1774 (1975).
37. T. Ishizaki, M. Watanabe, and N. Morishita, Br. J. Clin. Pharmacol., 7, 333 (1979).
38. A. W. Burg, Drug Metab. Rev., 4, 199 (1975).
39. H. H. Cornish and A. A. Christman, J. Biol. Chem., 228, 315 (1957).
40. K. L. Khanna, G. S. Rao, and H. H. Cornish, Toxicol. Appl. Pharmacol., 23, 720 (1972).
41. G. S. Rao, K. L. Khanna, and H. H. Cornish, Experientia, 29, 953 (1973).
42. M. J. Arnaud, Experientia, 32, 1238 (1976).
43. S. Sved, R. D. Hossie, and I. J. McGilveray, Res. Commun. Chem. Pathol. Pharmacol., 13, 185 (1976).
44. K. K. Midha, S. Sved, R. D. Hossie, and I. J. McGilveray, Biomed. Mass Spectrom., 4, 172 (1977).

45. A. Aldridge, J. V. Aranda, and A. H. Neims, Clin. Pharmacol. Ther., 25, 447 (1979).
46. A. Aldridge and A. H. Neims, Drug Metab. Disp., 7, 378 (1979).
47. F. L. S. Tse, K. H. Valia, D. W. Szeto, T. J. Raimondo, and B. Koplowitz, J. Pharm. Sci., 70, 395 (1981).
48. F. L. S. Tse and K. H. Valia, J. Pharm. Sci., 70, 579 (1981).
49. K. Nakano, S. P. Azzenza, and P. R. Brown, J. Chromatogr., 233, 51 (1982).
50. M. A. Evenson and B. L. Warren, Clin. Chem., 22, 851 (1976).
51. K. Nakatsu, J. A. Owen, and K. Scully, Clin. Biochem., 11, 148 (1978).
52. R. E. Hill, J. Chromatogr., 135, 419 (1977).
53. M. J. Cooper, B. L. Mirkin, and M. W. Anders, J. Chromatogr., 143, 324 (1977).
54. S. J. Soldin and J. G. Hill, Clin. Biochem., 10, 74 (1977).
55. M. A. Peat, T. A. Jennison, and D. M. Chinn, J. Anal. Toxicol., 1, 204 (1977).
56. F. Nielsen-Kudsk and A. K. Pedersen, Acta Pharmacol. Toxicol., 42, 298 (1978).
57. D. J. Popovich, E. T. Butts, and C. J. Lancaster, J. Liq. Chromatogr., 1, 469 (1978).
58. S. A. McKenzie, A. T. Edmunds, E. Baillie, and J. H. Meek, Arch. Dis. Childhood, 53, 322 (1978).
59. G. P. Butrimovitz and V. A. Raisys, Clin. Chem., 25, 1461 (1979).
60. B. R. Manno, J. E. Manno, and B. C. Hilman, J. Anal. Toxicol., 3, 81 (1979).
61. P. J. Naish, M. Cooke, and R. E. Chambers, J. Chromatogr., 163, 363 (1979).
62. L. A. Broussard, Clin. Chem., 27, 1931 (1981).
63. C.-N. Ou and V. L. Frawley, Clin. Chem., 28, 2157 (1982).

19
Nucleic Acid Constituents in Disease Processes

MONA ZAKARIA Hoffmann LaRoche, Nutley, New Jersey

   I. Introduction 365
  II. Extraction Procedures for Purine and Pyrimidine Analysis 366
     A. From cells 366
     B. From biological fluids 367
 III. Diseases Related to a Specific Enzymatic Disorder 367
     A. In purine metabolism 367
     B. In pyrimidine metabolism 372
     C. In general 373
 IV. Diseases Associated with Several Changes in Purine and Pyrimidine Metabolism 373
     A. Gout 373
     B. Neoplastic diseases 376
     C. Other diseases 383
  V. Conclusion 386
     References 386

I. INTRODUCTION

The biochemical importance of purines and pyrimidines stems from their role as nucleic acid precursors essential for cellular proliferation, and as enzyme cofactors, responsible for the transfer of chemical energy in metabolic and regulatory processes.

Alterations in the structures and levels of certain nucleotides, nucleosides, and bases in physiological fluids and cells have been associated with diseases of varying clinical severity. Specific enzymatic deficiencies cause some of the changes noted; for example, hypoxanthine-guanine phosphoribosyl transferase deficiency is responsible for the elevation of hypoxanthine, xanthine, and uric acid levels in the blood and urine of children with the Lesch-Nyhan syndrome. The changes which are observed in neoplastic disease states are not yet fully understood; they can, however, provide a clue in elucidating the exact pathogenesis of the illness.

High performance liquid chromatography (HPLC) is widely used in medicine today to diagnose diseases which result from errors in purine and pyrimidine metabolism. This technique has enabled the routine quantification of various nucleotides, nucleosides, and their bases in the biological fluids and cells of individuals with different diseases. Nucleic acid constituents have been investigated as markers not only to diagnose a particular illness, but also to assess the adequacy of treatment as well as the clinical response of the patient under therapy.

This chapter reviews various medical applications where HPLC has been used to identify changes in nucleic acid constituents typically associated with a certain illness. First summarized are the procedures currently used to extract purines and pyrimidines from biological fluids and cells. Next, diseases such as the Lesch-Nyhan syndrome (LNS) and severe combined immunodeficiency (SCID), which are associated with specific enzymatic disorders, will be discussed. Finally, diseases which may be attributed to more than one defect in nucleic acid metabolic pathways, such as neoplasms and the gout syndrome, will be reviewed. In all these cases, alterations in nucleotide, nucleoside, and base levels will be described together with the liquid chromatographic systems used in the assay procedure. Therapy and its effects on purine and pyrimidine compounds will also be mentioned.

II. EXTRACTION PROCEDURES FOR PURINE AND PYRIMIDINE ANALYSIS

A. From cells

In order to study the free nucleotide, nucleoside, and base content of a certain volume or number of cells, it is important to extract compounds of interest into a liquid medium. Desirable reagents in the extraction step are those that can:

1. Lyse the cell
2. Precipitate the protein (to stop the enzymatic degradation of the nucleotides as well as prevent the clogging of the chromatographic column)

3. Afford the highest recovery of the compounds of interest
4. Provide a neutral environment for the storage of these compounds

Perchloric acid (PCA) is commonly used to extract nucleotides from biological cells [1-7]. Normally, the resulting acidic supernatant is neutralized with KOH [2-6] or the amine-Freon solution [7,8].

B. From biological fluids

Protein removal is probably the most important step in the analysis of nucleosides and bases in physiological fluids. Conventionally, PCA and trichloroacetic acid (TCA) have been used. Recently however, serum [9-11], plasma [12], and urine [9,10] samples have been ultrafiltered through membrane cones which retain the high-molecular-weight proteins. This method is preferred since it does not alter the pH of the medium, dilute the sample, or interfere with the ultraviolet (UV) absorbance of sample constituents. Gehrke et al. [13] developed a different extraction procedure for the analysis of ribonucleosides in urine. The samples were passed through a boronate gel column. The ribonucleosides were retained on the column as cis-diol boronate complexes and subsequently eluted with 0.1 M formic acid (see Chap. 3).

III. DISEASES RELATED TO A SPECIFIC ENZYMATIC DISORDER

A. In purine metabolism

Purine metabolic disorders can have symptoms of varying severity, and may be due to enzymatic deficiencies. In the adult male, for instance, primary gout can be characterized by hyperuricemia, by arthritis, or by deposits of monosodium urate as tophae in the cartilage and calculi in the kidney. This disease is usually caused by a deficiency of xanthine oxidase. It is, in addition, often associated with a partial inborn deficiency of the enzyme hypoxanthine-guanine phosphoribosyl transferase (HGPRT). In prepubertal children with the Lesch-Nyhan syndrome, an X-linked disorder, HGPRT is almost completely or totally inactive. These children are mentally retarded and have a tendency toward self-multilzation. Death, induced by renal failure, occurs early.

Severe combined immunodeficiency (SCID) has also been associated with enzymatic deficiencies in purine metabolic pathways, namely, deficiencies of purine nucleoside phosphorylase (PNP) and adenosine deaminase (ADA).

These disorders are all characterized by abnormal levels of nucleosides and bases in biological fluids and cells. Early diagnosis and treatment are often of vital importance in reversing the disease process and providing symptomatic relief. The following section briefly describes the HPLC methodology developed to study and quantify such enzymatic aberrations.

1. Hypoxanthine-guanine phosphoribosyl transferase (HGPRT)

HPLC was used to examine the purine and pyrimidine content of normal and HGPRT-deficient skin fibroblasts obtained from normal and hyperuricemic patients [5]. This key enzyme in the purine salvage pathway ensures the conversion of hypoxanthine and guanine into inosine 5'-monophosphate (IMP) and guanosine 5'-monophosphate (GMP), respectively. When the activity of HGPRT is reduced, xanthine oxidase catalyzes the conversion of larger amounts of hypoxanthine into xanthine and uric acid. The study of HGPRT deficiency in cellular extracts necessitates a chromatographic method which can separate simultaneously several purine bases, nucleosides, and nucleotides. This separation was achieved by Bakay et al. [5] on a column packed with Aminex A-25 (17.5 + 2 µm, Bio-Rad, 0.2 × 50 cm). A gradient was generated from a lower ionic strength mixture of 0.08 M sodium borate and 0.05 M ammonium chloride, buffered at pH 9.1, to a higher ionic strength mixture of 0.01 M sodium borate and 0.5 M ammonium chloride at pH 9.0. The flow rate was 0.5 ml/min and the temperature 60°C. The separation of adenosine, guanosine, adenine, hypoxanthine, xanthine, adenosine 5'-monophosphate (AMP), GMP, IMP, xanthosine 5'-monophosphate (XMP), adenosine 5'-diphosphate (ADP), guanosine 5'-diphosphate (GDP), adenosine 5'-triphosphate (ATP), and guanosine 5'-triphosphate (GTP) required 2 hr.

This method was then used to investigate the metabolic fate of hypoxanthine in the skin fibroblasts of various individuals after the incubation of these cells in a hypoxanthine-enriched medium. HGPRT-deficient fibroblasts were found to utilize 86% less hypoxanthine than normal cells and hence to produce less ATP and GTP. Conversely, fibroblasts from a hyperuricemic patient with normal HGPRT, mental retardation, and a tendency toward self-mutilation utilized three times more hypoxanthine than the normal. This resulted in the overproduction of inosine, IMP, AMP, GMP, ADP, GDP, and ATP. GTP alone was decreased. The AMP/ATP, GMP/GTP, and AMP/inosine ratios were shown to differ substantially in these cells from those in normal controls. Specific changes in the levels of the various nucleotides were associated with each clinical condition (Fig. 1).

Nissinen [14] modified the above chromatographic parameters in order to include the analysis of xanthosine, and that of the pyrimidines cytosine, thymine, uracil, cytidine, thymidine, uridine, cytidine 5'-monophosphate (CMP), thymidine 5'-monophosphate (TMP), uridine 5'-monophosphate (UMP), cytidine 5'-diphosphate (CDP), uridine 5'-diphosphate (UDP), cytidine 5'-triphosphate (CTP), thymidine 5'-triphosphate (TTP), and uridine 5'-triphosphate (UTP). The separation of all purines and pyrimidines was accomplished in less than 3 hr on a longer column packed with the same resin (Aminex A-25, 0.18 × 70 cm). The low-ionic-strength eluent was used isocratically during 15 min prior to the generation of the linear gradient. The flow rate was 0.5 ml/min and the temperature 65°C.

Endogenous levels of the various bases, nucleosides, and nucleotides were then determined in the skin fibroblasts of a normal individual as well as in those of a subject with HGPRT deficiency. Hypoxanthine was not elevated in the HGPRT-deficient fibroblasts, probably due to its rapid conversion to xanthine and uric acid. IMP could not be detected in these cells. Levels of ATP, CTP, and UTP were lower than in normal fibroblasts [14].

Higher concentrations of UDP-glucose [15] and pyrimidine nucleotides [16,17] have also been reported in HGPRT-deficient lymphoblast lines.

2. Adenosine deaminase (ADA)

Since the 1972 report by Giblett et al. [18] which associated adenosine deaminase deficiency with severe combined immunodeficiency, a number of similar cases have been described [19]. In the absence of the deaminase, the conversion of adenosine and deoxyadenosine into inosine and deoxyinosine, respectively, does not occur. The accumulation of deoxyadenosine results in abnormally high dATP levels which are toxic to lymphoid cells [20].

The quantification of adenosine and deoxyadenosine in plasma and serum, after their HPLC separation, is used to diagnose this disorder and control the efficacy of the enzyme replacement therapy [21]. Such assays were developed by Koller et al. [20] and by Hartwick and Brown [22], who used a μBondapak C_{18} in combination with a 9/1 phosphate buffer/methanol eluent. No adenosine was detected in normal human serum, whereas it was measurable in the serum of patients with ADA deficiency (Fig. 2).

Another means of diagnosing this disorder is assaying the activity of ADA in erythrocytes. A chromatographic method capable of separating adenosine, inosine, and hypoxanthine from other constituents in red blood cells is needed for this purpose. Hartwick et al. [23] devised such a method and were thus able to assay for ADA in less than 6 minutes. The column was a μBondapak C_{18}, the eluent 14% methanol in 0.01 M KH_2PO_4, and the flow rate 2.0 ml/min. When adenosine was added to an aliquot of healthy erythrocyte lysate, the decrease in this substrate was monitored as well as the corresponding increase in inosine and hypoxanthine. In ADA-deficient individuals, the conversion of the added adenosine to inosine did not take place.

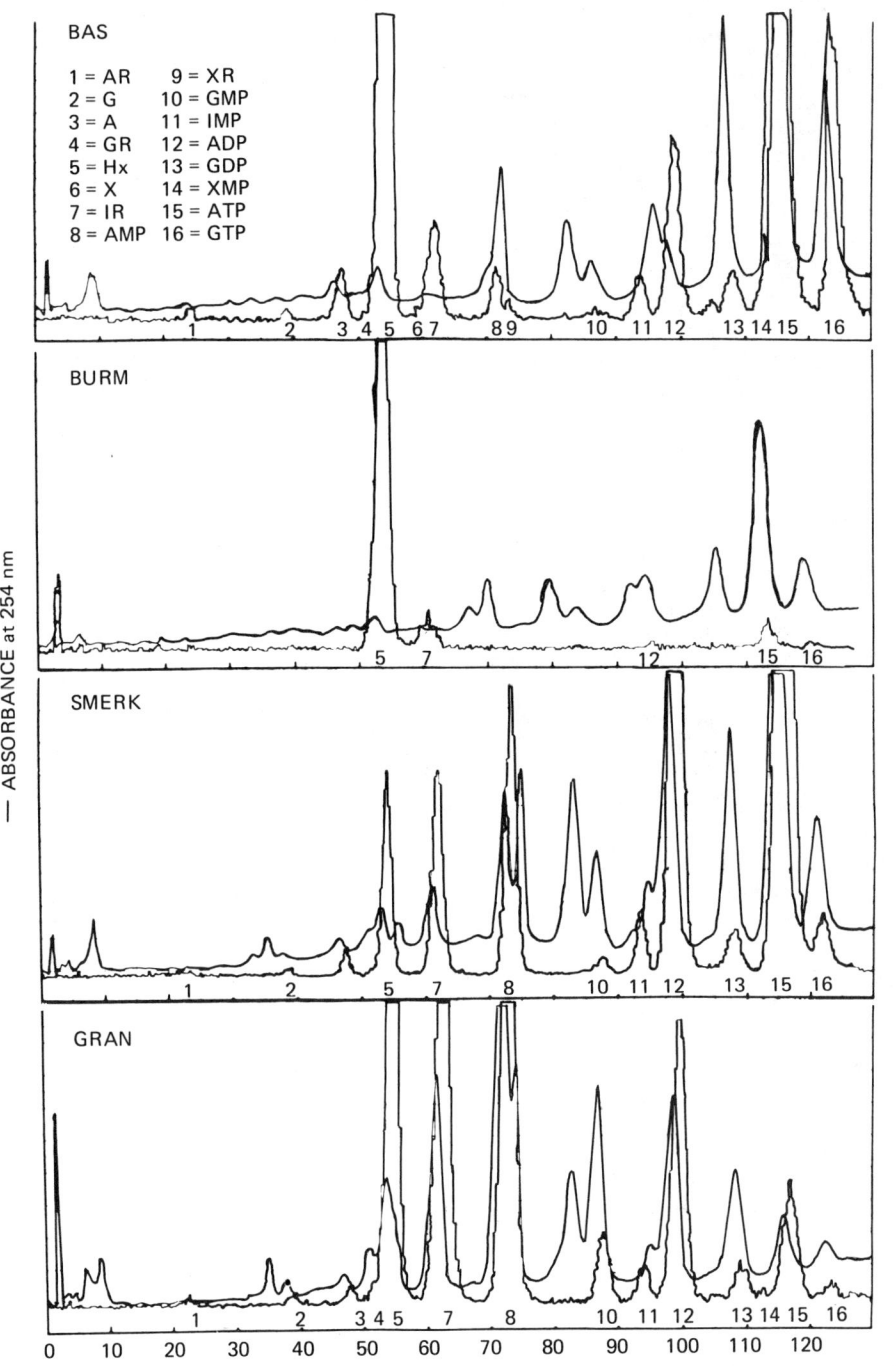

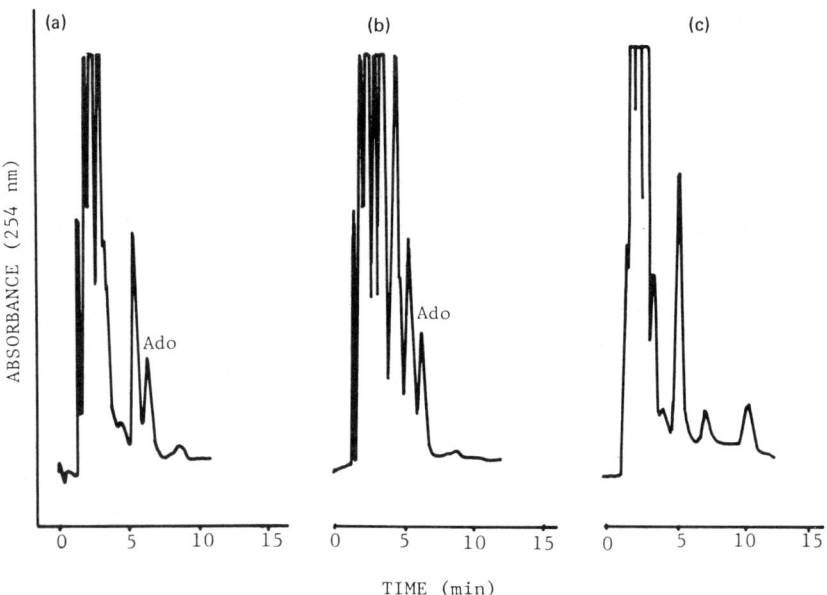

Figure 2. Samples (a) and (b), 50 µl of serum extract from two patients suffering from adenosine deaminase deficiency. 45 and 55 pmol are contained under the adenosine peak of samples (a) and (b), respectively. Sample (c) shows the injection of 50 µl of serum extract from a control patient. Integrator setting, 2; column packing, µBondapak C_{18}; temperature, ambient; detector sensitivity 0.02 aufs. Eluent, anhydrous methanol-0.007 F KH_2PO_4 (pH 5.8) (10:90). Flow rate 2.0 ml/min. (From Ref. 22.)

Figure 1. Chromatographic profiles of purine metabolites extracted with acid from cultured skin fibroblasts pulsed with ^{14}C hypoxanthine. Bas: 50-µl aliquot of extract from 7.4×10^6 cells of a normal control. Burm: 50-µl aliquot of extract from 3.5×10^6 cells of a gouty patient who has virtually no HPRT activity and clinically resembles a patient with the Lesch-Nyhan syndrome. However, he has no mental retardation. Smerk: 25-µl aliquot of extract from 3.3×10^6 cells of a patient with hyperuricemia who has a normal level of HPRT activity but who is mentally retarded and who has some tendency toward self-mutilation. Gran: 50-µl aliquot of extract from 5.3×10^6 cells of a patient with hyperuricemia who has a normal level of HPRT activity but who has clinical symptoms of the Lesch-Nyhan syndrome. (From Ref. 5.)

3. Purine nucleoside phosphorylase (PNP)

A few years after a deficiency of adenosine deaminase was implicated in severe combined immunodeficiency, Giblett et al. [24] reported a case of severe, selective cellular immune deficiency in which there was a deficiency of another enzyme in the erythrocytes of purine metabolism, namely, purine nucleoside phosphorylase. In the lymphocytes, the enzyme deficiency and the defective immunity were confined to the T-cells, leaving the B-cells unimpaired. Other patients with T-cell selective defective immunity and PNP deficiency have since been described [19,25].

Halfpenny and Brown [26] recently optimized an assay for PNP in erythrocytes using inosine as substrate. Xanthine oxidase was added to the incubation medium to prevent the accumulation of hypoxanthine, which would inhibit the forward reaction of PNP. Using a Partisil 5-ODS (0.4 × 25 cm, 5 μm), an eluent of 3% methanol in 0.02 F KH_2PO_4 buffer at pH 4.2, and a flow rate of 2.0 ml/min, uric acid, hypoxanthine, xanthine, and inosine could be resolved in 7 min with no interference from other compounds. The method is simple and provides a rapid diagnosis of the clinical condition.

In addition to the purine metabolic anomalies which result from PNP deficiency, Cohen et al. [27] and Van Gennip et al. [19] found increased levels of pyrimidine metabolites. They reported elevated excretion of orotic acid in the urine of patients with PNP deficiency. Van Gennip et al. [19] devised a method to separate orotic acid and orotidine from other urinary constituents. The system consisted of a μBondapak amino anion exchanger connected to a μBondapak C_{18} reversed-phase column, a 5 mM sodium acetate buffer at pH 3.5, and a flow rate of 1 ml/min. The effluent of the first column was switched to the reversed phase after 13 min, i.e., just prior to the elution of orotidine and orotic acid. Using these conditions, they observed that allopurinol administration to PNP-deficient patients increased the excretion of orotidine, whereas supplementing the diet with uridine and deoxycytidine increased orotic aciduria. Enzyme replacement with irradiated red cells was the only therapy which successfully reduced the excretion of orotic acid to control levles [19].

B. In pyrimidine metabolism

1. Orotic acid phosphoribosyl transferase (OPRT) and orotidine decarboxylase (ODC) deficiencies

The accumulation and urinary excretion of orotic acid, commonly termed orotic aciduria, are characteristic of deficiencies in OPRT and ODC, enzymes which lead to the formation of UMP. This disorder is hereditary. Afflicted children suffer from pyrimidine deprivation; growth is abnormal and megaloblastic anemia occurs. Using their assay method for pyrimidines in urine, Van Gennip et al. [19] observed increased levels of orotic acid in the urine of OPRT/ODC-deficient patients.

2. Ornithine transcarbamoylase (OTC)

The deficiency of OTC, unlike that of OPRT/ODC, leads to increased pyrimidine synthesis. Indeed, Van Gennip et al. [19] noted the increased excretion of orotic acid, uridine, and uracil in the urine of a patient with OTC deficiency.

C. In general

The enhanced activity of alkaline phosphatase has been observed in patients with osteogenic sarcoma, parathyroid adenoma or carcinoma, and cancer metastatic to the bone [28]. Abnormal levels of this enzyme have been found in the serum of patients with viral hepatitis and cirrhosis, as well as in diseases which do not involve the liver or bone—stage I or II Hodgkins' disease, myeloid metaplasia, congestive heart failure, and intraabdominal infections [28].

The enhanced activity of serum acid phosphatase has been observed in a variety of diseases, such as the carcinoma of the prostate, Gaucher's disease, and breast carcinoma [28].

Krstulovic et al. [28] developed a reversed-phase method to assay alkaline and acid phosphatase in serum. The column was an RP-8 and the mobile phase a 1:9 methanol:phosphate buffer (pH 5.5) mixture at a flow rate of 1.5 ml/min. The method could separate, in 7 min, the substrate AMP and product adenosine from other serum constituents. Interfering 5'-nucleotidase was inhibited by addition of Ni^{2+}. The assay required buffering the incubation medium to a pH of 9.8 for alkaline phosphatase, and 4.8 for acid phosphatase. Larger increases in the activity of alkaline phosphatase could be demonstrated in the serum of patients with cirrhosis and hepatitis.

IV. DISEASES ASSOCIATED WITH SEVERAL CHANGES IN PURINE AND PYRIMIDINE METABOLISM

A. Gout

Gout is a syndrome which may result from widely different biochemical anomalies. The earliest association of hyperuricaemia with gout dates back to 1776, when Scheele discovered uric acid in a kidney stone [29]. This compound was later found in the tophae of patients with gout. When Emile Fischer finally established its structure, in 1898, its relationship to purine bases became apparent [29]. Purine metabolism has since been extensively studied in patients with gout.

McBurney and Gibson [30] investigated purine and pyrimidine metabolism before and after therapy in subjects with gout and renal failure. The nucleoside and base contents of biological fluid samples were assayed by reversed-phase liquid chromatography. The stationary phase was a μBondapak C_{18} and the mobile phase a linear gradient, initiated after a 6-min delay period, between a 10 mg/l KH_2PO_4 buffer (pH 5.8) and methanol. The volume of the gradient chamber was 72 ml and the flow rate 1 ml/min. Under these conditions, uric acid, creatinine, pseudouridine, hypoxanthine, xanthine, oxypurinol, and N-methyl-2-pyridone-5-carboxamide were resolved in 20 min. Elevated concentrations of uric acid, hypoxanthine, and xanthine were observed in the plasma of gouty patients when compared to normal controls. As a result of defective elimination, levels of creatinine, pseudouridine, and N-methyl-2-pyridone-5-carboxamide were also increased in the blood and plasma of patients with renal failure, and levels of hypoxanthine and xanthine were decreased in the urine (Fig. 3) [30].

The hypoxanthine analog allopurinol, when administered to hyperuricaemic patients with gout and renal failure, inhibits xanthine oxidase and uses PRPP to form the allopurinol nucleotide. This reduces uric acid levels in blood, kidney, and urine. Plasma chromatograms show that oxypurinol, an allopurinol metabolite, does not have a rapid blood clearance, and that gout patients with renal failure exhibit higher levels of oxypurinol in their plasma than those with normal renal function. This is of consequence since the long-term accumulation of oxypurinol in the circulation results in bone marrow depression [30] and inhibition of orotidylate decarboxylase with mild to severe orotic aciduria. Elevated concentrations of orotic acid and orotidine have been observed by Mrochek et al. [31] in the urine of hyperuricaemic patient on allopurinol. The sample was chromatographed on an anion-exchange column (0.62 × 150 cm) filled with Bio-Rad A15 resin. Elution with a 0.015 to 6 M ammonium acetate-acetic acid buffer (pH 4.28) separated the compounds of interest in 42 hr. Temperature was ambient during the first 11.5 hr, and was increased thereafter to 60°C [31].

Its adverse effects make it important to monitor oxypurinol in patients on allopurinol therapy. This may be done rapidly using the assay developed by Taylor et al. [32] for the quantification of allopurinol, oxypurinol, hypoxanthine, thymine, and thymidine in plasma. A μBondapak C_{18} was used with 0.05 M ammonium acetate at a pH of 5.0. The flow rate was 2 ml/min and the temperature 23.5°C. Total analysis time was 25 min [32].

Gout associated with malignant disease states is due to an increased rate of cellular turnover with catabolism of nucleic acid purines to uric acid. The production of uric acid in amounts which exceed its solubility in the urine (2000 mg/l at pH 7) results in its precipitation as crystals in the kidney. Renal blood flow and glomerular filtration

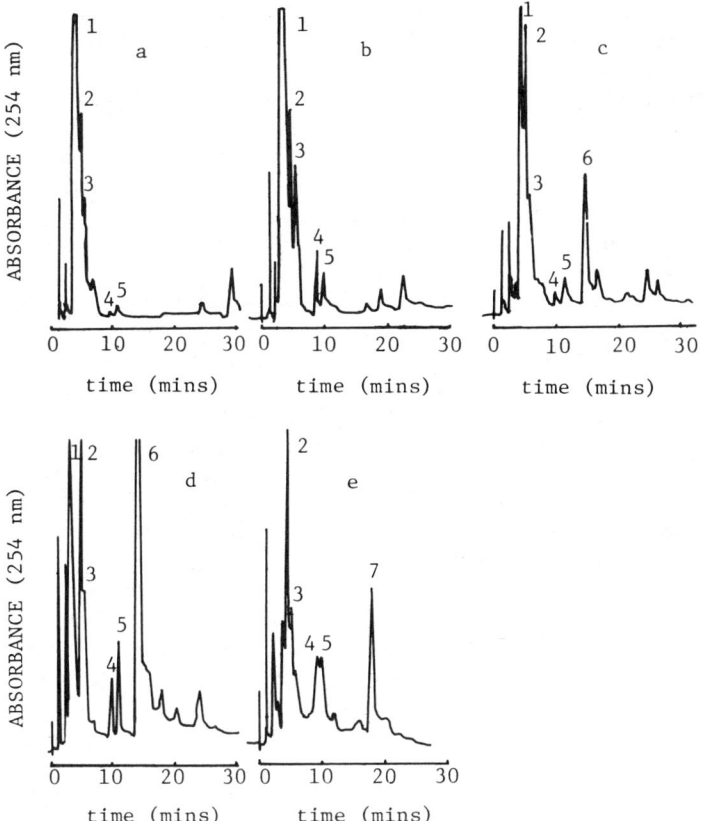

Figure 3. Representative chromatograms of plasma extracts from the following subjects: (a) normal, (b) gouty, (c) gouty on allopurinol, (d) gout and renal failure on allopurinol, (e) renal failure. Peak identification is as follows: (1) uric acid, (2) creatinine, (3) pseudo-uridine, (4) hypoxanthine, (5) xanthine, (6) oxypurinol, (7) N-methyl-2-pyridone-5-carboxamide. (From Ref. 30.)

are decreased, leading, in severe cases, to renal failure. The administration of effective antineoplastic agents enhances the gravity of the problem. The prophylactic use of allopurinol prior to the initiation of chemotherapy reduces the incidence of renal failure in all patients except those with clinically aggressive, rapidly dividing lymphomas [33].

Hande et al. [33], who quantified uric acid, hypoxanthine, and xanthine in urine, elucidated the reason for the renal failure in these patients. Urine extracts were assayed by RPLC on a μBondapak C_{18} and with a 0.05 M ammonium phosphate (pH 4.5) eluent. The method separated xanthine and hypoxanthine clearly from allopurinol, oxipurinol, and uric acid with no interference from caffeine or other methylated xanthines. After chemotherapy, a 2.2-fold increase in mean total daily excretion of uric acid, a 6.6-fold increase in that of hypoxanthine, and 6.9-fold increase in that of xanthine were observed in patients with rapidly growing, chemotherapy-sensitive lymphomas on allopurinol. Six of 11 patients excreted xanthine in concentrations which exceeded its solubility in urine at pH 5 (50 mg/l), or 7 (150 mg/l). Renal failure in these patients was attributed to xanthine nephropathy [33].

Even when its use may precipitate xanthine in patients' kidneys, allopurinol must be maintained at levels as high as 300-600 mg/day to reduce the uric acid nephropathy [33]. Administering antineoplastic drugs at lower doses, which cause a more gradual breakdown of the tumor mass, would reduce the severity of the nephropathy. However, this might also seriously compromise the possibility of complete remission. The problem, so far, remains unsolved. Patients are well hydrated (4-5 liters of water per day) for 4 to 5 days prior to and following chemotherapy. Hemodialysis is performed when renal failure still occurs [33]. This procedure requires constant monitoring since the characteristics of the dialysis membrane may change during the operation as a result of occlusion by adsorbing particulates. Knudson et al. [9] developed a HPLC method to study the change with time of purine and pyrimidine concentrations in the physiological fluids of patients under treatment. Samples of hemodialysate, serum, and urine were assayed by reversed-phase liquid chromatography (Fig. 4). The column was a μBondapak C_{18} (0.4 × 60 cm). A 90-min concave gradient was generated between a 0.025 M acetic acid solution titrated with sodium hydroxide to a pH value of 4.5, and a 0.2 M solution of acetic acid in methanol. The flow rate was 1.0 ml/min. Orotic acid, creatinine, uracil, uric acid, hypoxanthine, xanthine, allopurinol, 1-methylxanthine, and adenosine were separated in 45 min. The serum levels of these metabolites decreased during hemodialysis. Final concentrations were nearly half the initial concentrations, with the greatest decrease at the beginning of the 6-hr procedure. This assay can be used routinely in clinics and hospitals to determine the efficiency of hemodialysis processes.

B. Neoplastic diseases

Alterations in the levels and composition of purine and pyrimidine compounds have been observed in individuals with neoplastic diseases, along with changes in the activity of related enzymes [28,34-36]. Hartwick and coworkers [37,38] first established the system which separated the bases as well as major and modified nucleosides from other constituents of human serum. Gradient elution was used with a

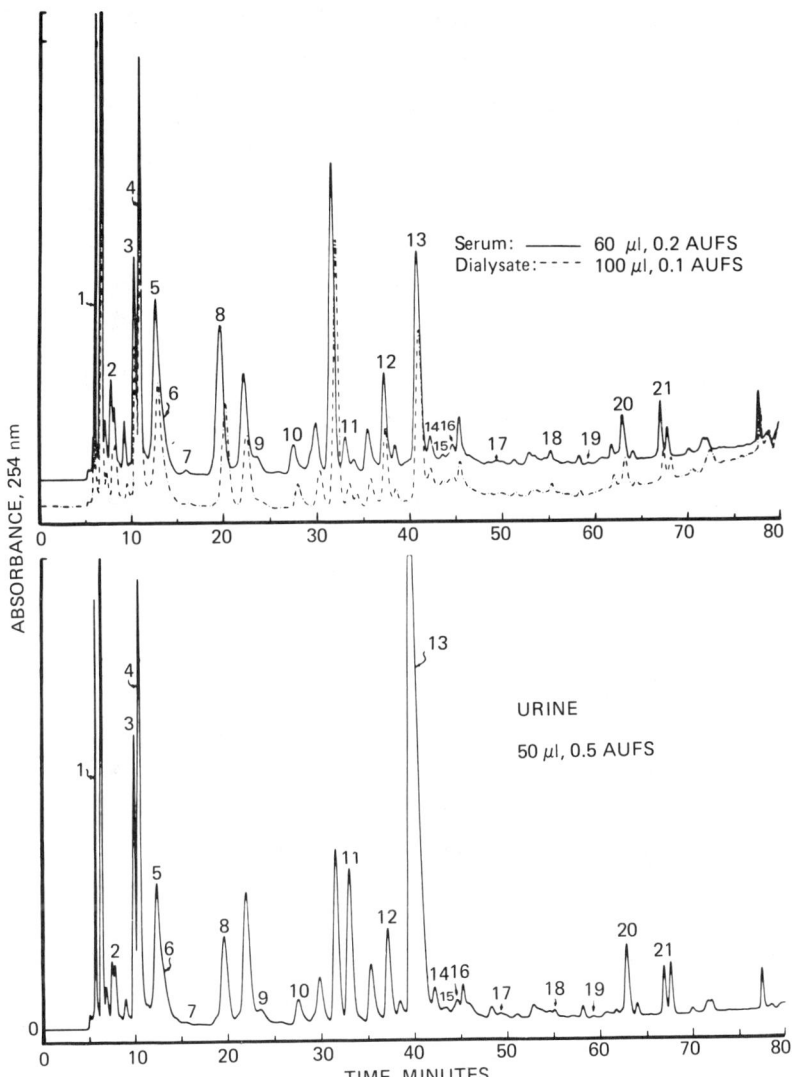

Figure 4. Liquid chromatograms for dialysate, serum, and urine samples simultaneously from a uremic male patient. Peak identities: 1, aconitic acid; 2, orotic acid; 3, 1-methylnicotinamide; 4, creatinine; 5, uracil; 6, nicotinic acid; 7, tyrosine, 8, uric acid; 9, hypoxanthine; 10, xanthine; 11, allopurinol; 12, 1,6-dihydro-1-methyl-6-oxonicotinamide; 13, hippuric acid; 18, theophylline; 19, syringic acid; 20, caffeine; 21, indole-3-acetic acid. (From Ref. 9. © 1978, Clinical Chemistry.)

Table 1 Purine and Pyrimidine Compounds Identified in Normal Human Serum[a]

Compound	Sex	X (μmol/liter)[b]	Range (2 S.D.)
Creatinine	F	63.4 (8.7)	46.0-80.8
Uric A	F	171 (30)	110-232
Tyr	N.S.[c]	62.2 (16.3)	29.6-94.8
Hyp	N.S.	7.16 (2.81)	1.56-12.8
Urd (E)	N.S.	3.17 (1.11)	0.951-5.39
Xan (E)	N.S.	2.62 (1.04)	0.542-4.70
Ino (H)	N.S.	5.62 (2.87)	0.0-11.4
Guo (I)	N.S.	0.881 (0.515)	0.0-1.98
Hipp A. (K)	N.S.	0.613 (0.477)	0.0-1.57
Trp (L)	N.S.	13.7 (3.57)	6.63-20.8
Dietary compounds			
Thb (M)	–	–	0.0-6.35
Caf (O)	–	–	0.0-12.2

[a] Sera were processed by membrane centrifuge cones (nominal molecular weight 25,000). Means represent 31 donors (17 males, 14 females). N.A. = Not available.
[b] Standard deviation of triplicate analyses are given in parentheses.
[c] No significant sex-related differences.
Source: Ref. 39.

μBondapak C_{18} column. The low-strength was a 0.02 M KH_2PO_4 buffer at a pH of 5.6 and the high-strength eluent, a 3/2 methanol/water mixture. The gradient slope was 0.69%/min and the flow rate 1.5 ml/min. Uric acid, hypoxanthine, uridine, xanthine, inosine, and guanosine levels were determined in normal serum (Table 1) and plasma (Table 2). 1-Methylinosine and N^2-methylguanosine were present in the serum of individuals with malignant lung cancer and breast cancer with metastasis to the bone (Fig. 5) [39,41]. Inosine, guanosine, adenosine, hypoxanthine, and xanthine were elevated in the plasma of patients with acute lymphocytic leukemia (ALL) on chemotherapy (Fig. 6). Inosine performed as a suitable marker for predicting the prognosis of the disease. The highest inosine levels corresponded to leukemic subjects whose condition deteriorated severely with time, whereas patients with inosine levels closer to normal continued in remission [40].

Gehrke and coworkers [42,43] investigated the use of methylated nucleosides as markers in neoplasms associated with enhanced tRNA methylase activity. Several methylated compounds were separated on a μBondapak C_{18}/Porasil (0.4 × 30 cm). The eluent consisted of 6% methanol in 0.01 M ammonium phosphate (pH 5.1) and the flow rate was 1.0 ml/min. The nucleosides 1-methyladenosine, 7-methylguanosine, guanosine, 1-methylinosine, 1-methylguanosine, N^2-methylguanosine, adenosine, and N^2,N^2-dimethylguanosine were quantified in normal human urine (Table 3) [43]. It was observed that 1-methylinosine and N^2,N^2-dimethylguanosine were increased in the urine of patients with leukemia (Fig. 7) and breast cancer [42], as well as N^1-methylguanosine and N^2-methylguanosine in urine of patients with colon cancer [43].

De Abreu et al. [44] used the anion-exchange mode to compare the nucleotide content of leukemic blast pools with that of normal lymphocytes. NAD, cyclic AMP, NADP, uridinediphosphoglucose, as well as the mono-, di-, and triphosphates of adenosine, cytidine, guanosine,

Table 2 Normal Levels of UV-Absorbing Constituents in Plasma (in μmol/liter)

	Means ± S.D.	Range
Creatinine	55.2 ± 15.0	29.5-84.8
Uric acid	191 ± 42.2	110-292
L-Tyrosine	72.2 ± 22.7	30.2-120
L-Phenylalanine	70.8 ± 29.5	27.6-120
L-Kynurenine	3.79 ± 1.14	1.84-5.96
L-Tryptophan	7.49 ± 2.38[a]	3.47-13.0[a]
Inosine	0.604 ± 0.453	0.102-1.81
Guanosine	0.258 ± 0.174	0.106-0.618
Hypoxanthine	3.98 ± 1.42[b]	2.36-7.34[b]
Xanthine	0.827 ± 0.434	0.270-2.01
Uridine	3.66 ± 1.24	2.02-6.23

[a] Free L-tryptophan levels reported.
[b] Values of hypoxanthine are strongly dependent on the time at which the samples are processed.
Source: Ref. 40.

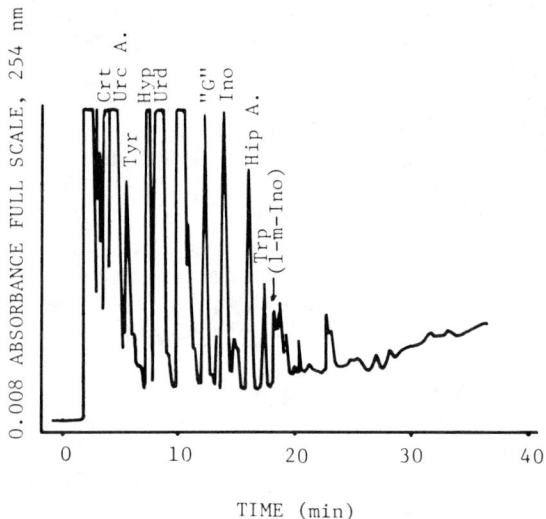

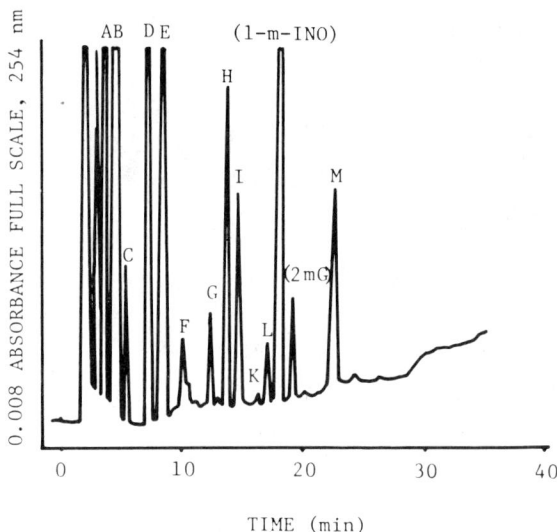

Figure 5. Top: chromatogram of a serum filtrate from a 46-year-old male patient with lung cancer. Peak in parenthesis represents tentative identification by retention times. Bottom: serum profile of a postoperative breast cancer patient with metastasis of the bone. (From Ref. 39.)

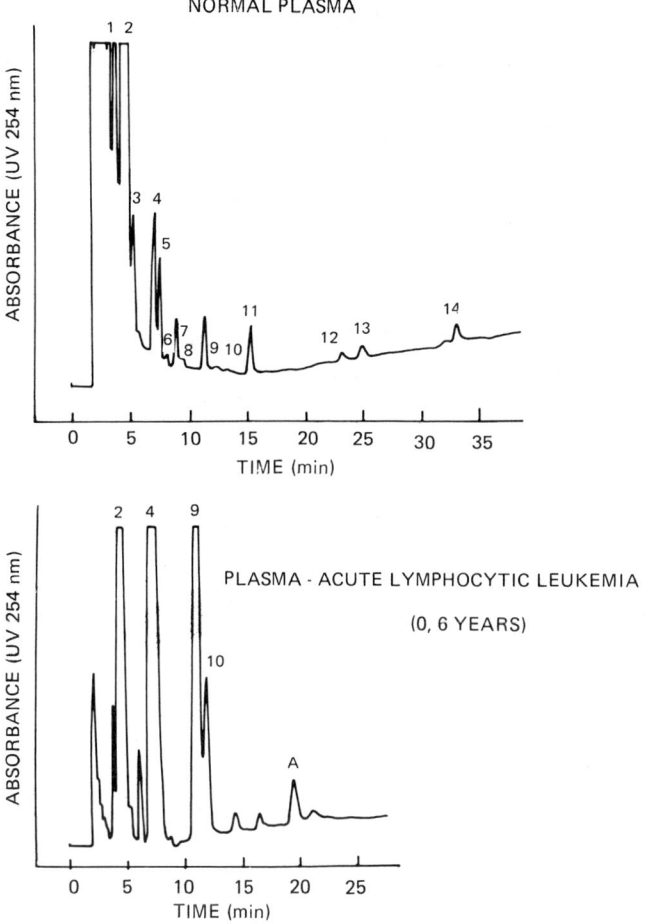

Figure 6. Top: chromatogram of a plasma sample obtained from a normal individual. Injection volume: 40 µl. Integrator attenuation: 8. UV detector at 254 nm. Column: Whatman ODS-3 (10 µm). Guard column: Whatman, Co-Pell ODS packing material. Eluents: A: 0.02 mol/liter KH_2PO_4, pH 5.7; B: 3:2 methanol-water. Gradient: linear, 0-40% B in 35 min. Flow rate: 1.5 ml/min. Peak identification: 1, creatinine; 2, uric acid; 3, L-tyrosine; 4, hypoxanthine; 5, uridine; 6, xanthine; 7, L-phenylalanine; 8, L-kynurenine; 9, inosine; 10, guanosine; 11, L-tryptophan; 12, theobromine; 13, paraxanthine; 14, caffeine. Bottom: chromatogram of a plasma sample obtained from an ALL patient. Same chromatographic conditions and peak identification numbers as in Fig. 1. A, adenosine. (From Ref. 40.)

Table 3 HPLC Analysis for Urinary Nucleosides[a]

Nucleoside[b]	Mean (nmol/ml)	S.D.
ψ	225.3	4.07
m^1A	15.18	0.54
m^7G[c]	6.72	0.31
m^1I	10.26	0.35
m^1G	5.69	0.16
m^2G	5.52	0.23
A	2.52	0.15
m_2^2G	11.37	0.82

[a] Each value is the average of four independent runs with four different affinity columns and a pooled urine control.
[b] ψ, pseudouridine; m^1A, 1-methyladenosine; m^7, 7-methylguanosine; m^1I, 1-methylinosine; m^1G, 1-methylguanosine; m^2G, N^2-methylguanosine; A, adenosine; m_2^2G, N^2,N^2-dimethylguanosine.
[c] Identity based on retention time only; needs further confirmation by other methods.
Source: Ref. 43.© 1977 Clinical Chemistry.

inosine, uridine, and xanthosine were separated in 100 min. The chromatographic conditions were based on the method originally devised by Brown [45] and coworkers [46,47]. The column was a Partisil-10 SAX column (0.46 × 25 cm, 10 μm, Whatman) and the eluent a pH/concentration ternary solvent gradient. One of the mobile phase components was a solution of 2% acetonitrile in water, the second, 2% acetonitrile in 0.05 M KH_2PO_4 (pH 3.35), and the third, 2% acetonitrile in 0.25 M KH_2PO_4, 0.05 M KCl (pH 5.25). The flow rate was 1.3 ml/min. AMP, ADP, ATP, GMP, GDP, GTP, CDP, CTP, UDP, and UTP were quantified (Table 4). Significantly higher levels of ATP and UTP were observed in the leukemic blasts of untreated ALL patients. (Fig. 8).

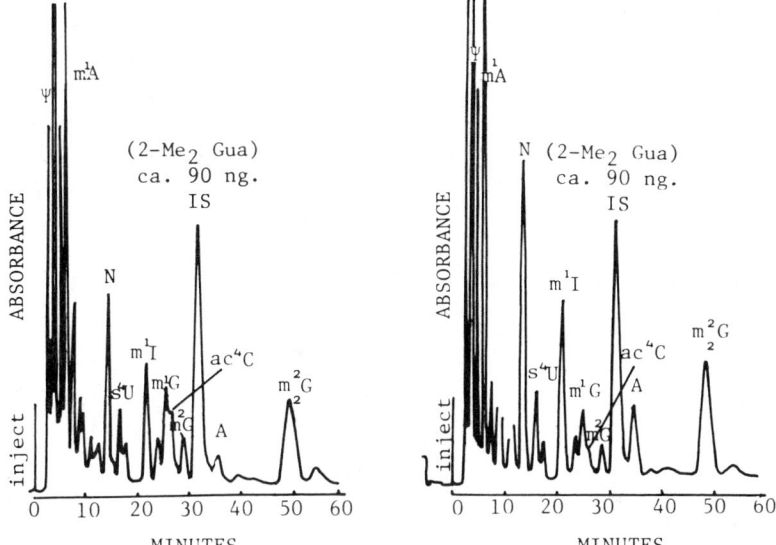

Figure 7. Left: reversed-phase HPLC isocratic separation of nucleosides in control urine. Right: reversed-phase HPLC isocratic separation of nucleosides in leukemia urine. Conditions: column, μBondapak C_{18} (4 × 300 mm). Buffer, 0.01 M $NH_4H_2PO_4$ (pH 5.10) with 6% methanol. Flow rate, 1.0 ml/min. Detector, 254 nm, 0.01 aufs. Temperature, 24°C. (From Ref. 42.)

Increased urinary levels of cyclic CMP, cyclic GMP, and cyclic AMP, as well as leukocytic levels of cyclic CMP were reported in patients with leukemia [48]. Elevations of these cyclic nucleotides and especially cyclic CMP were more pronounced in acute leukemia, whether myeloblastic, monoblastic, or lymphoblastic, than in chronic leukemia cases, thus indicating that cyclic CMP is a suitable marker for the disease. Other methods for the separation of cyclic ribonucleotides have been reported by anion-exchange [1] and by reversed-phase [49,50] liquid chromatography. The adaptability of such methods to the analysis of urine and leukocyte extracts has not yet been demonstrated.

C. Other diseases

1. Hypoxia and ischemia

Purine nucleoside and base levels were found to be changed in the physiological fluids of patients with cardiac malfunction. Harmsen et al. [51] demonstrated the marked elevation of hypoxanthine in the

Table 4 Nucleotide Concentration in Normal Peripheral Blood Lymphocytes

Nucleotide concentration (pmol/10^6 cells; mean ± S.D.)

	Normal lymphocytes
	PB (n = 21)[a]
AMP	25 ± 11
ADP	53 ± 40
ATP	400 ± 140
GMP	7 ± 5
GDP	22 ± 10
GTP	128 ± 47
CDP	12 ± 12
CTP	73 ± 44
UDP	14 ± 12
UTP	102 ± 31

[a] n = number of individuals.
Source: Ref. 44.

blood of patients with ischemic heart disease. Their chromatographic system consisted of a μBondapak C_{18} (0.4 × 30 cm) and an eluent of 10% methanol in 0.01 M ammonium phosphate (pH 5.5) at a flow rate of 1 ml/min. Uric acid, uracil, uridine, hypoxanthine, xanthine, xanthosine, inosine, guanosine, adenine, and adenosine were separated in 20 min. Similarly, Harkness et al. [52] reported increased concentrations of hypoxanthine and xanthine after hypoxia, especially in the cerebrospinal fluid. The results are easily understood in view of the excessive ATP breakdown in the heart and other muscles during and after hypoxia and ischemia. The chromatographic assays were performed on a ODS Hypersil-5 μm column (Shandon Southern Products, U.K.) eluted with 1% methanol in 0.01 M KH_2PO_4 (pH 6.5) at a flow rate of 1 ml/min.

2. *Hermansky-Pudlak syndrome*

The purine nucleotide content of storage pool-deficient platelets, obtained from patients with the Hermansky-Pudlak syndrome, was investigated by Rao et al. [53] using HPLC. AMP, ADP, GDP, and ATP were separated for this study in less than 15 min on an AS-Pellionex-SAX column (1.7 mm × 1 m) using a 0.15 M KH_2PO_4 buffer (pH 3.3).

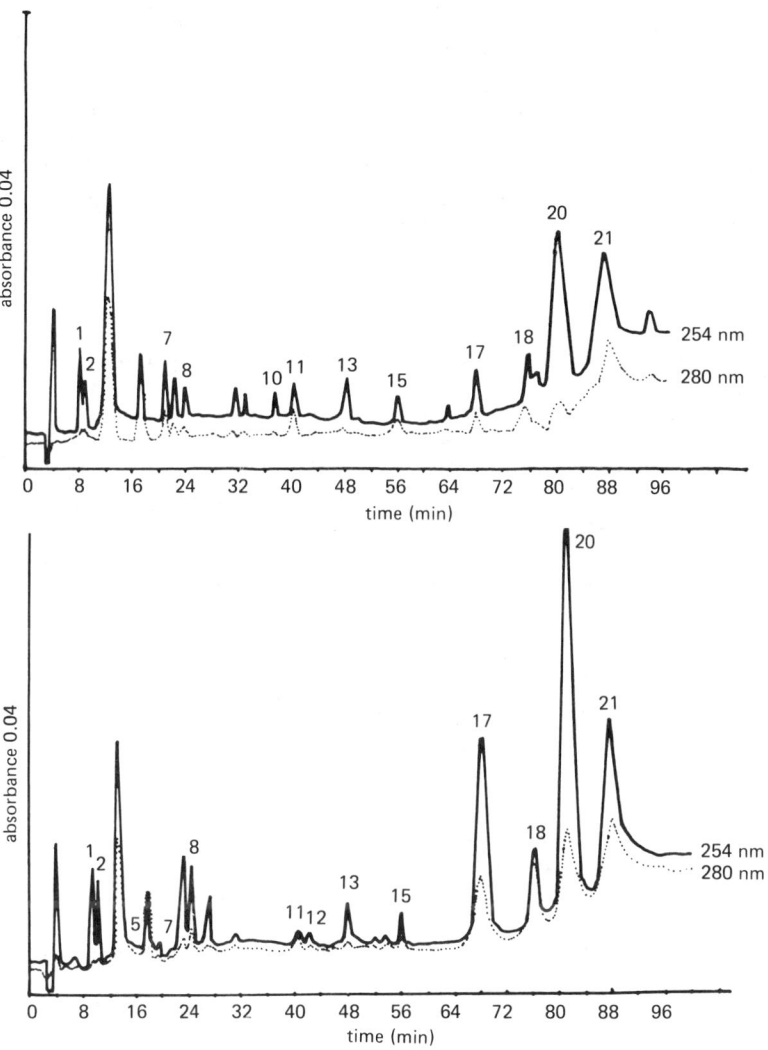

Figure 8. Top: HPLC profile of nucleotides in a cell extract from normal peripheral blood lymphocytes. Bottom: HPLC profile of nucleotides in a cell extract from lymphoblasts from a patient with non-B, non-T ALL. Retention times in minutes are as follows: (1) cyclic AMP, 8.9; (2) NAD, 10.0; (3) CMP, 12.0; (4) UMP, 15.1; (5) AMP, 16.6; (6) IMP, 17.6; (7) GMP, 21.4; (8) UDPG, 24.8; (9) XMP, 30.8; (10) NADP$^+$, 37.4; (11) CDP, 39.2; (12) UDP, 42.0; (13) ADP, 47.7; (14) IDP, 50.2; (15) GDP, 56.8; (16) XDP, 63.4; (17) UTP, 68.0; (18) CTP, 75.1; (19) ITP, 77.7; (20) ATP, 80.8; (21) GTP, 86.9; (22) XTP, 94.4. (From Ref. 44.)

The inlet pressure was 350 psi and the operating temperature 50°C. ADP levels in the Hermansky-Pudlak platelets were lower than those in normals. This finding is typical of and characterizes the disease condition. ATP levels in these platelets were normal or slightly higher than in the controls, yielding higher ATP/ADP ratios. The method did not detect the heterozygous deficiency state, but is valuable for characterizing variations in the nucleotide pools of platelets from patients with various storage pool diseases.

V. CONCLUSION

The biochemical importance of purine and pyrimidine compounds and their metabolic implication in a variety of disease states has been demonstrated using high performance liquid chromatography. A number of these compounds may be possible markers for certain disorders. Nucleic acid constituents are affected in individuals with other genetic and oncogenic illnesses which have not been mentioned in this chapter. The increasing use of HPLC in medical research will help identify these changes when they do occur, and will prove valuable in monitoring the efficiency of the therapy and the course of the illness.

REFERENCES

1. E. H. Edelson, J. G. Lawless, C. T. Wehrand, and S. R. Abbott, J. Chromatogr., 174, 409 (1979).
2. E. J. Ritter and L. M. Bruce, Biochem. Med., 21, 16 (1979).
3. C. Garret and D. V. Santi, Anal. Biochem., 99, 268 (1979).
4. W. Plunkett, J. A. Benvenuto, D. J. Stewart, and T. L. Loo, Cancer Treatment Rep., 63, 415 (1979).
5. B. Bakay, E. Nissinen, and L. Sweetman, Anal. Biochem., 86, 65 (1978).
6. A. Floridi, C. A. Palmerini, and C. Fini, J. Chromatogr., 138 (1977).
7. M. B. Cohen, J. Maybaum, and W. Sadee, J. Chromatogr., 198, 435 (1980).
8. J. X. Khym, Clin. Chem., 21, 1245 (1975).
9. E. J. Knudson, Y. C. Lau, H. Veening, and D. A. Dayton, Clin. Chem., 24, 686 (1978).
10. W. Voelter, K. Zech, P. Arnold, and G. Ludwig, J. Chromatogr., 199, 345 (1980).
11. R. A. Hartwick, D. Van Haverbeke, M. McKeag, and P. R. Brown, J. Liq. Chromatogr., 2, 725 (1979).
12. M. Zakaria and P. R. Brown, Anal. Biochem., 120, 25 (1982).
13. C. W. Gehrke, K. C. Kuo, G. E. Davis, R. D. Suits, T. P. Waalkes, and E. Borek, J. Chromatogr., 150, 455 (1978).

14. E. Nissinen, Anal. Biochem., 106, 497 (1980).
15. G. Nuki, K. H. Astrin, D. P. Brenton, M. K. Cruikshank, J. Lever, and J. E. Seegmiller, Purine and Pyrimidine Metabolism, Elsevier, Amsterdam, 1977, p. 127.
16. D. P. Brenton, K. H. Astrin, M. K. Cruikshank, and J. E. Seegmiller, Chem. Med., 17, 231 (1977).
17. G. Nuki, K. Astrin, D. Brenton, M. K. Cruikshank, J. Lever, and J. E. Seegmiller, Am. J. Human Genet., 25, 56A (1973).
18. E. R. Giblett, J. E. Anderson, F. Cohen, B. Pollara, and H. J. Meuwissen, Lancet, ii, 1067 (1972).
19. A. H. Van Gennip, J. Grift, P. K. De Bree, B. J. M. Zegers, J. W. Stoop, and S. K. Wadman, Clin. Chim. Acta, 93, 419 (1979).
20. C. A. Koller, P.L. Stetson, L. D. Nichamin, and B. S. Mitchell, Biochem. Med., 24, 179 (1980).
21. S. H. Polmar, R. C. Stern, A. L. Schwartz, E. M. Wetzler, P. A. Chase, and R. Hirschhorn, N. Engl. J. Med., 295, 1337 (1976).
22. R. A. Hartwick and P. R. Brown, J. Chromatogr., 143, 383 (1977).
23. R. A. Hartwick, A. Jeffries, and P. R. Brown, J. Chromatogr. Sci., 16, 427 (1978).
24. E. R. Giblett, A. J. Amman, D. W. Wasa, R. Sandman, and L. K, Diamond, Lancet, i, 1010 (1975).
25. J. W. Stoop, B. J. M. Zegers, G. F. M. Hendrickx, L. H. Siegenbeek van Heukelom, G. E. J. Staal, P. K. De Bree, S. K. Wadman, and R. E. Ballieux, N. Engl. J. Med., 296, 651 (1977).
26. A. P. Halfpenny and P. R. Brown, J. Chromatogr., 199, 275 (1980).
27. A. Cohen, G. E.J. Staal, A. J. Amman, and D. W. Martin Jr., J. Clin. Invest., 60, 491 (1977).
28. A. M. Krstulovic, R. A. Hartwick, and P. R. Brown, J. Chromatogr. Biomed. Appl., 163, 19 (1979).
29. J. B. Wyngaarden, in Harrisson's Principles of Internal Medicine, 7th ed., vol. I. McGraw-Hill Book Company, 1974, p. 607.
30. A. McBurney and T. Gibson, Clin. Chim. Acta, 102, 19 (1980).
31. J. E. Mrochek, W. C. Butts, W. T. Rainey, Jr., and C. A. Burtis, Clin. Chem., 17, 72 (1971).
32. G. A. Taylor, P. J. Dady, and K. R. Harrap, J. Chromatogr., 183, 421 (1980).
33. K. R. Hande, C. V. Hixson, and B. A. Chabner, Cancer Res., 41, 2273 (1981).
34. M. J. Cronklyn and R. Silber, Leukemic Res., 6, 203 (1982).
35. E. Samuel, C. H. Chung, N. Scher, B. Rosenzweig, and R. Silber, Blood, 55, 618 (1980).
36. R. K. Robins, Nucleosides and Nucleotides, II, 35 (1982).

37. R. A. Hartwick and P. R. Brown, J. Chromatogr., *126*, 679 (1976).
38. R. A. Hartwick, S. P. Assenza, and P. R. Brown, J. Chromatogr., *186*, 647 (1979).
39. R. A. Hartwick, A. M. Krstulovic, and P. R. Brown, J. Chromatogr., *186*, 659 (1979).
40. M. Zakaria, P. R. Brown, M. P. Farnes, and B. E. Barker, Clin. Chem. Acta, *126*, 69 (1982).
41. A. M. Krstulovic, R. A. Hartwick, and P. R. Brown, Clin. Chem. Acta, *97*, 159 (1979).
42. C. W. Gehrke, K. C. Kuo, G. E. Davis, R. D. Suits, T. P. Waalkes, and E. Borek, J. Chromatogr., *150*, 455 (1978).
43. G. E. Davis, R. D. Suits, K. C. Kuo, C. W. Gehrke, T. P. Waalkes, and E. Borek, Clin. Chem., *23*, 1427 (1979).
44. R. A. De Abreu, J. M. Van Baal, J. A. J. M. Bakkesen, C. H. M. M. De Bruyn, and E. D. A. M. Schretlen, J. Chromatogr. Biomed. Appl., *227*, 45 (1982).
45. P. R. Brown, J. Chromatogr., *52*, 257 (1970).
46. R. A. Hartwick and P. R. Brown, J. Chromatogr., *112*, 651 (1975).
47. M. McKeag and P. R. Brown, J. Chromatogr., *52*, 253 (1978).
48. J. Scavennec, Y. Carcassone, J.-A. Gastaut, A. Blanc, and H. L. Cailla, Cancer Res., *41*, 3222 (1981).
49. A. M. Krstulovic, R. A. Hartwick, and P. R. Brown, Clin. Chem., *25*, 235 (1979).
50. S. P. Assenza, J. Chromatogr. Biomed. Appl., *272*, 373 (1983).
51. E. Harmsen, J. W. De Jong, and P. W. Serruys, Clin. Chim. Acta, *115*, 73 (1981).
52. R. A. Harkness, R. J. Simmonds, M. C. O'Connor, and A. D. B. Webster, Biochem. Soc. Trans., *7*, 1021 (1979).
53. G. H. R. Rao, J. G. White, A. A. Jachimowicz, and C. J. Wiktop, J. Lab. Clin. Med., *84*, 839 (1974).

Selected Recent Publications

Sample Preparation

Olempska-Beer, Z. and Freese, E. B., Anal. Biochem. *140*, 236 (1984).

Ion Pairing

Perrone, P. A. and Brown, P. R., in Ion Pair Chromatography, Marcel Dekker, Inc., New York (in press).
Perrone, P. A. and Brown, P. R., J. Chromatogr. (in press).
Perrone, P. A. and Brown, P. R., J. Chromatogr. Biomed. App. *307*, 53 (1984).

Nucleic Acids

Demidou, V. V. and Potaman, V. N., J. Chromatogr. *285*, 135 (1984).
Graham, G. J., J. Chromatog. Libr. *22B*, 345 (1983).

Oligonucleotides

Delort, A. M., Derbyshire, R., Duplaa, A. M., Guy, A. A., Molko, D., and Teoule, R., J. Chromatogr. *283*, 462 (1984).

Demidov, V. V. and Potaman, V. N., J. Chromatogr. 285, 135 (1984).
Pearson, J. D. and Regnier, F. E., J. Chrolmatogr. 255, 137 (1983).

Nucleotides

Bedford, G. K. and Chiong, M. A., J. Chromatogr. Biomed. Applications 305, 183 (1984).
Floyd, R. A., Arch. Biochem. Biophys. 225, 263 (1983).
Folley, L. S., Power, S. D., and Poyton, R. O., J. Chromatogr. 281, 199 (1983).
Fruse, E., Olempska-Beer, Z., and Eisenberg, M., J. Chromatogr. 284, 125 (1984).
Haseltine, W. A., Franklin, W., and Lippke, J. A., Environmental Health Perspectives, 48, 29 (1983).
Hull-Ryde, E. A., et al., J. Chromatogr. 285, 411 (1983).
Linz, U., J. Chromatogr. 260, 161 (1983).
Reiss, P. D., Zuurendonk, P. F., and Veech, R. L., Anal. Biochem. 162 (1984).
VanBuren, C., et al., Transplantation, 36, 350 (1983).

Nucleosides and Their Bases

Assenza, S. P. and Brown, P. R., J. Chromatogr. 282, 477 (1983).
Assenza, S. P. and Brown, P. R., J. Chromatogr. 289, 355 (1984).
Assenza, S. P. and Brown, P. R., J. Chromatogr. Biomed. Applications, 277, 305 (1983).
Assenza, S. P. and Brown, P. R., Sep. Purif. Methods, 12, 177 (1983).
Boulieu, R., Baltassat, C., and Gonnet, J. Chromatogr. Biomed. Applications, 307, 469 (1984).
Brown, P. R., Cancer Investigations, 1(5), 439 (1983).
Brown, P. R., Zakaria, M., and Grushka, E., Anal. Chem. 55, 475 (1983).
Perrone, P. A. and Brown, P. R., Comp. Biochem. and Physiol. (in press).
Ramos, D. L. and Sholfstall, A. L., J. Chromatogr. 261, 83 (1983).

Cyclic Nucleotides

Assenza, S. P. and Brown, P. R., J. Chromatogr. Biomed. Applications, 272, 383 (1983).

Nucleotide Coenzymes

Entsch, B. and Sim, R. G., Anal. Biochem., *133*, 401 (1983).

Diseases

Gewirtz, H., Brantigan, D. L., Olsson, R. A., Brown, P. R., and Most, A. S., Circ. Res. *53*, 42 (1983).

Scoble, H. A., Brown, P. R., and Fasching, J. L., Anal. Chim. Acta, *150*, 171 (1983).

Scoble, H. A., Zakaria, M., Brown, P. R., and Martin, Computers and Biomed. Research, *16*, 300 (1983).

Index

Absorbance ratios:
  in peak identification, 82-85, 92
Acid-citrate-dextrose:
  in plasma preparation, 41, 43
Acid phosphatase, 372
  HPLC assay for, 295
Accuracy, 108
Acyclovir, 320, 321, 330, 333
Adenine, 258, 262, 308
  arabinoside, 320, 321, 330, 331, 334
  compounds, 260, 308
  nucleotides, 223-225, 227, 229, 236
    as coenzymes, 304-306
    HPLC analysis of, 308, 309
Adenosine, 76, 77, 249, 251 255, 259, 260, 262
  deaminase, 294, 295, 367, 369 371
    HPLC assay for, 295
  in ADA deficiency, 369, 371
  in assay of ADA, 394

[Adenosine]
  in leukemic plasma, 378
  in phosphatase assay, 295
Adenyl cyclase, 267, 271, 275
  HPLC assay for, 272, 295
Adsorption chromatography, 50
  (see also LSC)
Affinity chromatography, 182, 183
  in sample preparation, 40
Alkaline phosphatase, 295, 299, 372
  HPLC assay for, 295
Allopurinol, 251, 258, 320, 321, 329, 330, 333, 374-376
Analytical errors, 108
  determinate, 108
  indeterminate, 108
Anion exchange, 51, 73
  for antimetabolites, 333
  for oligonucleotides, 182, 202-203
  for tRNAs, 182

[Anion exchange]
  of cyclic AMP, 271
  of nucleotides, 67, 69, 72
    73, 216-225, 240
  of nucleotides, nucleosides,
    and bases, 76
Anticoagulants, 41, 249
Antimetabolites, 317-337
  purine, 317, 318, 320, 321,
    328-331
  pyrimidine, 317-319, 324-327
Assays:
  of enzymes, 285-302
Atomic absorption:
  in peak detection, 155
Azathioprine, 320, 321, 328,
    329, 333
ATP metabolites:
  by RPLC, 44, 275, 308, 309
Automation, 62
5-Azacytidine, 318, 319, 326
5-Aza-2'Deoxycytidine, 318

Bacteria:
  nucleotides in, 216, 230, 236
Band broadening, 58
  (see also Zone Spreading
    Band Spreading)
Band spreading, 58
Base pairing, 23
Biochemical markers, 248, 366, 378
Biogenetic engineering, 5
Biosynthesis, 8, 9, 10, 13
  DeNovo, 8, 9, 10
  of purines, 287
  of pyrimidines, 287, 288
  Salvage pathway, 8, 11
Blood:
  nucleotides in, 223-226, 229
    232
Blood cells:
  sample preparation of, 41,
    42, 366, 367
Blood fluids, 248-253, 367
  sample preparation of, 41-44
  (see also Serum and Plasma)

Borate complexes, 41, 259
Boronate gel, 259, 260
Brain:
  nucleotides in, 236

Caffeine, 45, 251, 258, 341-347,
    350-361
  by microbore, 131-134
Calculations:
  of enzyme activity, 299, 300
Calibration methods, 104-107
  external standardization, 106
  internal standardization, 105
  normalization, 105
  response factors, 104, 105
  standard addition, 106, 107
Capacity factor, 54, 55, 77, 78
  definition of, 54
  determination of, 55
  optimal, 55
Catabolism:
  of nucleotides, 11, 12
Cation exchange, 51, 73, 333
  in antimetabolite analysis, 333
Chromatographic mode:
  choice of, 73
Chromatography:
  description of, 49, 50
Ciliates:
  nucleotides in, 236
Cochromatography:
  in peak identification, 82, 84,
    89
Coenzyme A, 305-307
  precursors, 228, 236
Coenzyme nucleotides, 303-315
  FAD, 20
  FADH, 21
  NAD^+, 21
  NADH, 21
  $NADP^+$, 21
  NADPH, 21
  chemistry of, 304, 306, 307
  functions of, 303, 304, 306, 307
  HPLC of, 308-313
  structures of, 304

Column packings:
  in HPLC of oligonucleotides, 200-207
Column parameters, 76
Column switching, 21, 78
  for separation of nucleotides, nucleosides and bases, 235
Columns:
  microbore, 113-134
  short, 64
Cost savings:
  of microbore HPLC, 118
Cyclic AMP, 222, 227, 231, 232, 236, 267-281
Cyclic CMP, 272, 273, 277-280
Cyclic GMP, 269, 270, 272, 273, 275-281
Cyclic IMP, 272, 273, 277-280
Cyclic UMP, 272, 273, 277-280
Cytosine, 258, 318, 319, 323-325, 332-334
  arabinoside, 318, 319, 325, 326, 332, 334

Deoxyadenosine, 251, 255, 262
  in ADA deficiency, 369
Deoxyribonucleotides, 19, 218
Deproteinization:
  matrix effects, 34, 37-39
  solute effects, 34, 39
  techniques, 32-36
    organic solvents, 33, 34
    precolumns, 36
    salts, 35
    strong acids, 33
    temperature, 35
    ultrafiltration, 35, 36
Detection:
  of antimetabolites, 322, 323
  of cyclic nucleotides, 271
  electrochemical:
    of nicotinamide coenzymes, 310, 311

[Detection]
  of oligonucleotides, 207
  limits, 109
Detectors, 82-90, 91-95, 123-127, 140-157
  absorbance, 143-148
    fixed wavelength and filter photometers, 144-146
    variable wavelength, 147, 148
    rapid scanning, 147, 148
  characteristics, 140-143
  combined, 154
  electrochemical, 310
    in peak identification, 93, 94
    in peak detection, 149, 150
  fluorescence:
    derivatization with, 151
    description of, 151
    emission spectra, 92, 150, 151
    excitation spectra, 92, 150, 151
    in peak detection, 150-152
    in peak identification, 92, 94
    operation, 151
    synchronous scanning technique, 92
  radioactivity, 152-154
  reaction:
    chemical derivatization for: 194
    immobilized enzymes, 94, 157
    in peak identification, 93, 94
    post column, 94, 156, 157
  refractive index, 148
Direct injection:
  in antimetabolite analysis, 332
Diseases, 367-386
  nucleic acid constituents in, 365
DNA, 95, 182, 188-190, 206
Drug metabolism, 356
  of theophylline, 357, 358

[Drug metabolism]
 of caffeine, 360
Drugs, 317-337, 339, 363
 effectiveness, 5
 monitoring, 5
 toxicity, 5

Eddy diffusion, 58
Efficiency, 57, 58
 of columns, 57, 58
 of protein removal, 36
Elution, 76
 gradient, 67, 74, 76, 77
 isocratic, 76, 77
Energy:
 charge, 7, 162-165
 metabolism, 6-7
Enzymatic peak shifts:
 in peak identification, 82-84, 86, 89
Enzyme:
 assays, 285-302
  definition of, 286
 deficiencies, 366, 367
Enzymes:
 activities of, 5, 299, 300
 functions of, 286
 optimal temperatures, 293
 PH, 291-293
 units for activity, 293
 velocity of reactions, 288-291
Equations:
 of chromatographic terms, 54
Ethylenediaminetetraacetate (EDTA);
 in plasma preparation, 42, 51
Evaluation of results, 108-110
Exclusion chromatography, 50, 54, 55, 73
 (see also molecular sieving
  Gel Filtration
  Gel Permeation)
 description of, 54
 in sample preparation, 39

Extraction:
 in antimetabolite analysis, 323
 procedures:
  from cells, 366, 367
  from physiological fluids, 367
Flavin coenzymes, 305-307
 by ion exchange chromatography, 311
Flow programming, 78
5-Fluorocytosine, 318, 319, 327
5-Fluoroacil, 251, 318, 319, 324, 325, 332, 333
Fluoropurines, 218, 227, 230, 233
Ftorafur, 318, 319, 324
Fungi:
 nucleotide pools in, 217, 236
Gout, 366, 367, 373, 374
Gradient shope:
 effect of, 67, 74
Guanine, 78, 249
Guanosine, 249, 252, 257
 in leukemic plasma, 378
 in normal plasma, 379
Guanylate cyclase, 295
Guidelines:
 for chromatographic optimization, 73

Hemodialysates:
 sample preparation, 45
Heparin:
 in plasma preparation, 41, 44
High efficiency microbore separations:
 of oligonucleotides, 129
 of dimers of ribonucleotides, 129
High performance size exclusion (HPSEC), 187-189, 191
High speed microbore separations, 116, 117, 130, 131
 of ribonucleosides, 131, 135
 of xanthines, 131-134
 of theophylline, 131, 133

History:
  of nucleic acid research, 1-5
HPLC:
  analysis:
    steps involved, 71
  enzyme assays, 285-302
    advantages of, 300
  of antimetabolites, 322-328
  of cyclic nucleotides, 270-281
    requirements of, 270
    advantages of, 270
  of nucleic acids, 181-190
  of nucleosides and bases, 247-263
  of nucleotide coenzymes, 303-312
  of nucleotides, 215-242
  of oligonucleotide, 200-206
  of purine antimetabolites, 317-333
  of xanthines, 339-360
Hydrolysis:
  of nucleic acids, 14, 206
Hyperuricemia, 374
Hypoxanthine, 76, 78, 249, 251, 253, 255, 258, 274, 276
  in HGPRTase assay, 298
  in ischemic heart disease, 383, 384
  in leukemic plasma, 373
  in normal plasma, 378
  in normal serum, 378
  in PNPase assay, 288
  in xanthine oxidase assay, 295
Hypoxanthine guanine phosphoribosyl transferase, 294, 366, 367, 368, 369
Hypoxia, 383, 384
Hydroxylapatite, 182, 188, 191

Iodoxuridine, 325, 332
Immunodeficiencies, 12, 288

Inductively coupled plasma emission spectrometers (ICP):
  in peak detection, 155, 156
  in peak identification, 94
Infrared spectrometers, 94
Inhibition:
  of nucleic acid biosynthesis, 13
Injectors, 61
Inosine, 76, 249, 251, 252, 255, 257, 258
  in ADA deficiency, 369
  in leukemic plasma, 378
  in normal plasma, 379
  in normal serum, 378
  in PNPase assay, 288, 295
Instrumentation, 60-63
  automation, 62, 63
  for nucleotide analyses, 239-240
  injectors, 61
  pumps, 62
Integrators:
  electronic, 100, 104
    A/P conversion, 101
    baselines, 102
    method of reporting, 103
    peak initiation, 102
    peak recognition, 101
    peak termination, 102
    sampling rate, 102
    setting of thresholds, 102
Interconversion, 12
Interferences:
  in HPLC analysis of methylxanthines, 341, 345, 356
Ion exchange, 4, 21, 163, 184, 185, 191
  advantages of, 51
  in sample preparation, 40
  limitations of, 51
  of antimetabolites, 333, 334
  of methylxanthines, 341, 343
  of nucleotides, 240

Ion pairing, 21, 73, 78, 349, 356-359
  models, 171-173
  of antimetabolites, 333
  of isomeric monoribonucleotides, 167
  of linear and cyclic nucleotides, 242, 277, 280, 281
  of methylxanthines, 356-359
  of oligonucleotides, 209, 210
  of purine and pyrimidine bases, 169
Ionic strength:
  effect on separations, 64, 67, 68
Isocratic elution:
  of cyclic AMP, 271
Isoenzymes:
  separation of, 299

Kinetics:
  of enzymes, 288, 289
Km, 289, 290, 291
  measurment of, 290

Lesch Nyhan syndrome, 288, 299, 366, 367, 371
Leukemia, 318, 378, 379, 382, 383
Leukocytes:
  nucleotides in, 236
Liquid-Liquid chromatography (LLC), 50, 53
  (see also Partition), 50
  normal phase, 53
  reversed phase, 53
Liquid-solid chromatography (LSC), 50, 51
  (see also Adsorption), 50
Liver:
  nucleotides in, 236
  tissue, 46
Longitudinal diffusion, 58

Mammalian cell cultures, 217-219, 227, 230
Mass spectrometers:
  in peak detection, 154
  in peak identification, 94, 95
Mercaptopurines, 238, 320, 321, 328, 329, 332, 333
Metabolism, 3
  of bases, 8-12
  of nucleic acids, 13, 14
  of nucleosides, 8-12
  of nucleotides, 8-12
Metal ions:
  as ion pairing agents, 170
Methylxanthines, 339-363, 376
  analyses of, 340
  functions of, 339
  HPLC analysis of, 340-357
  sample preparation, 340
Michaelis-Menten Equation, 288, 289, 290
Microbore design requirements, 120-121
  detector cell volumes, 120
  detector sensitivity, 121
  injection volumes, 120
Microbore instrumentation, 121-127
  column design, 127
  detectors, 123, 124
  injectors, 122
  pumping systems, 121, 122
Microbore HPLC, 144
  performance potential of, 115-117
Miniaturization, 113
Mixed beds, 21
  columns, 78
Mobile phase:
  effects, 24-26
    ionic strength, 25
    organic modifiers, 24
    PH, 24
  ion exclusion, 161, 162
  ion pairing, 161, 174

Index   399

[Mobile phase]
  ion suppression, 161, 162, 175
  parameters, 64-67
  techniques, 161
Molecular sieving, 54
Multiple identification techniques,
    82, 84-89
Muscle:
  nucleotides in, 236

Neoplasms, 366
Nicotinamide coenzymes, 304-306
  HPLC of, 309-311
Nicotinate phosphoribosyltrans-
    ferase:
  HPLC assay for, 295
Novel microbore packings, 119
Nuclear magnetic resonance
    (NMR):
  in peak identification, 95
Nucleic acids, 3-8, 13, 14, 181-
    191, 286, 318
  (see also DNA, 4, 17
    RNA, 4, 17
  biosynthesis of, 13
  composition of, 20
  hydrolysates of, 4, 206
  hydrolysis of, 14
  inhibition of biosynthesis,
    13
  metabolism of, 13-14
Nucleosides:
  in normal plasma, 379
  in normal serum, 378
  modified, 248, 254, 256, 258-
    263, 376-379, 382
  selected analysis of, 251, 254,
    259, 260
Nucleosides and bases, 247-
    263
  separation of, 248-251, 373-
    386
Nucleotides, 4
  biosynthesis of,

[Nucleotides]
  by anion exchange, 67, 69,
    72, 73, 216-225, 240
  by reversed phase, 227-234,
    240, 242, 281
  catabolism of, 11, 12
  cyclic, 5, 7, 20, 227, 236,
    267-281
    analyses of, 270, 277, 278,
      279, 280, 281, 384
    functions of, 269
    in leukemic urine, 383
    structure of, 267
  functions of, 6, 7
  in Hermansky-Pudlak syn-
    drome, 384, 385
  in HGPRTase deficiencies,
    368, 369
  in leukemic blast cells,
    382
  in lymphocytes, 284
  interconversion of, 12
  metabolism of, 3
  structures of, 4, 6

Olignucleotides, 182-184, 195-
    213
  chromatographic properties
    of, 196
  detection of, 207
  identification of, 207-209
    for relative base composition,
      208, 209
    for sequences, 208
  preparation of, 196, 197
Open column chromatography:
  of oligonucleotides, 198
Operating conditions, 63
  mobile phase, 64-67
  stationary phase, 63, 64
  temperature, 67
Optimization, 70-78
  of chromatographic conditions,
    58

Organic solvents:
  effect on separations, 64, 67, 69
  for deproteinization, 33, 35
Orotic acid, 372, 374, 376
Orotic aciduria, 247, 372, 374
Orotic acid phosphoribusyl transferase, 372
Orotidine, 372
Orotidine decarboxylase, 372
Oxypurinol, 251, 258, 329, 330, 333, 374, 376

Paraxanthine, 257, 258
Peak areas:
  in peak measurements, 99, 100
Peak height ratios:
  in peak identification, 87, 88
Peak heights:
  in peak measurements, 99, 100
Peak identification, 82-95, 207-209
Peak measurements, 99, 100
Peak purity, 83, 84
Peak recognition, 101
Perchloric acid:
  for deproteinization, 33
Periodate/methylamine:
  to remove nucleotides, 218, 239
PH:
  effect on enzyme activity, 291-293
  effect on separations, 64, 66
Pharmacokinetics:
  of caffeine, 360, 361
  of theophylline, 359
  of theophylline and caffeine, 361
Phosphodiesterase:
  HPLC analysis of, 295
Physiological fluids, 247-263, 367

Plasma:
  profiles of, 248-255, 375, 381
Platelets:
  nucleotides in, 236
Precision, 108
Precolumns:
  for sample preparation, 36
Precipitation:
  in antimetabolite analysis, 323
Problem solving, 70-78
Profiles:
  of plasma, 248-255, 375, 381
  of saliva, 254-257
  of serum, 84-90, 248-251, 377, 380
  of urine, 258, 259, 377, 383
Protected nucleotides, 197
  separation of, 204
Pseudouridine, 259, 260, 262
Pumps, 62
  of antimetabolites, 321, 322
Purine antimetabolites, 317-334
  structures of, 319, 320
Purine nucleoside physohorylase, 367, 372
  HPLC analysis of, 295
Purine bases, 3, 5
  biosynthesis of, 8-10
  interconversion of, 12
  metabolism of, 8-12
  nucleosides and nucleotides, 217, 218, 223, 229, 235
Purine drugs:
  uses of, 318
Purines (see also individual purines), 17-19
  metabolites of, 258-263
  properties of, 17-19
  structures of, 17, 18
Pyrimidine antimetabolites, 317-334
  structure of, 318-320
Pyrimidines (see also individual pyrimidines), 17, 19
  metabolites of, 258-263
  properties of, 17, 18, 19
  structure of 17, 18

Pyridine coenzymes, 304-306
  (see also Nicotinamide coenzymes)
Pyrimidine drugs:
  uses of, 318
Pyrimidine metabolites:
  in PNPase deficiency, 372
  in OTC deficiency, 373
Pyrimidine-5-nucleotidase, 295

Radioactive labels:
  for antimetabolites, 322
Radionuclides, 238
Rate theory, 58
Recombinant DNA research, 195
Recovery studies:
  in sample preparation, 36-39
Recycle microbore separations, 129
Recycling, 78
Relationships:
  structure-retention, 23
Repeatability, 108
Reproducibility, 108
Requirements:
  of sample preparation, 31, 32
Resistance to mass transfer, 58
Resolution, 56, 57
  definition of, 56
Retention time, 54, 77
  adjusted, 54
  definition of, 54
  hold-up time, 54
Reversed phase, 21, 53, 73, 187, 188, 191
  advantages of, 53
  description of, 53
  in nucleoside and base analyses
  in nucleotide analyses, 227-234, 240-242
  in olignucleotide separations, 203-204

[Reversed phase]
  of adenines, 271
  of antimetabolites, 332, 333
  of coenzymes, 308-312
  of cyclic AMP, 271
  of methylated xanthines, 344-347, 350-355
  mechanisms of retention, 22
Riboflavin, 311, 312
RNA, 182, 184, 185, 187-189
  mRNA, 188-189
  tRNA, 182, 183, 185-187
RPC-5 chromatography, 183-187
RPC-5 packings:
  in olignucleotide separations, 198-200

Saliva, 254-257
  sample preparation, 45
"Salting Out":
  for deproteinization, 35
Sample preparation, 31-48, 238, 239
  antimetabolites, 323-332
  biological samples, 366, 367
  for chromatography:
    by exclusion, 39, 40
    by ion exchange, 40
    by affinity, 40, 41
  for enzyme activity, 294
  for methylxanthines, 340
  for nucleotides in biological samples, 238, 239
  for nucleosides, 259
Selective adsorption:
  in sample preparation, 40, 41
Selectivity, 56, 77
  definition of, 56
Sensitivity, 109
Separation methods:
  of oligonucleotides, 198, 210
Serum profiles:
  of nucleosides and bases, 84-90, 248-251, 377, 380

Silanophilic interaction, 22
Size exclusion:
 in oligonucleotide separations, 204-206
Soap chromatography, 162, 165
 of adenine and guanine nucleotides, 165
Solvophobicity, 22
Sonication:
 for sample preparation, 46
Stacking:
 vertical base, 23
Stationary phase, 63, 64
 effects on separation, 63, 64
 parameters, 76
Statistics, 109
Step-gradient:
 for cyclic AMP, 275
Surface modified packings:
 for oligonucleotides, 210
Synthesis:
 of oligonucleotides, 196, 197

Temperature:
 effect on enzyme activity, 293
 effect on separations, 67-69
 for deproteinization, 35
Temperature programming, 67, 68
Terminology, 54
Theobromine, 257
 by microbore columns, 117
Theophylline, 45, 257, 339, 341-347, 350-161
 by microbore, 117, 131, 133
Theoretical plates, 57, 58
 height equivalent to a theoretical plate (HETP), 58
 number of (N), 57, 58
6-Thioguanine, 320, 321, 328, 329
Thymidine, 255, 374
Thymine, 255, 374
Tissues, 227-229, 231

[Tissues]
 nucleotide pools in, 219-222
 sample preparation, 46
Trematodes:
 nucleotides in, 236
Trichloracetic acid:
 for deproteinization, 33
Types:
 of HPLC separation, 50

Ultrafiltration:
 for deproteinization, 35, 36
 in antimetabolite analysis, 323, 332
Uracil, 258, 262, 318, 319, 323-325, 332-334, 367
Uracil arabinoside, 325, 326, 334
Uric acid, 3, 11-12, 249, 251-253, 376
 in Lesch-Nyhan syndrome, 288
 in normal plasma, 379
 in normal serum, 378
 in xanthine oxidase assay, 295
Uridine, 249, 251-258
Urine, 258-263
 nucleotides in, 226-229
 profiles of, 258-259, 377, 383
 sample preparation, 45
UV detector rapid scanning:
 in peak identification, 84, 85
UV identification techniques, 84
 absorbance ratios, 83-85, 92
 cochromatography, 83
 enzymatic peak shifts, 83, 84, 86, 89
 retention characteristics, 82, 83
 UV scanning spectra, 83-85, 89
UV spectra:
 in peak identification, 82, 84, 86, 89, 91

Van Deemter equation, 58, 59, 68